Advances in Intelligent Systems and Computing

Volume 951

Series Editor

Janusz Kacprzyk, Systems Research Institute, Polish Academy of Sciences, Warsaw, Poland

Advisory Editors

Nikhil R. Pal, Indian Statistical Institute, Kolkata, India
Rafael Bello Perez, Faculty of Mathematics, Physics and Computing, Universidad Central de Las Villas, Santa Clara, Cuba
Emilio S. Corchado, University of Salamanca, Salamanca, Spain
Hani Hagras, Electronic Engineering, University of Essex, Colchester, UK
László T. Kóczy, Department of Automation, Széchenyi István University, Gyor, Hungary
Vladik Kreinovich, Department of Computer Science, University of Texas at El Paso, El Paso, TX, USA
Chin-Teng Lin, Department of Electrical Engineering, National Chiao Tung University, Hsinchu, Taiwan
Jie Lu, Faculty of Engineering and Information Technology, University of Technology Sydney, Sydney, NSW, Australia
Patricia Melin, Graduate Program of Computer Science, Tijuana Institute of Technology, Tijuana, Mexico
Nadia Nedjah, Department of Electronics Engineering, University of Rio de Janeiro, Rio de Janeiro, Brazil
Ngoc Thanh Nguyen, Faculty of Computer Science and Management, Wrocław University of Technology, Wrocław, Poland
Jun Wang, Department of Mechanical and Automation Engineering, The Chinese University of Hong Kong, Shatin, Hong Kong

The series "Advances in Intelligent Systems and Computing" contains publications on theory, applications, and design methods of Intelligent Systems and Intelligent Computing. Virtually all disciplines such as engineering, natural sciences, computer and information science, ICT, economics, business, e-commerce, environment, healthcare, life science are covered. The list of topics spans all the areas of modern intelligent systems and computing such as: computational intelligence, soft computing including neural networks, fuzzy systems, evolutionary computing and the fusion of these paradigms, social intelligence, ambient intelligence, computational neuroscience, artificial life, virtual worlds and society, cognitive science and systems, Perception and Vision, DNA and immune based systems, self-organizing and adaptive systems, e-Learning and teaching, human-centered and human-centric computing, recommender systems, intelligent control, robotics and mechatronics including human-machine teaming, knowledge-based paradigms, learning paradigms, machine ethics, intelligent data analysis, knowledge management, intelligent agents, intelligent decision making and support, intelligent network security, trust management, interactive entertainment, Web intelligence and multimedia.

The publications within "Advances in Intelligent Systems and Computing" are primarily proceedings of important conferences, symposia and congresses. They cover significant recent developments in the field, both of a foundational and applicable character. An important characteristic feature of the series is the short publication time and world-wide distribution. This permits a rapid and broad dissemination of research results.

**** Indexing: The books of this series are submitted to ISI Proceedings, EI-Compendex, DBLP, SCOPUS, Google Scholar and Springerlink ****

More information about this series at http://www.springer.com/series/11156

Francisco Martínez Álvarez ·
Alicia Troncoso Lora · José António Sáez Muñoz ·
Héctor Quintián · Emilio Corchado
Editors

International Joint Conference: 12th International Conference on Computational Intelligence in Security for Information Systems (CISIS 2019) and 10th International Conference on EUropean Transnational Education (ICEUTE 2019)

Seville, Spain, May 13–15, 2019, Proceedings

Editors
Francisco Martínez Álvarez
Data Science and Big Data Lab
Pablo de Olavide University
Seville, Spain

Alicia Troncoso Lora
Data Science and Big Data Lab
Pablo de Olavide University
Seville, Spain

José António Sáez Muñoz
University of Salamanca
Salamanca, Spain

Héctor Quintián
Department of Industrial Engineering
University of A Coruña
A Coruña, Spain

Emilio Corchado
University of Salamanca
Salamanca, Spain

ISSN 2194-5357 ISSN 2194-5365 (electronic)
Advances in Intelligent Systems and Computing
ISBN 978-3-030-20004-6 ISBN 978-3-030-20005-3 (eBook)
https://doi.org/10.1007/978-3-030-20005-3

© Springer Nature Switzerland AG 2020
This work is subject to copyright. All rights are reserved by the Publisher, whether the whole or part of the material is concerned, specifically the rights of translation, reprinting, reuse of illustrations, recitation, broadcasting, reproduction on microfilms or in any other physical way, and transmission or information storage and retrieval, electronic adaptation, computer software, or by similar or dissimilar methodology now known or hereafter developed.
The use of general descriptive names, registered names, trademarks, service marks, etc. in this publication does not imply, even in the absence of a specific statement, that such names are exempt from the relevant protective laws and regulations and therefore free for general use.
The publisher, the authors and the editors are safe to assume that the advice and information in this book are believed to be true and accurate at the date of publication. Neither the publisher nor the authors or the editors give a warranty, expressed or implied, with respect to the material contained herein or for any errors or omissions that may have been made. The publisher remains neutral with regard to jurisdictional claims in published maps and institutional affiliations.

This Springer imprint is published by the registered company Springer Nature Switzerland AG
The registered company address is: Gewerbestrasse 11, 6330 Cham, Switzerland

Preface

This volume of Advances in Intelligent and Soft Computing contains accepted papers presented at CISIS 2019 and ICEUTE 2019, both conferences held in the beautiful and historic city of Seville (Spain), in May 2019.

The aim of the twelfth CISIS 2019 conference is to offer a meeting opportunity for academic and industry-related researchers belonging to the various, vast communities of computational intelligence, information security, and data mining. The need for intelligent, flexible behavior by large, complex systems, especially in mission-critical domains, is intended to be the catalyst and the aggregation stimulus for the overall event.

After a thorough peer review process, the CISIS 2019 International Program Committee selected 20 papers which are published in these conference proceedings achieving an acceptance rate of 30%. In this relevant edition, a special emphasis was put on the organization of special sessions. One special session was organized related to relevant topics as: From the least to the least: cryptographic and data analytics solutions to fulfil least minimum privilege and endorse least minimum effort in information systems.

In the case of 10th ICEUTE 2019, the International Program Committee selected 15 papers, which are published in these conference proceedings. Two special sessions were organized related to relevant topics as: Looking for Camelot: New Approaches to Asses Competencies; Innovation in Computer Science Education.

The selection of papers was extremely rigorous in order to maintain the high quality of the conference, and we would like to thank the members of the Program Committees for their hard work in the reviewing process. This is a crucial process to the creation of a high standard conference, and the CISIS and ICEUTE conferences would not exist without their help.

CISIS 2019 and ICEUTE 2019 conferences enjoyed outstanding keynote speeches by distinguished guest speakers: Prof. Dieu Tien Bui (University of South-Eastern Norway, Norway), Prof. Juan Manuel Corchado (University of Salamanca, Spain) and Prof. Julien Jacques (University of Lyon, France).

CISIS 2019 special edition, as a follow-up of the conference, we anticipate further publication of selected papers in a special issue, in the prestigious Logic Journal of the IGPL (Oxford Academic).

Particular thanks go as well to the conference main sponsors, Startup Ole, IEEE SMC Spanish Chapter, the International Federation for Computational Logic, who jointly contributed in an active and constructive manner to the success of this initiative.

We would like to thank all the special session organizers, contributing authors, as well as the members of the Program Committees and the Local Organizing Committee for their hard and highly valuable work. Their work has helped to contribute to the success of the CISIS 2019 and ICEUTE 2019 events.

May 2019

Francisco Martínez Álvarez
Alicia Troncoso Lora
José António Sáez Muñoz
Héctor Quintián
Emilio Corchado

CISIS 2019

Organization

General Chairs

Francisco Martínez Álvarez	Pablo de Olavide University, Spain
Alicia Troncoso Lora	Pablo de Olavide University, Spain
Emilio Corchado	University of Salamanca, Spain

International Advisory Committee

Ajith Abraham	Machine Intelligence Research Labs—MIR Labs, Europe
Michael Gabbay	King's College London, UK
Antonio Bahamonde	University of Oviedo at Gijón, Spain

Program Committee Chairs

Emilio Corchado	University of Salamanca, Spain
Francisco Martínez Álvarez	Pablo de Olavide University, Spain
Alicia Troncoso Lora	Pablo de Olavide University, Spain
Héctor Quintián	University of A Coruña, Spain

Program Committee

Adolfo R. De Soto	University of Leon, Spain
Agustin Martin Muñoz	CSIC, Spain
Alberto Peinado	Universidad de Malaga, Spain
Amparo Fuster-Sabater	CSIC, Spain

Ana I. González-Tablas	University Carlos III de Madrid, Spain
Andreea Vescan	Babes-Bolyai University, Cluj-Napoca, Romania
Angel Arroyo	University of Burgos, Spain
Angel Martin Del Rey	University of Salamanca, Spain
Antonio J. Tomeu-Hardasmal	University of Cadiz, Spain
Camelia Serban	Babes-Bolyai University, Cluj-Napoca, Romania
Carlos Cambra	University of Burgos, Spain
Carlos Pereira	ISEC, Portugal
Carmen Benavides	University of León, Spain
Cosmin Sabo	Technical University of Cluj-Napoca, Romania
Cristina Alcaraz	University of Malaga, Spain
David Alvarez Leon	University of León, Spain
David Arroyo	CSIC, Spain
Eduardo Solteiro Pires	UTAD University, Portugal
Enrique Onieva	University of Deusto, Spain
Fernando Tricas	University of Zaragoza, Spain
Francisco Martínez Álvarez	Pablo de Olavide University, Spain
Francisco Rodríguez-Sedano	University of León, Spain
Guillermo Morales-Luna	CINVESTAV-IPN, Mexico
Héctor Quintián	University of A Coruña, Spain
Hugo Scolnik	University of Buenos Aires, Argentina
Ioana Zelina	Technical University of Cluj-Napoca, North University Center in Baia Mare, Romania
Isaac Agudo	University of Malaga, Spain
Isaias Garcia	University of León, Spain
Javier Areitio	University of Deusto, Spain
Jesús Díaz-Verdejo	University of Granada, Spain
Joan Borrell	Universitat Autònoma de Barcelona, Spain
Jose A. Onieva	University of Malaga, Spain
José F. Torres	Pablo de Olavide University, Spain
Jose Luis Calvo-Rolle	University of A Coruña, Spain
Jose Luis Imana	Polytechnic University of Madrid, Spain
José Luis Casteleiro-Roca	University of A Coruña, Spain
Jose M. Molina	University Carlos III de Madrid, Spain
Jose Manuel Gonzalez-Cava	Universidad de La Laguna, Spain
Jose Manuel Lopez-Guede	Basque Country University, Spain
Josep Ferrer	Universitat de les Illes Balears, Spain
Juan Jesús Barbarán	University of Granada, Spain
Juan Pedro Hecht	University of Buenos Aires, Argentina
Leocadio G. Casado	University of Almeria, Spain
Lidia Sánchez-González	University of León, Spain
Luis Hernandez Encinas	ITEFI, Spain
Manuel Grana	University of Basque Country, Spain
Michal Choras	ITTI Ltd., Poland

Paulo Moura Oliveira	UTAD University, Portugal
Paulo Novais	University of Minho, Portugal
Petrica Pop	Technical University of Cluj-Napoca, North University Center at Baia Mare, Romania
Rafael Alvarez	University of Alicante, Spain
Rafael Corchuelo	University of Seville, Spain
Ramón-Ángel Fernández-Díaz	University of León, Spain
Raúl Durán	University of Alcalá, Spain
Robert Burduk	Wroclaw University of Technology, Poland
Roman Senkerik	TBU in Zlin, Czechia
Salvador Alcaraz	Miguel Hernandez University, Spain
Sorin Stratulat	Université de Lorraine, France
Castejon Limas	University of León, Spain
Urko Zurutuza	Mondragon University, Spain
Vicente Matellan	University of León, Spain
Wenjian Luo	University of Science and Technology of China, China
Zuzana Kominkova Oplatkova	Tomas Bata University in Zlin, Czechia

Special Sessions

From the Least to the Least: Cryptographic and Data Analytics Solutions to Fulfil Least Minimum Privilege and Endorse Least Minimum Effort in Information Systems

Program Committee

Víctor Gayoso Martínez (Organizer)	CSIC, Spain
Agustín Martín (Organizer)	CSIC, Spain
Alberto Peinado	Universidad de Malaga, Spain
Amalia Orúe	CSIC, Spain
Araceli Queiruga	Universidad de Salamanca, Spain
David Arroyo	CSIC, Spain
Luis Hernández	Institute of Physical and Information Technologies (ITEFI), Spain
Sara Cardell	UNICAMP Universidade Estadual de Campinas, Brazil
Slobodan Petrovic	Gjøvik University College, Norway

CISIS 2019 Organizing Committee

Francisco Martínez Álvarez	Pablo de Olavide University, Spain
Alicia Troncoso Lora	Pablo de Olavide University, Spain
José F. Torres Maldonado	Pablo de Olavide University, Spain
David Gutiérrez-Avilés	Pablo de Olavide University, Spain
Rubén Pérez Chacón	Pablo de Olavide University, Spain
Ricardo L. Talavera Llames	Pablo de Olavide University, Spain
Federico Divina	Pablo de Olavide University, Spain
Gualberto Asencio Cortés	Pablo de Olavide University, Spain
Miguel García-Torres	Pablo de Olavide University, Spain
Cristina Rubio-Escudero	University of Seville, Spain
María Martínez-Ballesteros	University of Seville, Spain
Álvaro Herrero	University of Burgos, Spain
José Antonio Sáez-Muñoz	University of Salamanca, Spain
Héctor Quintián	University of A Coruña, Spain
Emilio Corchado	University of Salamanca, Spain

ICEUTE 2019

Organization

General Chairs

Francisco Martínez Álvarez	Pablo de Olavide University, Spain
Alicia Troncoso Lora	Pablo de Olavide University, Spain
Emilio Corchado	University of Salamanca, Spain

Program Committee Chairs

Emilio Corchado	University of Salamanca, Spain
Francisco Martínez Álvarez	Pablo de Olavide University, Spain
Alicia Troncoso Lora	Pablo de Olavide University, Spain
Héctor Quintián	University of A Coruña, Spain

Program Committee

Antonio Morales Esteban	University of Seville, Spain
Carlos Cambra	University of Burgos, Spain
Damián Cerero	University of Seville, Spain
Federico Divina	Pablo de Olavide University, Spain
Francisco Gómez-Vela	Pablo de Olavide University, Spain
Francisco Martínez Álvarez	Pablo de Olavide University, Spain
Héctor Quintián	University of A Coruña, Spain
José F. Torres	Pablo de Olavide University, Spain
José Lázaro Amaro Mellado	University of Seville, Spain
Jose Manuel Gonzalez-Cava	University of La Laguna, Spain
Khawaja Asim	Pakistan Institute of Engineering and Applied Sciences (PIEAS), Pakistan
Miguel García-Torres	Universidad Pablo de Olavide, Spain

Special Sessions

Looking for Camelot: New Approaches to Asses Competencies

Program Committee

Araceli Queiruga-Dios (Organizer)	University of Salamanca, Spain
Agustín Martín Muñoz	CSIC, Spain
Ángel Martín del Rey	University of Salamanca, Spain
Ascensión Hernández Encinas	University of Salamanca, Spain
Cristina M. R. Caridade	Coimbra Superior Institute of Engineering, Portugal
Daniela Richtarikova	SjF STU, Slovakia
Deolinda Rasteiro	Coimbra Superior Institute of Engineering, Portugal
Fatih Yilmaz	Ankara Hacı Bayram Veli University, Turkey
Ion Mierlus-Mazilu	Technical University of Civil Engineering of Bucharest, Bucharest
José Javier Rodriguez Santos	University of Salamanca, Spain
José María Chamoso Sánchez	University of Salamanca, Spain
Juan J. Bullon Perez	University of Salamanca, Spain
Luis Hernandez Encinas	Institute of Physical and Information Technologies (ITEFI), Spain
María Jesús Sánchez Santos	University of Salamanca, Spain
María José Cáceres García	University of Salamanca, Spain
Marie Demlova	Czech Technical University in Prague, Czechia
Michael Carr	DIT, Ireland
Snezhana Gocheva-Ilieva	University of Plovdiv Paisii Hilendarski, Bulgaria
Víctor Gayoso Martínez	CSIC, Spain

Innovation in Computer Science Education

Program Committee

Cristina Rubio-Escudero (Organizer)	University of Seville, Spain
María Martínez-Ballesteros (Organizer)	University of Seville, Spain
José C. Riquelme (Organizer)	University of Seville, Spain
Damián Cerero	University of Seville, Spain
Francisco Martínez Álvarez	Universidad Pablo de Olavide, Spain
Héctor Quintián	University of A Coruña, Spain

José F. Torres	Pablo de Olavide University, Spain
Samuel Conesa	Pablo de Olavide University, Spain

ICEUTE 2019 Organizing Committee

Francisco Martínez Álvarez	Pablo de Olavide University, Spain
Alicia Troncoso Lora	Pablo de Olavide University, Spain
José F. Torres Maldonado	Pablo de Olavide University, Spain
David Gutiérrez-Avilés	Pablo de Olavide University, Spain
Rubén Pérez Chacón	Pablo de Olavide University, Spain
Ricardo L. Talavera Llames	Pablo de Olavide University, Spain
Federico Divina	Pablo de Olavide University, Spain
Gualberto Asencio Cortés	Pablo de Olavide University, Spain
Miguel García-Torres	Pablo de Olavide University, Spain
Cristina Rubio-Escudero	University of Seville, Spain
María Martínez-Ballesteros	University of Seville, Spain
Álvaro Herrero	University of Burgos, Spain
José Antonio Sáez-Muñoz	University of Salamanca, Spain
Héctor Quintián	University of A Coruña, Spain
Emilio Corchado	University of Salamanca, Spain

Contents

Special Session: From the Least to the Least: Cryptographic and Data Analytics Solutions to Fulfill Least Minimum Privilege and Endorse Least Minimum Effort in Information Systems

Special Session: Looking for Camelot: New Approaches to Asses Competencies

Special Session: Innovation in Computer Science Education

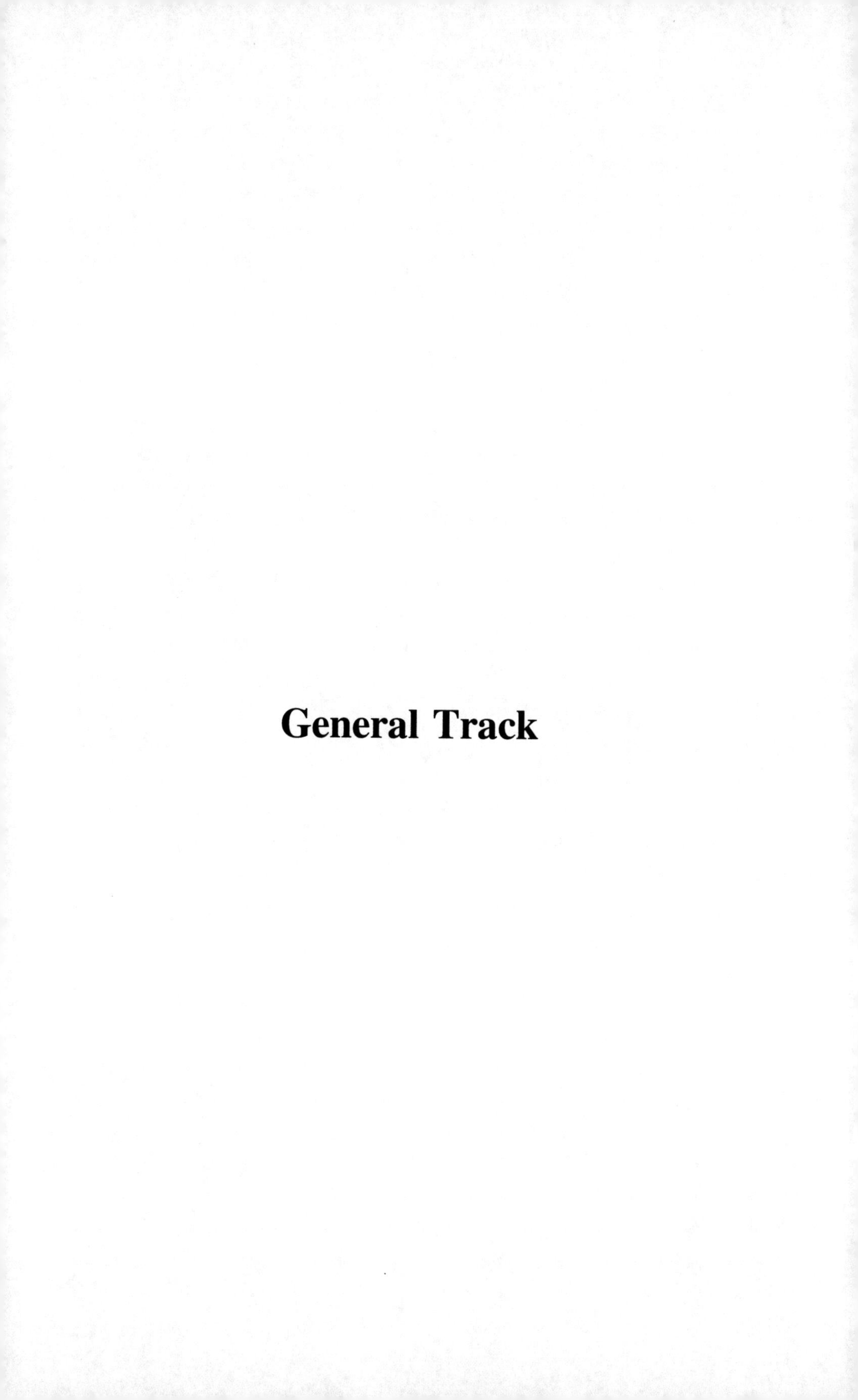

General Track

Perturbing Convolutional Feature Maps with Histogram of Oriented Gradients for Face Liveness Detection

Yasar Abbas Ur Rehman(✉), Lai-Man Po, Mengyang Liu, Zijie Zou, and Weifeng Ou

Department of Electronic Engineering, City University of Hong Kong, Kowloon, Hong Kong
{yaurehman2-c,mengyaliu7-c,zijiezou2-c, weifengou2-c}@my.cityu.edu.hk, eelmpo@cityu.edu.hk

Abstract. Face anti-spoofing in unconstrained environment is one of the key issues in face biometric based authentication and security applications. To minimize the false alarms in face anti-spoofing tests, this paper proposes a novel approach to learn perturbed feature maps by perturbing the convolutional feature maps with Histogram of Oriented Gradients (HOG) features. The perturbed feature maps are learned simultaneously during training of Convolution Neural Network (CNN) for face anti-spoofing, in an end-to-end fashion. Extensive experiments are performed on state-of-the-art face anti-spoofing databases, like OULU-NPU, CASIA-FASD and Replay-Attack, in both intra-database and cross-database scenarios. Experimental results indicate that the proposed framework perform significantly better compare to previous state-of-the-art approaches in both intra-database and cross-database face anti-spoofing scenarios.

Keywords: Convolution Neural Networks · Face liveness detection · Histogram of Oriented Gradients

1 Introduction

Preserving the privacy of individuals using their biometric traits has been a prevailing issue in biometric based user authentication systems. Among various biometric traits like fingerprint, palm vein and iris; face biometric based user authentication systems have been widely used in mobile applications, boarder security, and automatic transactions [1]. However, face biometric based user authentication systems are highly vulnerable to face spoofing attacks, also known as Presentation Attacks (PA). These face PA vary from a simple photographic image to complex and expensive 3D realistic Mask. Due to surge in the availability of high-end cameras, printers and display screens, and the easy access to user face information from social media like Facebook, twitter, WeChat and Instagram etc., face PA could be easily developed with limited resources. As such, it has become indispensable to develop a robust face anti-spoofing system that can detect and classify a range of face PA with high accuracy and low tolerance [2].

© Springer Nature Switzerland AG 2020
F. Martínez Álvarez et al. (Eds.): CISIS 2019/ICEUTE 2019, AISC 951, pp. 3–13, 2020.
https://doi.org/10.1007/978-3-030-20005-3_1

To detect face PA, a wide range of face anti-spoofing (also known as face liveness detection) algorithms and systems have been proposed in literature. These face liveness detection algorithms and systems can be divided into two broader categories, i.e. Hand-crafted features based liveness detection systems [3] and deep Convolution Neural Networks (CNN) [4] based face liveness detection systems. Although, the performance of face liveness detection methods using CNN in intra-database scenarios has been quite remarkable [5], it only imply an upper bound on the performance of face liveness detection method under evaluation. Further, improving the performance of face liveness detection in cross-database domain is still an open research problem [6]. To make matters worse, current face anti-spoofing databases possesses low diversity, number of samples and environmental scenarios. Most face anti-spoofing techniques proposed so far have focused only on liveness detection (binary classification) problem without paying much attention to engineering the hidden layers of CNN classifiers. Recently, CNN classifiers have shown to improve its performance when trained with an additional domain knowledge data [7–9].

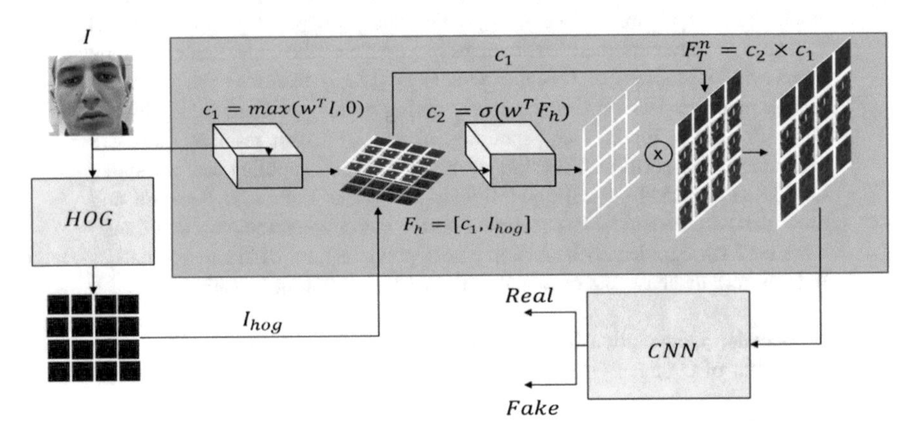

Fig. 1. Proposed approach for perturbing convolutional feature maps with HOG features for face liveness detection

Motivating by these facts, in this paper, we propose a novel approach for face liveness detection by perturbing the hidden convolutional layers of CNN with Histogram of Oriented Gradient (HOG) features. As shown in Fig. 1, before feeding an RGB face image to the CNN classifier, the HOG features of face image are computed first. The HOG features in the form of a 2-dimensional matrix is then concatenated with each output feature map of a candidate convolutional layer of CNN. Afterwards, the concatenated HOG features and convolutional feature maps are passed through a convolution layer to compute a weight matrix. The computed weight matrix is then multiplied with the convolution feature maps of candidate convolution layer to generate perturbed feature maps. These perturbed feature maps are then passed through the rest of the layers in the CNN for face liveness detection.

The main contribution of this paper can be summarized as follows:

- We propose a novel approach for face liveness detection by perturbing the convolutional feature maps with HOG features.
- Further, we propose an end-to-end scheme to simultaneously generate perturbed feature maps and supervise a CNN network using the generated perturbed feature maps for face liveness detection.
- We provide a comprehensive experimental evaluation of the proposed system, evaluating various activation functions for generating perturbing weight matrix and their effect on face liveness detection in general. Further, we evaluate the performance of the proposed system in both intra-database and cross-database face liveness detection scenarios on challenging face anti-spoofing databases like OULU [10], CASIA [10] and Replay-Attack [11].

The rest of this paper has been organized as follows. In Sect. 2, we review the state-of-the-art work done in face liveness detection. In Sect. 3, we discuss our proposed method for face liveness detection. In Sect. 4, we provide experimental results and discussions. Finally in Sect. 5, we conclude the paper with a conclusion and possible future research directions.

2 Literature Review

The remarkable success of CNNs in various computer vision tasks, including classifying face spoofing attacks, have provided a way to categorize face anti-spoofing classifiers into two broader domains, i.e. Hand-crafted features based classifiers and learnable or dynamic features based classifiers. Hand-crafted features based classifiers utilize liveness cues in a face image such as motion [11–13], texture [14–16] and spectral energy contents [17]. Approaches that utilized texture information for face anti-spoofing applications were proved to be quite robust in detection of different types of PA. On the other hand, learnable or dynamic feature based classifiers, like CNNs, learn features directly from the raw data fed to it. Therefore, the features learn by CNN classifiers are dynamic and represent a wide range of patterns in the data as compared to fixed feature based classifiers [18].

In [4], a CNN architecture was proposed for the combined detection of fingerprint, iris and face liveness detection. Although, the authors obtained benchmark accuracy for face liveness detection; no cross-database results for face liveness detection were reported. In [19], an LBP-net, that combine LBP feature maps with CNN, was proposed for classification of live face and PA. In [20], face-depth estimation and individual facial regions like eyes, mouth, nose and eyebrows were utilized using a two stream CNN and an SVM with score level-fusion for face liveness detection. In [21], facial depth maps were obtained from kinetic sensors and combined with texture features extracted from CNN for face liveness detection. In [22], a spatio-temporal representation was learned from the input face sequences by utilizing an LSTM (Long Short Term Memory)-CNN.

In [23], spatio-temporal representation was obtained from the energy representation of each color channel in RGB face images. This spatio-temporal representation of face images was then fed to the CNN classifier for face liveness detection. Recently, in [24], a combination of hand-crafted features like LBP and deep features from CNN were utilized for face liveness detection. To overcome the high-dimensionality problem, a PCA was utilized first before feeding the combined features to an SVM classifier for the detection of PA. Similarly, in [25], the discriminative feature maps at different layers of VGG-Net's [26] were obtained first using PCA. These discriminative feature maps were then utilized to train an SVM classifier for face liveness detection.

3 Methodology

3.1 Generating Perturbed Feature Maps

Let suppose an input face image is represented as I, and a corresponding HOG features in the form of 2D matrix is represented as I_{hog}. Additionally, let the output feature maps of the candidate convolution layer is represented as c_i, where i represents the position of the candidate convolutional layer in a CNN. In this work, we select $i = 1$, (the first layer of the proposed CNN). For generating the perturbed feature maps, we first concatenated each feature maps of c_1 layer with hog features I_{hog}. Assuming there are n feature maps in a candidate convolutional layer, this process generate a $2n$ dimensional hybrid tensor F_h.

$$F_h^n = [[c_1(1), I_{hog}], \ldots, [c_1(n), I_{hog}]] \tag{1}$$

Each nth element of the hybrid tensor F_h is then passed through a convolution layer having a single shared weight matrix w^T and a sigmoid activation. We represent the output of this convolution layer as c_2.

$$c_2 = \sigma([w^T * [c_1(1), I_{hog}], \ldots, w^T * [c_1(n), I_{hog}]]) \tag{2}$$

In Eq. (2), * represents convolutional operation. After obtaining the feature maps from convolutional layer c_2, we calculate the Hadamard product of convolutional feature maps c_1 and c_2 to obtain the perturbed feature maps F_n^T using Eq. (3).

$$F_T^n = c_2 \circ c_1 \tag{3}$$

The perturbed feature maps F_n^T encapsulating both convolutional features and HOG features are passed through the rest of convolution layers in the CNN in the usual fashion for predicting the input face image being live or PA. Figure 2 shows an example of face samples from OULU-NPU face anti-spoofing database and corresponding class activation maps from layer c_1 and corresponding perturbed class activation maps. A quick comparison between the class activation maps from layer c_1 and the perturbed class activation maps indicated that the perturbed class activation maps boost the response of the convolutional feature maps in the discriminative regions selected by convolution layer c_1.

3.2 CNN Architecture

The proposed CNN has been shown in Table 1. The input to the proposed CNN is real world face image I and corresponding HOG features I_{hog}. As shown in Table 1, the output from the convolution layer conv_d1 is concatenated with input face image HOG features I_{hog}. Further, to ease the flow of gradient across the CNN architecture, we further map the 3 convolution layers, following the perturbation layer, using 1×1 convolution layers. At the end of the CNN architecture, all the feature maps from convolution layers 4, 5, 6 and 7 are concatenated followed by global average pooling layer that average all the feature maps and provide an output vector, which was then fed to the fully-connected layer with 2-way soft-max activation. Since global average pooling has no parameter to learn, a direct relationship can be established between the convolution layers and output of soft-max. We further used a dropout of 0.2 after each max-pooling layer and regularization factor of 0.0005 in each convolution layer except for the perturbation layer c_2.

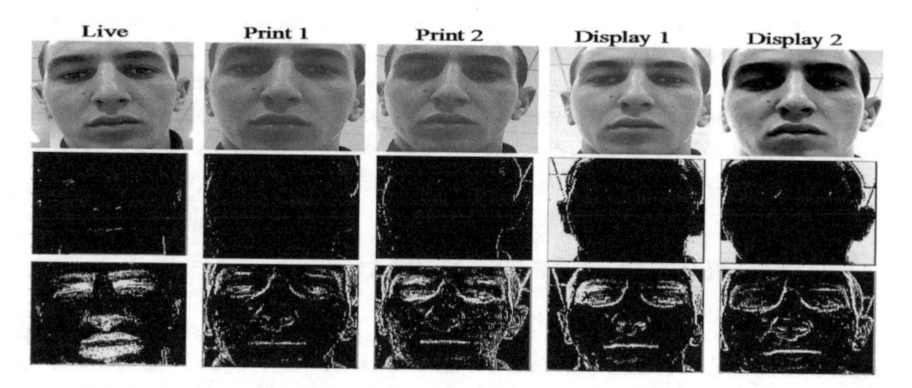

Fig. 2. Top row: Samples of face images from OULU face anti-spoofing database. Middle row: Corresponding convolutional layer c_1 feature maps and Bottom row: Corresponding perturbed feature maps F^n_T. The convolution feature maps and perturbed feature maps are represented in binary format by thresholding every pixel $p > 0.5$.

Table 1. Configuration of proposed CNN architecture

Layer name	Kernel size	Output channel	Input
Conv_d1	3×3	16	I
Perturbation layer	3×3	16	$[I_{hog}$, Conv_d1]
Conv_1	3×3	32	**Perturbation layer**
Max-pool_1	2×2	32	Conv_1
Conv_2	3×3	64	Max-pool_1
Max-pool_2	2×2	64	Conv_2
Conv_3	3×3	128	Max-pool_2

(*continued*)

Table 1. (*continued*)

Layer name	Kernel size	Output channel	Input
Max-pool_3	2 × 2	128	Conv_3
Conv_4	1 × 1	2	Max-pool_3
Conv_5	1 × 1	2	Conv_1
Conv_6	1 × 1	2	Conv_2
Conv_7	1 × 1	2	Conv_3
F1 = concatenate [Conv_4, Conv_5, Conv_6, Conv_7]			
Global average pooling	-	8	F1
Fc1	10	2	Global average pooling
2 way soft-max			

3.3 Training

We train the proposed CNN for a total of 20 epochs. The initial learning rate was set to 0.01, which was reduced by a factor of 0.1 after 10th and 15th epoch. The batch-size was set to 32. Before feeding the training data to the proposed CNN, samples in the training data were randomly shuffled. The proposed network took approximately 40 min to train on GTX 1080 GPU. Each epoch took approximately between 120 s to 128 s depending on the size of the input face image.

4 Experimental Results and Discussions

We first analyze various activation functions for generating perturbed feature maps weight matrix and their effect on face liveness detection performance. Afterwards, we present the analysis and discussion of the proposed system in intra-database and cross-database face liveness detection tests. Further we evaluated the performance of the proposed system using Half Total Error Rate (HTER), Equal Error Rate (EER), Bona Fide Presentation Classification Error Rate (BPCER), Attack Presentation Classification Error Rate (APCER) and Average Classification Error Rate (ACER). For intra-database evaluation, we utilized the BPCER, APCER and their average ACER metric. For cross-database evaluation, we utilized HTER value. Since HTER is threshold dependent, the threshold computed at EER point on the development set was used calculate the HTER on the database under consideration.

4.1 Effect of Activation Functions on Perturbed Feature Maps

We first analyzed the effect of utilizing various activation functions in generating perturbed feature maps for face liveness detection. Particularly, we evaluated the effect of three activation functions i.e., rectified linear unit (ReLU), exponential linear unit (ELU) and sigmoid activation function. Table 2 shows the performance of the final CNN classifier on face liveness detection using each activation function. It can be clearly seen in Table 2 that the use of sigmoid activation function provide best performance compared to using other activation functions. Further, we found that sigmoid activation function weight the pixels in the convolutional feature maps according to their magnitude.

4.2 Intra-database Face Liveness Detection

For the intra-database face liveness detection analysis, we utilized the BPCER, APCER and their average ACER metric. The operating threshold value for the test set was determined using the EER point on the development set. Table 3 shows the performance of the proposed system on intra-database face liveness detection tests. We represent the CNN classifiers without using the perturbation as "conventional" and the one using perturbation as "proposed" from now on. It can be seen in Table 3 that the proposed approach obtain a lowest ACER of 4.74%, 1.08% and 0.17% as compared to obtaining an ACER of 7.13%, 1.17% and 0.60% using conventional approach on OULU, CASIA and Replay-Attack face anti-spoofing databases. This indicated that the utilization of perturbed feature maps add new information to the following convolution layers of the CNN to leverage the classification decision of an input face image being considered as live or fake.

Table 2. Performance in (%) of using various activation functions in the perturbation layer

Activation	OULU (Development)			OULU (Test)		
	BPCER	APCER	ACER	BPCER	APCER	ACER
ReLU	6.33	1.59	3.96	8.97	3.43	6.20
ELU	6.14	1.51	3.83	9.10	3.18	6.14
Sigmoid	**5.63**	**1.39**	**3.51**	**7.40**	**2.07**	**4.74**

Table 3. Performance in (%) of conventional and proposed approach on intra-database face anti-spoofing tests

Method	OULU (Development)			OULU (Test)		
	BPCER	APCER	ACER	BPCER	APCER	ACER
Conventional	6.44	1.65	4.03	11.36	2.90	7.13
Proposed	**5.63**	**1.39**	**3.51**	**7.40**	**2.07**	**4.74**
				CASIA (Test)		
Conventional				2.63	0.79	1.71
Proposed				**1.60**	**0.58**	**1.08**
	Replay-attack (Development)			**Replay attack (Test)**		
Conventional	**1.23**	0.69	**0.96**	0.93	0.28	0.60
Proposed	2.35	**0.41**	1.38	**0.04**	**0.29**	**0.17**

4.3 Cross-Database Face Liveness Detection

For the cross-database analysis, we trained the face anti-spoofing system on one database and tested it on the other database. The operating threshold was determined by development set or the testing set of the face anti-spoofing database on which the face anti-spoofing system was trained on. Table 4 shows the performance of the conventional approach and the proposed approach on cross-database face anti-spoofing tests. It can be seen in Table 4, that the proposed method obtain better performance in cross-database tests when trained on OULU face anti-spoofing database as compared to conventional approach. When trained with OULU face anti-spoofing database, the proposed method obtain an all-time lower HTER of 9.62% and 10.31% on CASIA and Replay-Attack database as compared to the conventional approach.

On the other hand, the conventional approach obtain better performance when trained on CASIA face anti-spoofing database. Further, when trained on Replay-Attack database, we found that the conventional approach and the proposed approach obtain comparable performance. Further analyses of the CNN classifier trained with CASIA and Replay-Attack face anti-spoofing database in the Table 4 show higher percentages of HTER as compared to the CNN classifier trained with OULU face anti-spoofing database. This suggest that there is a certain bias in these database toward certain types of PA.

Table 4. HTER in (%) on cross-database tests

Training set	Database	Conventional	Proposed
OULU	CASIA	11.10	**9.62**
	Replay-Attack	17.83	**10.31**
CASIA	OULU	**18.32**	26.82
	Replay-Attack	**27.70**	31.54
Replay-Attack	OULU	25.88	**24.90**
	CASIA	**11.55**	11.76

4.4 Comparison with State-of-the-Art Approaches

We further compared the cross-database face anti-spoofing performance, in terms of % HTER, of the proposed method with state-of-the-art face anti-spoofing approaches in Table 5. It can be seen in Table 5 that the proposed method, when trained on OULU database, obtain an all-time lower HTER of 9.62% and 10.31% on CASIA and Replay Attack databases. However, when trained on CASIA database, we found that the proposed method obtain an HTER of 31.54% on Replay Attack database. The reason of this degradation in HTER is because the CASIA database was collected under uniform capturing (illumination) conditions compared to OULU and Replay Attack databases, which were collected under different capturing environments.

Table 5. Comparison of the proposed method and state-of-the-art face liveness detection methods. HTER in (%) on cross-database scenarios

Method	CASIA*	Replay-attack**
Pinto et al. [17]	50.0	34.4
Siddiqui et al. [27]	44.6	35.4
Boulkenafet et al. [15]	37.7	30.3
Li et al. [6]	36.0	27.4
Manjain et al. [28]	27.4	22.8
Proposed	**11.76**	**31.54**
Proposed†	**9.62**	**10.31**

* Train set: Replay-Attack
** Train set: CASIA
† Train set: OULU

5 Conclusion and Future Work

In this paper, we proposed a novel approach to face liveness detection by introducing perturbation in the convolutional feature maps. The proposed perturbation approach added only a minor parameter overhead in the CNN, while achieving significant performance improvement for face liveness detection task both in intra-database and cross-database scenarios. Further, we found that the proposed method, when trained on challenging face anti-spoofing database, like OULU, further improved the performance of face liveness detection in cross-database face anti-spoofing scenarios. This further demands that the face anti-spoofing databases should contain different variations in not only in imaging qualities and types of PA instruments, but also in environmental conditions to further facilitate the development of face anti-spoofing research.

Future work include evaluation of perturbing the convolutional layers of CNN using various hand-crafted features, such as Local Binary Patterns (LBP), Shearlet features and wavelet features. We believe that the information induced by these hand-crafted features in a CNN can further improve its robustness, efficiency and reliability in face liveness detection applications.

Acknowledgements. The work in this paper was supported by City University of Hong Kong under the research project with grant number 7004430.

References

1. Galbally, J., Marcel, S., Fierrez, J.: Biometric antispoofing methods: a survey in face recognition. IEEE Access **2**, 1530–1552 (2014)
2. Sepas-Moghaddam, A., Pereira, F., Correia, P.L.: Light field based face presentation attack detection: reviewing, benchmarking and one step further. IEEE Trans. Inf. Forensics Secur. **13**(7), 1696–1709 (2018)
3. Kim, W., Suh, S., Han, J.-J.: Face liveness detection from a single image via diffusion speed model. IEEE Trans. Image Process. **24**, 2456–2465 (2015)

4. Menotti, D., Chiachia, G., Pinto, A., Schwartz, W.R., Pedrini, H., Falcao, A.X., Rocha, A.: Deep representations for iris, face, and fingerprint spoofing detection. IEEE Trans. Inf. Forensics Secur. **10**, 864–879 (2015)
5. Rehman, Y.A.U., Po, L.M., Liu, M.: LiveNet: improving features generalization for face liveness detection using convolution neural networks. Expert Syst. Appl. **108**, 159–169 (2018). https://doi.org/10.1016/j.eswa.2018.05.004
6. Li, H., Li, W., Cao, H., Wang, S., Huang, F., Kot, A.C.: Unsupervised domain adaptation for face anti-spoofing. IEEE Trans. Inf. Forensics Secur. **13**, 1794–1809 (2018)
7. Yang, J., Lei, Z., Li, S.Z.: Learn convolutional neural network for face anti-spoofing (2014). arXiv Preprint: arXiv:1408.5601
8. Yang, J., Lei, Z., Yi, D., Li, S.Z.: Person-specific face antispoofing with subject domain adaptation. IEEE Trans. Inf. Forensics Secur. **10**, 797–809 (2015)
9. Li, L., Xia, Z., Hadid, A., Jiang, X., Roli, F., Feng, X.: Face presentation attack detection in learned color-liked space, pp. 1–13. arXiv:1810.13170v1
10. Boulkenafet, Z., Komulainen, J., Li, L., Feng, X., Hadid, A.: OULU-NPU: a mobile face presentation attack database with real-world variations. In: Proceedings of the 12th IEEE International Conference on Automatic Face and Gesture Recognition (FG 2017), pp. 612–618. IEEE (2017)
11. de Freitas Pereira, T., Komulainen, J., Anjos, A., De Martino, J.M., Hadid, A., Pietikäinen, M., Marcel, S.: Face liveness detection using dynamic texture. EURASIP J. Image Video Process. **2014**, 2 (2014)
12. Kim, Y., Yoo, J.-H., Choi, K.: A motion and similarity-based fake detection method for biometric face recognition systems. IEEE Trans. Consum. Electron. **57**, 756–762 (2011)
13. Anjos, A., Chakka, M.M., Marcel, S.: Motion-based counter-measures to photo attacks in face recognition. IET Biom. **3**, 147–158 (2013)
14. Wen, D., Han, H., Jain, A.K.: Face spoof detection with image distortion analysis. IEEE Trans. Inf. Forensics Secur. **10**, 746–761 (2015)
15. Boulkenafet, Z., Komulainen, J., Hadid, A.: Face spoofing detection using colour texture analysis. IEEE Trans. Inf. Forensics Secur. **11**, 1818–1830 (2016)
16. Gragnaniello, D., Poggi, G., Sansone, C., Verdoliva, L.: An investigation of local descriptors for biometric spoofing detection. IEEE Trans. Inf. Forensics Secur. **10**, 849–863 (2015)
17. Pinto, A., Pedrini, H., Schwartz, W.R., Rocha, A.: Face spoofing detection through visual codebooks of spectral temporal cubes. IEEE Trans. Image Process. **24**, 4726–4740 (2015)
18. Krizhevsky, A., Sutskever, I., Hinton, G.E.: ImageNet classification with deep convolutional neural networks. In: Advances in Neural Information Processing Systems, pp. 1097–1105 (2012)
19. de Souza, G.B., da Silva Santos, D.F., Pires, R.G., Marana, A.N., Papa, J.P.: Deep texture features for robust face spoofing detection. IEEE Trans. Circuits Syst. II Express Briefs **64**, 1397–1401 (2017). https://doi.org/10.1109/tcsii.2017.2764460
20. Atoum, Y., Liu, Y., Jourabloo, A., Liu, X.: Face anti-spoofing using patch and depth-based CNNs. In: Proceedings of the IEEE International Joint Conference on Biometrics (IJCB), pp. 319–328 (2017)
21. Wang, Y., Nian, F., Li, T., Meng, Z., Wang, K.: Robust face anti-spoofing with depth information. J. Vis. Commun. Image Represent. **49**, 332–337 (2017)
22. Xu, Z., Li, S., Deng, W.: Learning temporal features using LSTM-CNN architecture for face anti-spoofing. In: 2015 3rd IAPR Asian Conference on Pattern Recognition (ACPR), pp. 141–145 (2015)
23. Lakshminarayana, N.N., Narayan, N., Napp, N., Setlur, S., Govindaraju, V.: A discriminative spatio-temporal mapping of face for liveness detection. In: 2017 IEEE International Conference on Identity, Security and Behavior Analysis (ISBA), pp. 1–7 (2017)

24. Nguyen, D.T., Pham, T.D., Baek, N.R., Park, K.R.: Combining deep and handcrafted image features for presentation attack detection in face recognition systems using visible-light camera sensors. Sensors (Switzerland) **18** (2018). https://doi.org/10.3390/s18030699
25. Li, L., Feng, X., Boulkenafet, Z., Xia, Z., Li, M., Hadid, A.: An original face anti-spoofing approach using partial convolutional neural network. In: 2016 6th International Conference on Image Processing Theory Tools and Applications (IPTA), pp. 1–6 (2016)
26. Simonyan, K., Zisserman, A.: Very deep convolutional networks for large-scale image recognition (2014). arXiv Preprint: arXiv:1409.1556
27. Siddiqui, T.A., Bharadwaj, S., Dhamecha, T.I., Agarwal, A., Vatsa, M., Singh, R., Ratha, N.: Face anti-spoofing with multifeature videolet aggregation. In: 2016 23rd International Conference on Pattern Recognition (ICPR), pp. 1035–1040 (2016)
28. Manjani, I., Tariyal, S., Vatsa, M., Singh, R., Majumdar, A.: Detecting silicone mask based presentation attack via deep dictionary learning. IEEE Trans. Inf. Forensics Secur. **12**, 1713–1723 (2017)

Comparison of System Call Representations for Intrusion Detection

Sarah Wunderlich[1(✉)], Markus Ring[1], Dieter Landes[1], and Andreas Hotho[2]

[1] Coburg University of Applied Sciences and Arts, Coburg, Germany
sarah.wunderlich@hs-coburg.de
[2] Data Mining and Information Retrieval Group, University of Würzburg, Würzburg, Germany
http://www.hs-coburg.de/cidds

Abstract. Over the years, artificial neural networks have been applied successfully in many areas including IT security. Yet, neural networks can only process continuous input data. This is particularly challenging for security-related non-continuous data like system calls. This work focuses on four different options to preprocess sequences of system calls so that they can be processed by neural networks. These input options are based on one-hot encoding and learning word2vec and GloVe representations of system calls. As an additional option, we analyze if the mapping of system calls to their respective kernel modules is an adequate generalization step for (a) replacing system calls or (b) enhancing system call data with additional information regarding their context. However, when performing such preprocessing steps it is important to ensure that no relevant information is lost during the process. The overall objective of system call based intrusion detection is to categorize sequences of system calls as benign or malicious behavior. Therefore, this scenario is used to evaluate the different input options as a classification task. The results show that each of the four different methods is valid when preprocessing input data, but the use of kernel modules only is not recommended because too much information is being lost during the mapping process.

Keywords: HIDS · System calls · Intrusion detection · LSTMs

1 Introduction

In the age of digitization, many process flows involving personal or otherwise critical data have been digitized. Hence, it is more important than ever to protect data against unauthorized access and a lot of research on intrusion detection has been done over the past years. An obvious approach for this task is to monitor the system calls of different processes running on a host. Since standard application programs are running in user mode, they are not allowed to access the resources of an operating system on their own. Even for simple activities (like reading or writing a file), programs need to make a request to the operating system's kernel

© Springer Nature Switzerland AG 2020
F. Martínez Álvarez et al. (Eds.): CISIS 2019/ICEUTE 2019, AISC 951, pp. 14–24, 2020.
https://doi.org/10.1007/978-3-030-20005-3_2

in the form of system calls. Hence, if a program is exploited, every action the exploit takes will also be mirrored within the system call trace of the program. Consequently, system calls have widely been used as data source for security-critical events in intrusion detection [1–6]. A simplified approach is using the kernel modules to which the system calls belong rather than the system calls [7].

Problem Setting. The increasing interest in (deep) neural networks has also reached the area of IT security in recent years. Neural networks, however, can only process continuous input data. Yet, many security-related data like system calls are non-continuous which constitutes a major limitation to the application of neural networks in this area. In particular, neural networks cannot be directly applied to system calls as these do not conform to the expected input formats.

Objective. We intend to (pre–)process system calls in such a way that they can be analyzed through neural networks. Preprocessing can be accomplished in many ways. Complex structures like language demand a more sophisticated preprocessing step since, e.g., context information can play a decisive role for understanding a sentence. The field of natural language processing (NLP) has different approaches for inserting context into a word. Since logs of system calls are basically textual data where temporal ordering is relevant, we want to investigate the suitability of different approaches from NLP to transform system calls into continuous input data for neural networks.

Approach and Contribution. Benign and malicious behavior of processes can be analyzed by looking at sequences of system calls. Following this line, this work explores Long Short Term Memory Networks (LSTMs) for processing sequences of system calls. LSTMs are a type of neural networks which are able to process sequence data [8]. As mentioned above, preprocessing system calls such that they can be processed by neural networks is a crucial issue. In particular, it should be ensured that no relevant information is lost during the preprocessing step. We examine four preprocessing approaches from NLP, namely (1) one-hot encoding, (2) expanding the original network structure by an additional embedding layer and learning (3) word2vec and (4) GloVe representations of system calls prior to the classification task. Evaluation uses the ADFA-LD dataset [9].

As an alternative to using the original system calls, several generalizations might be applicable. For instance, system calls can be mapped to the kernel modules in which they are defined in. The four different representations used on the raw system call data are also applied to this kernel module representation. Because the respective kernel modules may contain additional information about the relationship between different system calls, a combination of system calls and kernel modules is also examined as input.

The paper's main contribution is a systematic evaluation of different preprocessing methods for system calls such that they can be used in LSTMs for detecting security critical events (malicious behavior). To the best of our knowledge, we are the first to test GloVe for system call representation and also to use a combination of system calls and kernel modules as input options.

Structure. The rest of this paper is organized as follows. Section 2 discusses related approaches for intrusion detection using system calls. Section 3 provides details on the four preprocessing approaches analyzed in this work. Section 4 presents available datasets and their properties. Experiments and results are presented in Sects. 5. Finally, Sect. 6 summarizes this work.

2 Related Work

This Section focuses on how sequences of system calls can be interpreted for intrusion detection and how the application of neural networks puts even more emphasis on the proper preprocessing of input data.

Starting with the work of Forrest et al. [2], the use of system call sequences for intrusion detection has gained considerable interest. Forrest et al. interpret system calls as categorical values and compare them with respect to equality or inequality. They use a sliding window across system call traces of benign programs to identify the relative positions of system calls to each other and store them in a database. If a program is run, its trace is compared to the relative positions from the database. Accumulated deviations between the current trace and database data which exceed a threshold may indicate an attack.

Ever since a lot of research has been based on the use of system call sequences for intrusion detection. During the last decade, methods shifted from Hidden Markov Models [10–12] and Support Vector Machines [13,14] towards various kinds of neural networks [1,4,15]. As neural networks can only process numerical data, a lot of effort was put into the meaningful transformation of system call sequences into numerical representations. Popular methods from NLP were applied to represent the traces, since temporal ordering of input data is also very important for analyzing language.

Creech et al. [1] create semantic models based on context-free grammars. Similar to Forrest [2], they create a database of normal behavior, but then use counting to generate input features for a decision engine. The database is built by forming a dictionary containing every contiguous trace by using multiple sliding window sizes, denoting the resulting traces as words. Those words (dictionary entries) are again combined to form phrases of length one to five. Finally, the counts of actual occurrences of those phrases are fed into different decision engines, e.g. a neural network called Extreme Learning Machine (ELM). Xie et al. [16], however, found that learning the dictionary is extremely time consuming, in particular an entire week for the ADFA-LD dataset.

Murtaza et al. [7] aim at reducing the execution time of anomaly detectors by representing system call traces as sequences of kernel module interactions. They map each single system call to its corresponding kernel module, thus reducing the range of input values from approximately 300 system calls to eight kernel modules. This approach achieves similar or fewer false alarms and is much faster. Yet, the original seven kernel modules of a 32 bit Unix system (architecture, file systems, inter process communication, kernel, memory management, networking, and security) need to be extended by another kernel module (unknown) to capture all possible system calls, since the system call table the authors use does not comprise every system call within the ADFA-LD.

While neural networks are widely used as classifiers, they can also be employed to extract a meaningful representation of the input data. Two of the most popular approaches up to date are word2vec [17] and GloVe [18]. Both methods are heavily used in NLP as they can extract a vector representation of words regarding their context of use. Variants of word2vec have already been used by Ring et al. [19] for network-based intrusion detection.

Following a similar idea, Kim et al. [4] combine an additional embedding layer with a LSTM-based sequence predictor. The authors state that through using an additional fully-connected layer in front of a LSTM layer, the sequence fed into the neural network becomes embedded to continuous space in a first step. Since the authors use this LSTM-based approach for sequence prediction on the ADFA-LD dataset as well, we adopt the idea of adding an additional embedding layer as one potential (built-in) preprocessing approach. The single system calls become embedded before reaching the LSTM layer.

3 Comparison of System Call Representations

The overall purpose of this work is to compare four different methods of system call representation such that neural networks can process them. Since isolated system calls do not contain any clue regarding their intent (benign or malicious), it is necessary to take their relationships into account by analyzing sequences of system calls.

Consequently, a simple LSTM classifier is used for comparing the four input methods. The LSTM classifier used, in its original definition, consists of an input layer, a LSTM layer, a fully connected (FC) layer and an output layer. Figure 1 on the left shows the used network structure for three of the four methods being compared. The right side shows the network structure for the fourth method.

The main contribution is the comparison of system call representation techniques, in particular widely used methods from NLP. The use of a LSTM for classification is a mere tool to show the effect different input methods have on the intent of a sequence (benign or malicious). Again, our approach should not be seen as a solution to intrusion detection, but rather as a systematic comparison of different input possibilities.

Like Murtaza et al. [7], we map system calls to their corresponding kernel modules. Since ADFA-LD was recorded on an Ubuntu 11.04 32 bit operating system, all system calls can be assigned to their respective kernel modules by looking into the system call implementation in the source files of kernel 2.6.35. Thus, in contrast to [7], no unknown kernel module is needed.

With this, three forms of input to a neural network emerge for a comprehensive comparison: Using (1) only the system call representation, (2) only the kernel module representation or (3) both combined. Consequently, this paper compares twelve different input variants of four categories as discussed below.

One-Hot Encoding. One-hot encoding is the typical approach for feeding categorical data into a neural network. Each system call is represented with a vector in which every position represents a specific system call. Hence, the size of the

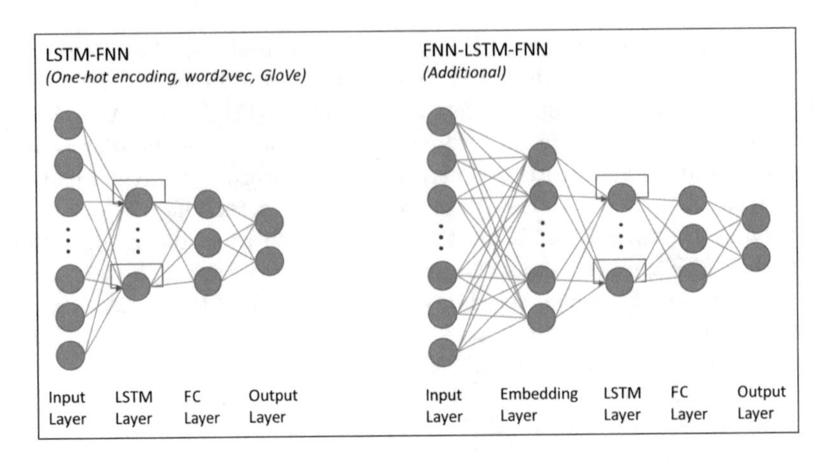

Fig. 1. Network structures used.

input vector equals the number of different system calls. A particular system call is mapped to a binary vector of all zeros except for a single one at the appropriate position for the corresponding call. This method of representation will be tested using three forms: (1) The system calls in one-hot encoding. (2) The mapped kernel modules in one-hot encoding. (3) Both one-hot vectors (the system call vector of size 341 and the module vector of size 7) concatenated. Any of these input techniques will be referred as *One-hot* in the following.

Additional Embedding Layer. This category accords with *one-hot encoding*, except for an additional (fully-connected) embedding layer between input and LSTM layers (see right part of Fig. 1). Thus, the input will be embedded to a continuous vector representation using the embedding layer in front of the LSTM layer. This approach is inspired by a similar network structure for sequence prediction [4]. As before, the following three forms of input will be used: (1) The system calls in one-hot encoding. (2) The mapped kernel modules in one-hot encoding. (3) Both one-hot vectors (the system call vector of size 341 and the module vector of size 7) concatenated. These input techniques using the additional embedding layer, will be referred to as *Additional* in this work.

Word2vec. One-hot encoding has still many meaningful applications and is very popular due to its simplicity, but cannot cope with more complex structures like natural language text since the meaning of words may depend on their context. Mikolov et al. [17] presented an approach now known as word2vec which generates word vectors based on the context in which they are used. Word vectors may be used following two approaches, namely the Continuous Bag-of-Words model (CBOW) and the Skip-Gram model. CBOW learns to predict target words from given context (e.g. the context being 'Molly is already at' and the target word being 'home'). The Skip-Gram model works the other way around, predicting context from a target word. The basic idea of word2vec is to train a neural network, discard the model, but use the weights of the fully trained hidden layer as word vectors.

Since system calls may also vary in their intent (benign or malicious) given their context around them, we adapt this NLP approach using the CBOW model. Consequently, this category consists of the following three input methods: (1) The system calls in word2vec representation. (2) The mapped kernel modules in word2vec representation. (3) The concatenation of system calls in word2vec representation and one-hot encoded kernel modules. This input category will be referred to as *word2vec* in this work.

GloVe. GloVe [18] is a count-based model that also learns a vector representation of words regarding their context. For GloVe, context amounts to a co-occurrence matrix, thus including word statistics into their model. GloVe is then trained with the non-zero entries of that co-occurrence matrix. As with word2vec, this category consists of the three input methods: (1) The system calls in GloVe representation. (2) The mapped kernel modules in GloVe representation. (3) The concatenation of system calls in GloVe representation and one-hot encoded kernel modules. These three forms of input will be referred to as *GloVe* in the remaining chapters.

4 Data

Three well-known datasets, namely the DARPA [20], UNM [21] and ADFA-LD [9], are very popular for evaluating and comparing host-based intrusion detection systems. A typical system call sent to the kernel consists of its name, parameters and a return value. Since number and type of the transfer values vary among system calls, many researchers focus on analyzing only the temporal ordering of system calls, i.e. their names, but neglecting parameters to reduce complexity [1,2,16,22]. Hence, popular datasets like the UNM and ADFA-LD only take the system calls names into account.

The DARPA and UNM datasets are already more than 15 years old, but still in use due to a lack of better alternatives. However, they are also heavily criticised due to their age and lack of complexity [23]. Creech et al. [9] recorded the ADFA-LD dataset in order to replace the outdated DARPA (KDD) collection. Hence we focus solely on the newer ADFA-LD set recorded in 2013.

The ADFA-LD dataset consists of three parts named *Training_Data_Master*, *Attack_Data_Master* and *Validation_Data_Master*, which contain files with system call traces of processes. Table 1 shows the number of system calls and traces contained in the three dataset partitions together with the type of behavior.

Table 1. Number of system calls and traces in ADFA-LD.

	System calls	Traces	Label of traces
Training	308077	833	Benign
Attack	317388	746	Malicious
Validation	2122085	4372	Benign

Only the *Attack_Data_Master* contains exploited processes. In order to use the ADFA-LD dataset for classification, the attack subset needs to be split, mixing one half with the training and the other half with the validation subset. Since training and attack subsets encompass roughly the same number of system calls, the new combined training dataset is skewed with respect to its class label after splitting the attack subset. The new set contains around twice as many benign sequences as malicious sequences. This is even worse in the validation data set since splitting results in a combination of the original roughly 2000000 system calls from benign traces and approximately 150000 system calls from malicious traces. If such a skewed model classifies every input as benign, an accuracy of 93% will be achieved even without any attacks detected. Working with these unbalanced sets demands two adaptations: (1) To avoid a resulting model which tends to classify sequences as benign, the new training set is balanced by using data point duplication in the less represented class. This technique is called random oversampling [24]. (2) Since the new combined validation dataset is extremely unbalanced, accuracy is no clear indicator towards the quality of the sequence classifier. Therefore, true positive rates and false positive rates (TPR and FPR) are used as evaluation criteria.

Training the LSTM model using a whole process trace file does not make much sense due to the nature of attacks. An intrusion detection system should not wait until an exploit reaches its end, but rather intervene early, e.g. by stopping the exploited process. A typical approach (see Sect. 2) is to use sliding windows to split the dataset into smaller sequences for training, attaching the corresponding label of the trace. A process trace is divided into smaller sequences that can be processed without waiting for the process to end. In this work, a sliding window size of 20 is used due to the structure of the ADFA-LD. The *Attack_Data_Master* consists of traces of exploited processes. Yet, not every partial sequence exhibits malicious behavior, but it still would be labeled by the overall intent of the process (malicious). For example, a process may start normally and an attack could exploit a vulnerability near the end of the trace. So, to reduce the risk of learning malicious labels for benign sequences we refrain from using typical smaller window sizes like 5, 6 or 11 as used in [2].

5 Experiments

5.1 Experiment Setup

Table 2 shows the parameters used for the neural networks. For a fair comparison, the same overall parameters are used in all settings. Embedding sizes are generally set to a size of 8 except when using kernel modules only. Since there are only seven different kernel modules compared to 341 different system calls or 348 system calls and kernel modules combined, using identical sizes for embedding, LSTM and fully-connected (FC) layer is not reasonable. Thus, an embedding size and FC size of 3 is used when working with kernel modules only. In Table 2, text embedding parameters like the vector size of word2vec or GloVe are indicated as TE (text embedding). Text embeddings are learned for 10 epochs.

Table 2. Parameterset (N.A. – not applicable).

Source	Method	TE size	TE window	Embed size	LSTM size	FC size
System calls	One-Hot	N.A	N.A	N.A	32	16
	Additional	N.A	N.A	8	32	16
	Word2vec	8	5	N.A	32	16
	GloVe	8	5	N.A	32	16
Kernel modules	One-Hot	N.A	N.A	N.A	5	3
	Additional	N.A	N.A	3	5	3
	Word2vec	3	5	N.A	5	3
	GloVe	3	5	N.A	5	3
Both	One-Hot	N.A	N.A	N.A	32	16
	Additional	N.A	N.A	8	32	16
	Word2vec	8	5	N.A	32	16
	GloVe	8	5	N.A	32	16

Dropout is a general strategy to avoid overfitting in neural networks [25], whereas peepholes are an optimization specifically for the timing in LSTM cells [26]. Here, dropout is set using a keeping probability of 0.8 for all experiments on the final FC layer, and peepholes in the LSTM layers are set to true.

As mentioned in Sect. 4, we use a sequence length of 20. For fair comparison, each model is trained the same amount of epochs (20). Naturally, the number of classes (benign, malicious) implies a size of 2 for the output layer.

5.2 Evaluation of Results

Table 3 shows the experiment results. Since the dataset is skewed in its class distribution, higher accuracy does not necessarily mean a better result. As explained above, true positive rates (TPR) and false positives rates (FPR) might be more meaningful. In our setting, a true positive is a correctly detected malicious sequence while a false positive is a normal sequence that is classified to be malicious. To reduce confounding effects, we conduct our experiments twice on the ADFA-LD dataset with different splits. Table 3 shows the mean results of both experiments.

There is always the risk that valuable information regarding the intent of a sequence (benign or malicious) is lost during preprocessing steps. In our case, this does not seems to be the case for all four transformation methods (one-hot, additional, word2vec or GloVe). Nevertheless, one-hot encoding, which is the most direct transformation of the non-continuous system calls, achieves the best results with a TPR of 0.95 and a FPR of 0.16.

Our results are in line with other intrusion detection based results on this dataset. The results of Creech et al. [1] show a TPR of approximately 90% at a

Table 3. Results: true positive rate/false positive rate (accuracy).

	One-hot	Additional	Word2vec	GloVe
System calls	0.95/0.16 (0.85)	0.90/0.16 (0.85)	0.92/0.17 (0.84)	0.79/0.14 (0.85)
Kernel modules	0.80/0.24 (0.77)	0.89/0.25 (0.75)	0.77/0.25 (0.76)	0.77/0.24 (0.77)
Both	0.95/0.16 (0.86)	0.93/0.17 (0.84)	0.91/0.16 (0.85)	0.87/0.16 (0.84)

FPR of around 15%. Xie et al. [16] achieve a TPR of 70% at a FPR of around 20%. It should however be noted, that those two approaches are anomaly-based (hence they can not be directly compared to our classification-based results), but even if the setting is not comparable, the similar range shows that the representations work quite well.

However, using kernel modules only results in a considerably lower TPR and higher FPR in comparison to our other approaches and is, thus, not advised to use in this classification setting. Surprisingly, enriching the system call data with their corresponding kernel modules does not yield better results. Also, neither the additional embedding layer nor the pre-learned NLP representations show better results than one-hot encoding without the additional embedding layer. This might be because of the nature of LSTM cells. Since LSTM cells automatically embed sequences of data, additional embedding prior to the LSTM might not be helpful. However, aside from using kernel modules only, the overall information regarding the intent of the sequences is being kept to a certain extent. So far, we used the same overall parameters for all approaches in order to have a fair comparison base. Better results could be achieved for each approach with parameter optimization. By and large, each of the four methods seem to be rewarding in their setting.

Which method from Table 3 is considered to be the best also heavily relies on the problem setting at hand. In a setting of intrusion detection through classification, we would argue that a small false positive rate is more important than a high true positive rate for two reasons. (1) Analyzing an alert with respect to it being a true or false positive is very expensive. (2) It may not even be necessary to get all malicious sequences, since it is sufficient to raise an alert on one of the malicious system call sequences within one exploited process trace. The latter again depends on the problem setting or rather how to handle an attack reported. For instance, in critical infrastructures it may be more important to capture every attack possible. In this case, analyzing potential false positives may be a necessary evil.

6 Conclusion

This work systematically compares different input methods for system call traces for intrusion detection. We use sequences of system calls, their mapped kernel modules or a combination of both as representation options. The three input options are combined with four different encodings, namely an one-hot vector, learning an embedding while training, and using word2vec or GloVe representations.

Results imply that working with kernel modules exclusively is not recommended for our setting, although they might still be helpful as supplementary information in other settings. One-hot encoding showed the best results. The other approaches, however, should not be discarded since they could be helpful for more sophisticated models for intrusion detection. Also we compare every approach with the same overall parameterset to achieve a fair comparison base. Better results in terms of TPR or FPR might be achieved by optimizing a specific approach.

With the results of this work in mind, future activities may focus on modeling normal behavior of programs based on system call sequences.

Acknowledgements. S.W. is funded by the Bavarian State Ministry of Science and Arts in the framework of the Centre Digitization.Bavaria (ZD.B). S.W. and M.R. are further supported by the BayWISS Consortium Digitization. Last but not least, we gratefully acknowledge the support of NVIDIA Corporation with the donation of the Titan Xp GPU used for this research.

References

1. Creech, G., Hu, J.: A semantic approach to host-based intrusion detection systems using contiguous and discontiguous system call patterns. IEEE Trans. Comput. **63**(4), 807–819 (2014)
2. Forrest, S., Hofmeyr, S.A., Somayaji, A., Longstaff, T.A.: A sense of self for unix processes. In: IEEE Symposium on Security and Privacy, pp. 120–128. IEEE (1996)
3. Hofmeyr, S.A., Forrest, S., Somayaji, A.: Intrusion detection using sequences of system calls. J. Comput. Secur. **6**(3), 151–180 (1998)
4. Kim, G., Yi, H., Lee, J., Paek, Y., Yoon, S.: LSTM-based system-call language modeling and robust ensemble method for designing host-based intrusion detection systems. arXiv preprint arXiv:1611.01726, pp. 1–12 (2016)
5. Kolosnjaji, B., Zarras, A., Webster, G., Eckert, C.: Deep learning for classification of malware system call sequences. In: Australasian Joint Conference on Artificial Intelligence (AI), pp. 137–149. Springer (2016)
6. Sharma, A., Pujari, A.K., Paliwal, K.K.: Intrusion detection using text processing techniques with a kernel based similarity measure. Comput. Secur. **26**(7–8), 488–495 (2007)
7. Murtaza, S.S., Khreich, W., Hamou-Lhadj, A., Gagnon, S.: A trace abstraction approach for host-based anomaly detection. In: IEEE Symposium on Computational Intelligence for Security and Defense Applications (CISDA), pp. 170–177. IEEE (2015)
8. Hochreiter, S., Schmidhuber, J.: Long short-term memory. Neural Comput. **9**(8), 1735–1780 (1997)

9. Creech, G., Hu, J.: Generation of a new IDS test dataset: time to retire the KDD collection. In: IEEE Wireless Communications and Networking Conference (WCNC), pp. 4487–4492. IEEE (2013)
10. Eskin, E., Lee, W., Stolfo, S.J.: Modeling system calls for intrusion detection with dynamic window sizes. In: DARPA Information Survivability Conference & Exposition II (DISCEX), vol. 1, pp. 165–175. IEEE (2001)
11. Hoang, X.D., Hu, J.: An efficient hidden markov model training scheme for anomaly intrusion detection of server applications based on system calls. In: IEEE International Conference on Networks (ICon), vol. 2, pp. 470–474. IEEE (2004)
12. Kosoresow, A.P., Hofmeyer, S.: Intrusion detection via system call traces. IEEE Softw. **14**(5), 35–42 (1997)
13. Eskin, E., Arnold, A., Prerau, M., Portnoy, L., Stolfo, S.: A geometric framework for unsupervised anomaly detection: detecting intrusions in unlabeled data. Appl. Data Min. Comput. Secur. **6**, 77–102 (2002)
14. Wang, Y., Wong, J., Miner, A.: Anomaly intrusion detection using one class SVM. In: IEEE SMC Information Assurance Workshop, pp. 358–364. IEEE (2004)
15. Chawla, A., Lee, B., Fallon, S., Jacob, P.: Host based intrusion detection system with combined CNN/RNN model. In: International Workshop on AI in Security, pp. 9–18 (2018)
16. Xie, M., Hu, J., Yu, X., Chang, E.: Evaluating host-based anomaly detection systems: application of the frequency-based algorithms to ADFA-LD. In: International Conference on Network and System Security, pp. 542–549. Springer (2014)
17. Mikolov, T., Sutskever, I., Chen, K., Corrado, G.S., Dean, J.: Distributed representations of words and phrases and their compositionality. In: Advances in Neural Information Processing Systems, pp. 3111–3119 (2013)
18. Pennington, J., Socher, R., Manning, C.: GloVe: global vectors for word representation. In: Conference on Empirical Methods in Natural Language Processing (EMNLP), pp. 1532–1543 (2014)
19. Ring, M., Landes, D., Dallmann, A., Hotho, A.: IP2Vec: learning similarities between IP adresses. In: Workshop on Data Mining for Cyber Security (DMCS), International Conference on Data Mining Workshops (ICDMW), pp. 657–666. IEEE (2017)
20. [Online] DARPA 1998/1999 Dataset. https://www.ll.mit.edu/r-d/datasets. Accessed 14 Mar 2019
21. [Online] UNM Dataset. https://www.cs.unm.edu/~immsec/systemcalls.htm. Accessed 14 Nov 2018
22. Warrender, C., Forrest, S., Pearlmutter, B.: Detecting intrusions using system calls: alternative data models. In: IEEE Symposium on Security and Privacy, pp. 133–145. IEEE (1999)
23. McHugh, J.: Testing intrusion detection systems: a critique of the 1998 and 1999 DARPA intrusion detection system evaluations as performed by Lincoln laboratory. ACM Trans. Inform. Syst. Secur. (TISSEC) **3**(4), 262–294 (2000)
24. He, H., Garcia, E.A.: Learning from imbalanced data. IEEE Trans. Knowl. Data Eng. **21**(9), 1263–1284 (2009)
25. Srivastava, N., Hinton, G., Krizhevsky, A., Sutskever, I., Salakhutdinov, R.: Dropout: a simple way to prevent neural networks from overfitting. J. Mach. Learn. Res. **15**(1), 1929–1958 (2014)
26. Gers, F.A., Schraudolph, N.N., Schmidhuber, J.: Learning precise timing with LSTM recurrent networks. J. Mach. Learn. Res. **3**, 115–143 (2002)

Efficient Verification of Security Protocols Time Properties Using SMT Solvers

Agnieszka M. Zbrzezny[1], Sabina Szymoniak[2(✉)], and Mirosław Kurkowski[3]

[1] Faculty of Mathematics and Computer Science, University of Warmia and Mazury in Olsztyn, Olsztyn, Poland
agnieszka.zbrzezny@matman.uwm.edu.pl

[2] Institute of Computer and Information Sciences, Częstochowa University of Technology, Częstochowa, Poland
sabina.szymoniak@icis.pcz.pl

[3] Institute of Computer Science, Cardinal St. Wyszynski University, Warsaw, Poland
m.kurkowski@uksw.edu.pl

Abstract. In this paper, we present a novel method for the verification of security protocols time properties using SMT-based bounded model checking (SMT-BMC). In our approach, we model protocol users' behaviours using networks of synchronized timed automata. Suitably specified correctness properties are defined as reachability property of some, chosen states in automata network. We consider most important time properties of protocols' executions using specially constructed time conditions. These are checked by a quantifier-free SMT encoding and SMT solver using BMC algorithms. In our work, we have also implemented the proposed method and evaluated it for four, well-known security protocols. We also compared our new SMT-based technique with the corresponding SAT-based approach.

1 Introduction

Every conscious and serious Internet user cannot imagine the reality of the computer network without the protocols that allow keeping good transmitted data security. The SSL or the TLS protocols are well known and widely used by almost all well-experienced networks users. A lot of people that surf the Internet, a lot of attention paid whether watched sites are marked by innocently looking little padlock that means that the site is secured by the TLS protocol. Significantly fewer people know that the heart, the most important fragment of these protocols, is a very short and simple scheme of some data secure exchanging. In the literature, these schemes are known as cryptographic or security protocols (SP). The properties of these schemes are most important from the open computer networks security point of view. It is obvious that if an SP is incorrectly constructed then all the implemented tools based on it are incorrect and insecure.

The project financed under the program of the Minister of Science and Higher Education under the name "Regional Initiative of Excellence" in 2019–2022 project number 020/RID/2018/19, the amount of financing 12,000,000 PLN.

© Springer Nature Switzerland AG 2020
F. Martínez Álvarez et al. (Eds.): CISIS 2019/ICEUTE 2019, AISC 951, pp. 25–35, 2020.
https://doi.org/10.1007/978-3-030-20005-3_3

Taking this into considerations, for many years there is a need for creating, and investigation of the different ways for the SP properties verification, especially widely understood correctness. From the nineties of the XX century, many methodologies and practically used tools have been developed and implemented in this case. From a practical (and methodological) point of view, it is obvious that the first attempts were checking SP properties by testing real or virtual environments. However, it is obvious that even many testing hours or years cannot answer the question whether the investigated system is correct or not. After long time testing, we can only say that so far system works correctly.

Another way of SP properties checking is using the formal, for example, deduction methods. It is known that since the creation of Hoare's logic, a huge number of logical systems have been introduced and helped in many different problems of software engineering. The case of security protocols is not isolated here. Remembering BAN logic and other systems [4,13] we have to say that this approach for SP verification was useful and successful.

A most important formal way for the SP properties verification are model checking techniques. Generally, in this case, an appropriate and adequate formal model of the protocol executions is created and searched. There are many concepts about how the model can be constructed and many algorithms how it can be searched. In the case of SP verification, we have to notice a few very useful and very well grounded approaches. First one is the well known the AVISPA tool [2], that uses HLPSL, a specially designed language for SP specification, and four independent subtools for checking SP properties. Another one is Scyther proposed by Cremers and Mauw [7]. These tools allow automatic investigations about SP properties especially their correctness. We can also annotate Kurkowski's solutions where protocols executions were modeled as work of synchronized automata networks [12].

Here, we have to annotate that none of these solutions can express or investigate SP time properties. In SP case it is very important problem because of adding into protocols schemes time tickets and lifetimes of messages. It is important to note that time modeling it is not an easy problem at all. There are several main ways of expressing time and time properties in formal computer systems models [1,3,9,11,17]. So far the main methods used for SP time properties verification are based on the translation to the SAT problem [12,19,23].

In the case of SP time modeling Penczek and Jakubowska have been introduced a concept how to use timed automata for modeling protocols executions and users' behaviors [10]. This approach allowed to express some important SP properties, for example, the existence of selected types of attacks upon protocols, taking into account delays in the network too. Due to the extension of the model with delays in the network, it became possible to consider the SP in a broader aspect of the influence of the message transmission time on the Intruder's capabilities. Next combining mentioned before solutions of Kurkowski, Penczek and Jakubowska these concepts were continued by Szymoniak et al. in [19,20].

SMT-based bounded model checking (BMC) consists in translating the existential model checking problem for a modal logic and for a model to the satisfiability modulo theory problem (SMT-problem) of a quantifier-free first-order formula. To the best of our knowledge, there is no work that considers SMT-based verification of security protocols.

In this paper, we make the following contributions. Firstly, we define and implement an SMT-based method for verification of security protocols correctness properties. The behaviours of security protocols are modelled using networks of communicating timed automata in order to verify them with SMT-based bounded model checking [22]. In our experiments as protocol correctness property, we investigate protocol resistance to a few kinds of attack (reply, and man-in-the-middle). Detail information about these properties can be found in [19].

Next, we report on the initial experimental evaluation of our new SMT method. To this aim, we use two four benchmarks: Needham-Schroeder Public Key (NSPK), Needham Schroeder Symmetric Key (NSSK), Woo Lam Pi (WLP) and Wide Mouth Frog (WMF) protocols.

2 Preliminaries

As we mentioned before, one of the SP executions modelling methods is a way via automata. In Genet's [8] and Monniaux's [14] works, the SP executions were modeled using automata. These automata accepted languages reflecting the users' activities including Intruder. Also in many verification tools automata were implemented. In Corin's works [5,6] timed automata from [1] were considered. The authors emphasized security protocols sensitivity to the passage of time. In this approach, the users' behaviors were presented by means of separate automata. Communication between users was appropriately synchronized. However, it was necessary to prepare a precise and very detailed protocol specification especially users' behavior.

In our approach, we build a synchronized network of timed automata. This network consists of two kinds of automata (automata of executions and automata of knowledge). First, of them are designed to present the next steps of the protocol execution. Second of them are designed to present the knowledge of users, which changes during the protocol execution. The prepared automata are properly synchronized to fully model the protocols executions. Precise definitions of mentioned types of automata were presented in [12,18].

Networks of Synchronized Timed Automata. Here we give definitions of a timed automaton and a network of synchronized timed automata. Next, we show how to model the executions of a protocol by the runs of a network of synchronized timed automata, where each timed automaton represents one component of the protocol. We also introduce automata representing messages transmitted during a protocol execution and automata representing users' knowledge. An appropriate synchronisation between the automata of the introduced network enables to model the executions of the protocol according to the knowledge acquired by its users.

We assume a finite set $\mathcal{X} = \{x_0, \ldots, x_{n-1}\}$ of variables, called *clocks*. A *clock valuation* is a total function $v : \mathcal{X} \mapsto \mathbb{R}$ that assigns to each clock x a non-negative real value $v(x)$. The set of all the clock valuations is denoted by $\mathbb{R}^{|\mathcal{X}|}$. For $X \subseteq \mathcal{X}$, the valuation $v' = v[X := 0]$ is defined as: $\forall x \in X$, $v'(x) = 0$ and $\forall x \in \mathcal{X} \setminus X$, $v'(x) = v(x)$. For $\delta \in \mathbb{R}$, $v + \delta$ denotes the valuation v'' such that $\forall x \in \mathcal{X}, v''(x) = v(x) + \delta$. Let $x \in \mathcal{X}$, $c \in \mathbb{N}$, and $\sim \in \{<, \leqslant, =, \geqslant, >\}$. *Clock constraints* over $\mathcal{X}$ are conjunctions of comparisons of a clock with a time constant from the set of non-negative natural numbers $\mathbb{N}$. The set $\mathcal{C}(\mathcal{X})$ of clock constraints over the set of clocks $\mathcal{X}$ is defined by the grammar: $\mathfrak{cc} := x \sim c \mid \mathfrak{cc} \wedge \mathfrak{cc}$. Let v be a clock valuation, and $\mathfrak{cc} \in \mathcal{C}(\mathcal{X})$. A clock valuation v satisfies a clock constraint $\mathfrak{cc}$, written as $v \models \mathfrak{cc}$, iff $\mathfrak{cc}$ evaluates to true using the clock values given by the valuation v.

Definition 1. *A* timed automaton *is a eight-tuple* $\mathcal{A} = (A, L, l^0, E, \mathcal{X}, Inv, AP, V)$, *where*

– A *is a finite set of* actions,

– L *is a finite set of* locations,

– $l^0 \in L$ *is the* initial location,

– $\mathcal{X}$ *is a finite set of clocks,*

– $Inv : L \mapsto \mathcal{C}(\mathcal{X})$ *is a state invariant function,*

– AP *is a set of atomic propositions,*

– $E \subseteq L \times A \times \mathcal{C}(\mathcal{X}) \times 2^{\mathcal{X}} \times L$ *is a* transition relation,

– $V : L \mapsto 2^{AP}$ *is a valuation function assigning to each location a set of atomic propositions true in this location.*

Each element e of E is denoted by $l \stackrel{a,\mathfrak{cc},X}{\longrightarrow} l'$, which represents a transition from the location l to the location l', executing the action a, with the set $X \subseteq \mathcal{X}$ of clocks to be reset, and with the clock condition $\mathfrak{cc} \in \mathcal{C}$ defining the enabling condition (guard) for e.

Given a transition $e = l \stackrel{a,\mathfrak{cc},X}{\longrightarrow} l'$, we write *source*$(e)$, *target*$(e)$, *action*$(e)$, *guard*$(e)$ and *reset*(e) for l, l', a, $\mathfrak{cc}$ and X, respectively. The clocks of a timed automaton allow to express the timing properties. An enabling condition constrains the execution of a transition without forcing it to be taken.

Several examples of timed automata used for modelling of protocols executions are given at the end of this section.

Definition 2. *Let $\mathcal{A} = (A, L, l^0, E, \mathcal{X}, Inv, AP, V)$ be a timed automaton, and v^0 a clock valuation such that $\forall x \in \mathcal{X}$, $v^0(x) = 0$. A* concrete model *for $\mathcal{A}$ is a tuple $\mathcal{M}_{\mathcal{A}} = (Q, q^0, \rightarrow, \mathcal{V})$, where $Q = L \times \mathbb{R}^n$ is a set of the concrete states, $q^0 = (l^0, v^0)$ is the initial state, $\rightarrow \subseteq Q \times Q$ is a total binary relation on Q defined by action and time transitions as follows. For $\sigma \in A$ and $\delta \in \mathbb{R}$,*

1. *Action transition: $(l, v) \stackrel{\sigma}{\rightarrow} (l', v')$ iff there is a transition $l \stackrel{\sigma,\mathfrak{cc},X}{\longrightarrow} l' \in E$ such that $v \models \mathfrak{cc} \wedge Inv(l)$ and $v' = v[X := 0]$ and $v' \models Inv(l')$,*
2. *Time transition: $(l, v) \stackrel{\delta}{\rightarrow} (l, v + \delta)$ iff $v \models Inv(l)$ and $v + \delta \models Inv(l)$.*

A valuation function $\mathcal{V} : Q \mapsto 2^{AP}$ is defined such that $\mathcal{V}((l, v)) = V(l)$ for all $(l, v) \in Q$.

Intuitively, a time transition does not change the location l of a concrete state, but it increases the clocks. An action successor corresponding to an action σ is executed when the guard $\mathfrak{cc}$ holds for v and the valuation v' obtained by resetting the clocks in $\mathcal{X}$. A *run* ρ of $\mathcal{A}$ is an infinite sequence of concrete states: $q_0 \xrightarrow{\delta_0} q_0 + \delta_0 \xrightarrow{\sigma_0} q_1 \xrightarrow{\delta_1} q_1 + \delta_1 \xrightarrow{\sigma_1} q_2 \xrightarrow{\delta_2} \ldots$ such that $q_i \in Q$, $\sigma_i \in A$, and $\delta_i \in \mathbb{R}$ for each $i \in \mathbb{N}$. Notice that our runs are *weakly monotonic*. This is because the definition of the run permits two consecutive actions to be performed one after the other without any time passing in between. A k-run ρ_k of $\mathcal{M}_\mathcal{A}$ is a finite sequence of transitions: of concrete states: $q_0 \xrightarrow{\delta_0} q_0 + \delta_0 \xrightarrow{\sigma_0} q_1 \xrightarrow{\delta_1} q_1 + \delta_1 \xrightarrow{\sigma_1} q_2 \xrightarrow{\delta_2} \ldots \xrightarrow{\sigma_k} q_k$.

Now, we are going to use networks of timed automata for modelling executions of the protocol as well as for modelling the knowledge of the participants.

Product of a Network of Timed Automata. A network of timed automata can be composed into a global (*product*) timed automaton [17] in the following way. The transitions of the timed automata that do not correspond to a shared action are interleaved, whereas the transitions labelled with a shared action are synchronised.

Let $n \in \mathbb{N}$, $I = \{1, \ldots, n\}$ be a non-empty and finite set of indices, $\{\mathcal{A}_i \mid i \in I\}$ be a family of timed automata $\mathcal{A}_i = (A_i, L_i, l^0{}_i, E_i, \mathcal{X}_i, Inv_i, AP_i, V_i)$ such that $\mathcal{X}_i \cap \mathcal{X}_j = \emptyset$ and $AP_i \cap AP_j = \emptyset$ for $i \neq j$. Moreover, let $I(\sigma) = \{i \in I \mid \sigma \in A_i\}$. The *parallel composition* of the family $\{\mathcal{A}_i \mid i \in I\}$ of timed automata is the timed automaton $\mathcal{A} = (A, L, l^0, E, \mathcal{X}, Inv, AP, V)$ such that $A = \prod_{i \in I}(A_i)$, $L = \prod_{i \in I} L_i$, $l^0 = (l^0_1, \ldots, l^0_n)$, $\mathcal{X} = \bigcup_{i \in I} \mathcal{X}_i$, $Inv(l_1, \ldots, l_n) = \bigwedge_{i=1}^{n} Inv_i(l_i)$, $AP = \bigcup_{i \in I} AP_i$, $V(l_1, \ldots, l_n) = \bigcup_{i=1}^{n} V_i(l_i)$, and a transition is defined as follows:

$$((l_1, \ldots, l_n), (\sigma_1, \ldots, \sigma_n), \bigwedge\nolimits_{i \in I} \mathfrak{cc}_i, \bigcup\nolimits_{i \in I} X_i, (l'_1, \ldots, l'_m)) \in E$$
$$\text{iff } (\forall i \in I))(l_i, \sigma_i, \mathfrak{cc}_i, X_i, l'_i) \in E_i), \text{ and } (\forall i \in I \setminus I(\sigma))l'_i = l_i.$$

We used a special set of time conditions including delays in the network. This set is defined by the grammar: $\mathbf{tc} ::= true \mid \tau_i + \tau_d - \tau_j \leq \mathcal{L}_\mathcal{F} \mid tc \wedge tc$.

Grammar defines three kinds of time conditions. First of them ($true$) is the simplest condition meaning that the step is not time dependent. By the second kind of time conditions ($\tau_i + \tau_d - \tau_j \leq \mathcal{L}_\mathcal{F}$) we express the dependence between the time of sending the message (τ_i), delay in the network (τ_d), time of timestamp's generation τ_j and the lifetime $\mathcal{L}_\mathcal{F}$. Using the third kind of time conditions it is possible to combining second time conditions into complex conditions using conjunctions.

An Example. As an example, we consider a synchronized network of timed automata for well-known Needham Schroeder Symmetric Key Protocol, NSSK for short. This protocol aims to a distribution of a new shared symmetric key by a trusted server and mutual authentification of honest users. The syntax of the timed version of NSSK protocol in Common Language is as follows:

α_1 $A \rightarrow S$: I_A, I_B, T_A
α_2 $S \rightarrow A$: $\{T_A, I_B, K_{AB}, \{I_A, K_{AB}\}_{K_{BS}}\}_{K_{AS}}$
α_3 $A \rightarrow B$: $\{I_A, K_{AB}\}_{K_{BS}}$
α_4 $B \rightarrow A$: $\{T_B\}_{K_{AB}}$
α_5 $A \rightarrow B$: $\{T_B\}_{K_{AB}}$

NSSK protocol consists of five steps. The user A tries to communicate with the user B. First, he must establish the session with the trusted server. The server generates a new symmetric key (in the second step) which will be shared between the users A and B. After the exchange this key between the honest users and a trusted server, they are authenticated with each other. In the Fig. 1 the automaton of execution for NSSK protocol was presented. This automaton consists of six states, which correspond to the protocol's execution. The transition to the next state is possible only when the timed condition imposed on the protocol's step is met. For example, on the first step of NSSK protocol was imposed following condition $\tau_1 + \mathcal{D}_1 - \tau_A \leqslant \mathcal{L}_f$. This condition presents the relationship between the time of sending a message τ_1, delay in the network τ_A and the lifetime $\mathcal{L}_f$. The next states should be considered in this same way.

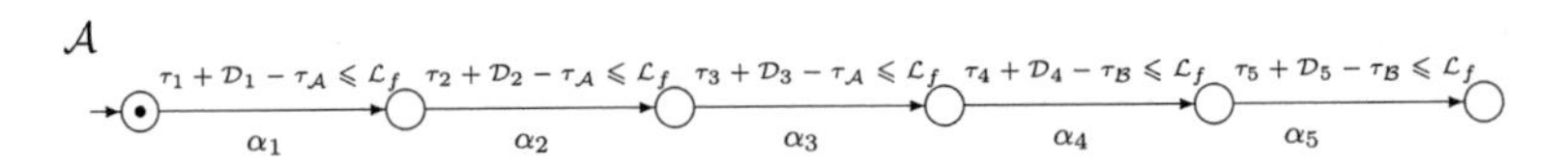

Fig. 1. The automaton of execution for NSSK protocol.

In the Fig. 2 the automata of knowledge for NSSK protocol was presented. There are eight automata which model the changes in users' knowledge during the protocol execution. For example, the first automaton models a change at user A's knowledge about the timestamp τ_A. In the first step, the user A generates this timestamp acquires knowledge about this timestamp. In the second step, he uses his knowledge because he gets back the timestamp τ_A from a trusted server. The next automata should be considered in this same way. The using of time conditions during the protocols execution is an important aspect of their security research. The Intruder's activity may involve the additional steps. The additional steps will also take additional time, which will affect the total duration of the step or protocol. Time conditions are therefore designed to protect the honest users from the Intruder's activity and to signal the unwanted actions as a result of prolonged step duration.

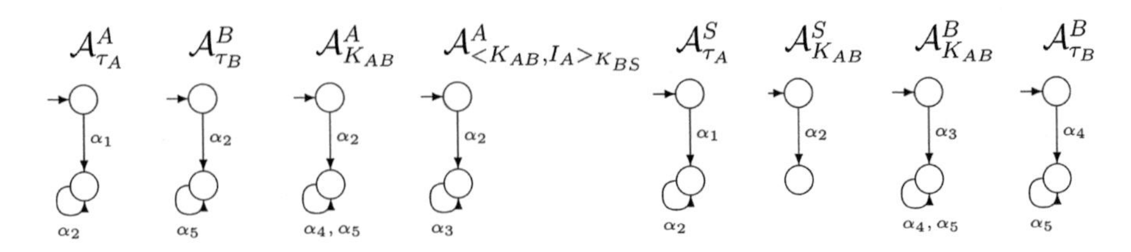

Fig. 2. The knowledge automata for NSSK protocol.

3 SMT-Based Verification

We decided to use as our first approach to reachability analysis the method based on the translation to an SMT problem (Satisfiability Modulo Theories Problem). The first step of this method is to encode actions and states (local and global) as vectors of individual variables, and hence to encode the transition relation by a quantifier-free first-order logic formula.

SMT-solvers extend the capabilities of SAT-solvers by allowing for first order formulae that can be defined over several built-in theories, and types other than booleans. Both, our new SMT-based reachability analysis and SAT-based reachability analysis are based on encoding presented in [24,25]. Let $\mathcal{A} = (A, L, l^0, E, \mathcal{X}, Inv, AP, V)$ be a TA that is a parallel composition of n timed automata, $\mathcal{M}_\mathcal{A} = (Q, q^0, \rightarrow, \mathcal{V})$ a concrete model for $\mathcal{A}$, and $k \in \mathbb{N}$. For a timed automaton $\mathcal{A}$ we encode its concrete model as follows. We assume that each state $s \in Q$ of $\mathcal{M}_\mathcal{A}$ is encoded by a valuation of a symbolic state $\overline{\mathbf{w}} = ((\mathtt{l}_1, \ldots, \mathtt{l}_n), (\mathtt{x}_1, \ldots, \mathtt{x}_{|\mathcal{X}|}))$ that consists of symbolic local states which a, symbolic clock valuations. Each symbolic local state $\mathtt{l}_i$ $(1 \leq i \leq n)$ is an individual variable ranging over the natural numbers. Each symbolic clock valuation $\mathtt{x}_i$ $(1 \leq i \leq |\mathcal{X}|)$ is an individual variable ranging over the real numbers.

Similarly, each action $\sigma \in A$ can be represented by a valuation of a symbolic action a_i that is an individual variable ranging over the natural numbers, each real number $\delta \in \mathbb{R}$ can be represented by a valuation of a symbolic number $\mathtt{b}$ that is an individual variable ranging over the real numbers.

In order to encode a k-run we use a finite sequence $(\overline{\mathbf{w}}_0, \ldots, \overline{\mathbf{w}}_k)$ of symbolic states that is called a *symbolic k-path*. Further, for a symbolic time passage $\mathtt{b}$, and for two symbolic states $\overline{\mathbf{w}} = ((\mathtt{l}_1, \ldots, \mathtt{l}_n), (\mathtt{x}_1, \ldots, \mathtt{x}_{|\mathcal{X}|}), \overline{\mathbf{w}}' = ((\mathtt{l}'_1, \ldots, \mathtt{l}'_n), (\mathtt{x}'_1, \ldots, \mathtt{x}'_{|\mathcal{X}|}),)$, we define the following propositional formulae:

- $\mathtt{I}(\overline{\mathbf{w}}) := (\bigwedge_{i=1}^{n} \mathtt{l}_i = l_1) \wedge (\bigwedge_{i=1}^{|\mathcal{X}|} \mathtt{x}_i = 0)$ is the formula that encodes the initial state $s^0 = ((l_1, \ldots, l_n), (0, \ldots, 0))$ of $\mathcal{M}_\mathcal{A}$.
- $\mathtt{P}_p(\overline{\mathbf{w}}) := \bigvee_{i=1}^{n} \bigvee_{l \in V_i[p]} (\mathtt{l}_i = l)$, is the formula that encodes the set of states of $\mathcal{M}_\mathcal{A}$ in which $p \in AP$ holds.
- $\mathtt{TT}(\overline{\mathbf{w}}, \mathtt{b}, \overline{\mathbf{w}}')$ is a formula that encodes the time transition relation of $\mathcal{M}_\mathcal{A}$.
- $\mathtt{AT}(\overline{\mathbf{w}}, \mathsf{a}, \overline{\mathbf{w}}')$ is a formula that encodes the action transition relation of $\mathcal{M}_\mathcal{A}$.

4 Experimental Results

In this section we provide experimental results, and discuss the performance of our SMT-based solution to verification of security protocols time properties. We used following protocols for our research: NSPK, NSSK (described in Sect. 2), WLP and WMF. NSPK[1] is one of the most significant asymmetric protocols. This protocol was proposed in [16]. NSPK ensure mutual authentication of two

[1] Syntax of NSPK timed version in Common Language is as follow: α_1 $A \rightarrow B$: $\{T_A, I_A\}_{K_B}$; $\alpha_2 B \rightarrow A$: $\{T_A, T_B\}_{K_A}$; α_3 $A \rightarrow B$: $\{T_B\}_{K_B}$.

users. WLP[2] protocol uses symmetric cryptography. This protocol was proposed in [21]. The purpose of this protocol is unilateral authentication with the trusted server. WMF[3] was proposed in [4]. This protocol uses symmetric cryptography with a trusted server too. The WMF task is to distribute the new session key between two users.

Performance Evaluation. We have performed our experimental results on a computer equipped with I7-8850U, 1.80 GHz x 8 processor, 8 GB of RAM, and the operating system Ubuntu Linux with the kernel 4.15.0-43. We compare our SMT-based bounded model checking (BMC) algorithm with the corresponding SAT-based BMC [24]. Our SMT-based algorithm is implemented as standalone program written in the programming language `C++`. For SMT-BMC we used the state of the art SMT-solvers Z3 [15] and Yices2 and for the SAT-BMC we used the state of the art SAT-solver MiniSAT. All the benchmarks together with an instruction how to reproduce our experimental results can be found at the web page https://tinyurl.com/bmc4sptp.

The experimental results show that the SMT-based approach in every case is better than SAT-based approach (Fig. 3).

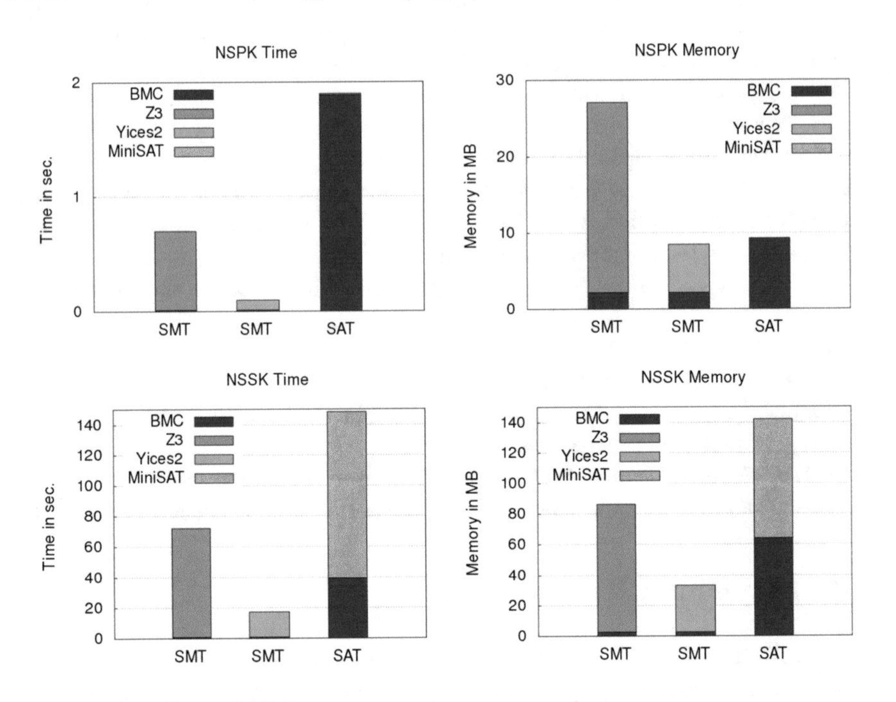

Fig. 3. NSPK and NSSK: an avarage total time and an avarage memory usage.

[2] Syntax of WLP timed version in Common Language is as follow: α_1 $A \rightarrow B$: I_A; α_2 $B \rightarrow A$: T_B; α_3 $A \rightarrow B$: $\{T_B\}_{K_{AS}}$; α_4 $B \rightarrow S$: $\{I_A, \{N_B\}_{K_{AS}}\}_{K_{BS}}$; α_5 $S \rightarrow B$: $\{T_B\}_{K_{BS}}$.

[3] Syntax of WMF timed version in Common Language is as follow: α_1 $A \rightarrow S$: $I_A, \{T_A, I_B, K_{AB}\}_{K_{AS}}$; α_2 $S \rightarrow B$: $\{T_S, I_A, K_{AB}\}_{K_{BS}}$.

In the experimental results for the NSPK, the main difference is in the translation of the model to SMT2 format and DIMACS format. The translation to SMT is quick (less than 0.01 s) while the translation to SAT takes 10 s. The SMT approach also consumes less memory than SAT approach. We have also compared two SMT-solvers. For NSPK, Yices2 has consumed less time and memory. The really interesting fact is that in this case SAT-solver needs less than 0.01 s to check satisfiability of the input formula.

In the next three benchmarks, again, SMT-based approach was much better, but here we can see the huge difference in the both approaches. The SAT-solver consumes more time and memory than two SMT-solvers. The same situation is with the translations.

The very interesting observation is Yices2 is much better than Z3 in the security protocols verification context (Fig. 4).

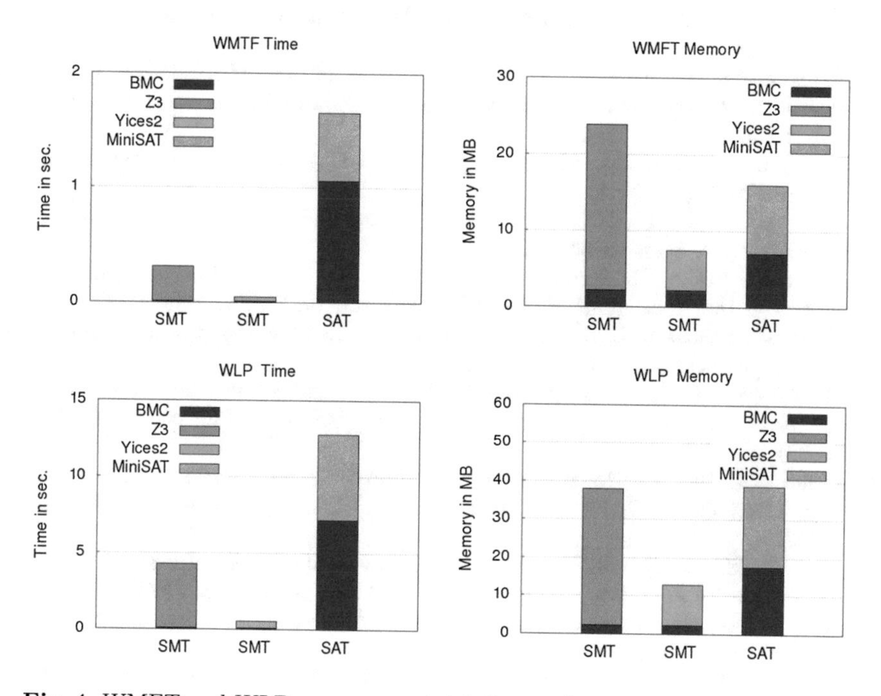

Fig. 4. WMFT and WLP: an average total time and an average memory usage.

5 Conclusions

The correct work of Timed Security Protocols depends on well-chosen values of their time parameters such as encryption, decryption and network delays times. For this reason, investigating of SPs timed properties is a very important problem. According to the difficulty of time modeling and checking time-sensitive systems properties, there is a need for investigating increasingly efficient modeling ways.

In the case of SPs timed properties, an investigation allows, for example, indicate the existence of an attack on the examined protocol and an indication of the value of time parameters determining its safety.

In this work, we have proposed, implemented, and experimentally evaluated the SMT-based BMC approach for the verification of security protocols time properties. For our experiments, we have considered protocols resistance on a few types of attacks. We have compared two methods: the new one with the old, SAT-based technique. The experimental results show that the SMT-based BMC in every case is better than SAT-based BMC. This is a novelty, in the case of SP verification, and a very interesting result.

References

1. Alur, R., Dill, D.L.: A theory of timed automata. Theor. Comput. Sci. **126**, 183–235 (1994)
2. Armando, A., et al.: The AVISPA tool for the automated validation of internet security protocols and applications. In: CAV 2005, Edinburgh, Scotland, UK, 6–10 July 2005, pp. 281–285 (2005)
3. Basin, D., Cremers, C., Meadows, C.: Model Checking Security Protocols, pp. 727–762. Springer (2018)
4. Burrows, M., Abadi, M., Needham, R.: A logic of authentication. ACM Trans. Comput. Syst. **8**(1), 18–36 (1990)
5. Corin, R., Etalle, S., Hartel, P.H., Mader, A.: Timed model checking of security protocols. In: Proceedings of the ACM Workshop on FMSE. ACM (2004)
6. Corin, R., Etalle, S., Hartel, P.H., Mader, A.: Timed analysis of security protocols. J. Comput. Secur. **15**(6), 619 (2007)
7. Cremers, C.J.F.: The Scyther tool: verification, falsification, and analysis of security protocols. In: CAV 2008, Princeton, NJ, USA, 7–14 July 2008, pp. 414–418 (2008)
8. Genet, T., Klay, F.: Rewriting for cryptographic protocol verification. In: Automated Deduction - CADE-17, Pittsburgh, PA, USA, 17–20 June 2000, pp. 271–290 (2000)
9. Hess, A., Mödersheim, S.: A typing result for stateful protocols. In: 31st IEEE Computer Security Foundations Symposium, CSF 2018, Oxford, pp. 374–388 (2018)
10. Jakubowska, G., Penczek, W.: Modelling and checking timed authentication of security protocols. Fundam. Inform. **79**(3–4), 363–378 (2007)
11. Koymans, R.: Specifying real-time properties with metric temporal logic. Real-Time Syst. **2**(4), 255–299 (1990)
12. Kurkowski, M., Penczek, W.: Applying timed automata to model checking of security protocols. In: Handbook of Finite State Based Models and Applications, pp. 223–254 (2012)
13. Kurkowski, M., Srebrny, M.: A quantifier-free first-order knowledge logic of authentication. Fundam. Inform. **72**(1–3), 263–282 (2006)
14. Monniaux, D.: Abstracting cryptographic protocols with tree automata. In: Static Analysis, SAS 1999, Venice, Italy, Proceedings, pp. 149–163 (1999)
15. De Moura, L., Bjørner, N.: Z3: an efficient SMT solver. In: Proceedings of (TACAS'2008). LNCS, vol. 4963, pp. 337–340. Springer (2008)
16. Needham, R.M., Schroeder, M.D.: Using encryption for authentication in large networks of computers. Commun. ACM **21**(12), 993–999 (1978)

17. Penczek, W., Pólrola, A.: Advances in Verification of Time Petri Nets and Timed Automata: A Temporal Logic Approach. Studies in Computational Intelligence, vol. 20. Springer (2006)
18. Szymoniak, S., Kurkowski, M., Piątkowski, J.: Timed models of security protocols including delays in the network. J. Appl. Math. Comput. Mech. **14**(3), 127–139 (2015)
19. Szymoniak, S., Siedlecka-Lamch, O., Kurkowski, M.: Timed analysis of security protocols. In: ISAT 2016 - Part II, pp. 53–63 (2016)
20. Szymoniak, S., Siedlecka-Lamch, O., Kurkowski, M.: On some time aspects in security protocols analysis. In: CN 2018, Proceedings, pp. 344–356 (2018)
21. Woo, T.Y.C., Lam, S.S.: A lesson on authentication protocol design. SIGOPS Oper. Syst. Rev. **28**(3), 24–37 (1994)
22. Wozna-Szczesniak, B., Zbrzezny, A.M., Zbrzezny, A.: SMT-based searching for k-quasi-optimal runs in weighted timed automata. Fundam. Inform. **152**(4), 411–433 (2017)
23. Zbrzezny, A.: Improvements in SAT-based reachability analysis for timed automata. Fundam. Inform. **60**(1–4), 417–434 (2004)
24. Zbrzezny, A.: SAT-based reachability checking for timed automata with diagonal constraints. Fundam. Inform. **67**(1–3), 303–322 (2005)
25. Zbrzezny, A.M., Wozna-Szczesniak, B., Zbrzezny, A.: SMT-based bounded model checking for weighted epistemic ECTL. In: EPIA 2015, Coimbra, Portugal, pp. 651–657 (2015)

A Framework of New Hybrid Features for Intelligent Detection of Zero Hour Phishing Websites

Thomas Nagunwa(✉), Syed Naqvi, Shereen Fouad, and Hanifa Shah

School of Computing and Digital Technology,
Birmingham City University, Birmingham, UK
thomas.nagunwa@mail.bcu.ac.uk

Abstract. Existing machine learning based approaches for detecting zero hour phishing websites have moderate accuracy and false alarm rates and rely heavily on limited types of features. Phishers are constantly learning their features and use sophisticated tools to adopt the features in phishing websites to evade detections. Therefore, there is a need for continuous discovery of new, robust and more diverse types of prediction features to improve resilience against detection evasions. This paper proposes a framework for predicting zero hour phishing websites by introducing new hybrid features with high prediction performances. Prediction performance of the features was investigated using eight machine learning algorithms in which Random Forest algorithm performed the best with accuracy and false negative rates of 98.45% and 0.73% respectively. It was found that domain registration information and webpage reputation types of features were strong predictors when compared to other feature types. On individual features, webpage reputation features were highly ranked in terms of feature importance weights. The prediction runtime per webpage measured at 7.63 s suggest that our approach has a potential for real time applications. Our framework is able to detect phishing websites hosted in either compromised or dedicated phishing domains.

Keywords: Phishing · Phishing webpage detection · Zero hour phishing website · Webpage features · Machine learning

1 Introduction

Phishing websites mimic their targeted legitimate websites to trick users to collect their Personal Identification Information (PII) for online frauds [1]. Today, many phishers use easily available sophisticated tools including phishing toolkits and fast flux networks to create and host large number of highly dynamic and quality phishing websites [2, 3]. Consequently, there has been a rapid growth of new and unknown (zero hour) phishing websites in recent years [4]. As 91% of all global security breaches begin with phishing attacks and phishing websites being the key player [5], effective detection of the websites is inevitable towards making cyber space safe.

Phishing website detection solutions are mainly based on blacklists and heuristics techniques. Blacklists extensively use human skills in maintaining records of the

© Springer Nature Switzerland AG 2020
F. Martínez Álvarez et al. (Eds.): CISIS 2019/ICEUTE 2019, AISC 951, pp. 36–46, 2020.
https://doi.org/10.1007/978-3-030-20005-3_4

databases therefore they lack real time intelligence to detect zero hour phishing websites [6, 7]. Heuristics solutions analyze distinctive webpage features using various algorithmic approaches to detect phishing webpages. Many of them have reported moderate accuracy and false alarm rates [7]. Phishers also have been learning their prediction features and adopt corresponding obfuscations in their phishing webpages to enhance detection evasion [7]. This is facilitated with the limited number and diversity of features used by most of the solutions. Therefore continuous discovery and adoption of new, robust and highly diversified features is vital in maintaining effective detection.

This paper proposes a framework of new hybrid features for real time prediction of zero hour phishing webpages using machine learning. A total of 31 features, of which 26 are novel, from five different types of features, the most diverse compared to previous works, are proposed. Features related to different URL components (FQDN[1], domain and path) are introduced to enable the framework to detect phishing websites hosted in either compromised or dedicated phishing domains. The framework for the implementation of the prediction process is designed and presented, in which three modules are introduced to pre-process webpages to improve accuracy and efficiency. Our webpage pre-processors include JavaScript form detector and URL redirections check modules which have never been used before. Eight machine learning classification algorithms are applied to evaluate the extracted features' data to develop a best performing prediction model. We also evaluate overheads of the prediction process by computing an average prediction runtime per a webpage.

The paper is organized as follows; Sect. 2 discusses related works, Sect. 3 introduces the framework's design while Sect. 4 describes the conducted experiments and their results. Finally, Sects. 5 and 6 provide discussions and conclusion respectively.

2 Related Works

Several studies have applied machine learning (ML) approaches for predicting zero hour phishing webpages. Generally, they have used different diversity of feature types, typically between one and four types of features. For instance, [8] extracted 1701 word and 40 natural language processing based features, all from the URL, for the prediction. By evaluating the features using seven classifiers, Random Forest produced the highest accuracy of 97.98%. Zuhair et al. [9] investigated and designed 58 predictive features in which 48 and 10 were webpage structure and URL related features respectively. Using SVM classifier, they evaluated the features and the resulting model produced false positives (FP) of 1.17% and false negatives (FN) of 0.03%. Li et al. [10] also developed 20 features of the same two types of features to create a fast real time prediction model. By combining Gradient Boosting DT, XGBoost and LightGBM algorithms, the model obtained an accuracy, misclassification rate and FN of 97.3%, 4.46% and 1.61% respectively. Jain et al. [11], similar to [10], did not use third party features to avoid network overheads so as to improve efficiency.

[1] Full Qualified Domain Name, also known as hostname of a webpage.

Studies including [12] used a hybrid of three types of features while others such as [1, 13, 14, 15] used four types of features. Mohammad et al. [12], for instance, developed 5, 9 and 3 features related to webpage structure, URL and domain registration information respectively. They applied deep neural network to develop a prediction model of an accuracy of 92.18%. Feng et al. [14] extracted 30 features related to webpage structure, URL, domain registration and webpage reputation and evaluated them against eight classifiers to develop the prediction model. A deep neural network algorithm obtained an optimal accuracy of 97.71%.

Generally, most of the reviewed studies scored accuracy and error rates of between 81% and 98.5%, and 0.43% and 18% respectively. We observed that only [13, 16] deployed webpage pre-processor in which a HTML form detector was implemented to filter out webpages without the HTML forms. The work [16] also filtered known blacklisted webpages using a computing intensive SHA1 hash value comparison method.

3 Framework Design

3.1 Architecture

We have designed a machine learning based framework of new hybrid of features to predict whether the user requested webpage, prompting for PII, is phishing or not. Figure 1 demonstrates the framework's architecture. Paths 1–3 and 4–7 represent modelling of the classifier and the prediction processes respectively. The framework consists of the following six modules;

PII Webpage Filtering. A webpage requested by a user is checked if it prompts PII by examining if the webpage contains at least one of the PII webpage phrases and a HTML form or a JavaScript popup. Most of the webpages use the form or popup to prompt and collect PII. The PII webpage phrase is a word such as *sign in* and *login* contained in a webpage that is related to purpose of the webpage in collecting specific PII. We identified 43 PII phrases (in English) that we found were commonly used by over 50 samples of English based legitimate webpages prompting PII that we collected before. We used simple lookups, for instance, searching for a HTML form tag *<form...> ...</form>* in combination with at least one of the phrases in a webpage, to reduce overheads. The role of the module is to filter out webpages which do not collect PII and therefore can never be phishing. This avoids misuse of computer and network resources for analyzing irrelevant webpages, thus optimizing users' web browsing experience as well as avoiding false positive errors due to positive prediction for webpages which do not prompt for PII.

Phishing Blacklist Check. A webpage's URL is checked if it exists in a PhishTank's phishing URL blacklist, one of the most reputable online databases of blacklisted phishing URLs. The module's role is to filter out webpages which have already been confirmed to be phishing, thus enhancing detection efficiency and reducing false negatives. We use a simple lookup of a URL in the blacklist to reduce overheads.

URL Redirections Check. We observed that a significant number of both phishing and legitimate webpages have their first visible URLs being redirected to other URLs when downloading the webpages. Redirections in phishing webpages may be for the reason of hiding the true identity of the actual URLs hosting the webpages. Our interest was to learn features of the final redirected URLs, as actual addresses of the hosts. The module's role is to detect existence of all common redirections and extract their final URLs. Types of URL redirections detected were client-end redirections (implemented in HTML Meta and JavaScript tags), server-end redirections and short to long URL conversions. By determining the final redirected URLs, we ensured that we have collected relevant URL feature data to improve accuracy and error rates.

Feature Data Extraction. In this module, data about the webpage features are extracted from the webpage as well as from third party services such as search engines.

Training a Classifier. A classifier builds a prediction model from the training dataset.

Prediction Analysis. The prediction model analyses features' data extracted from a new webpage and generates a prediction result.

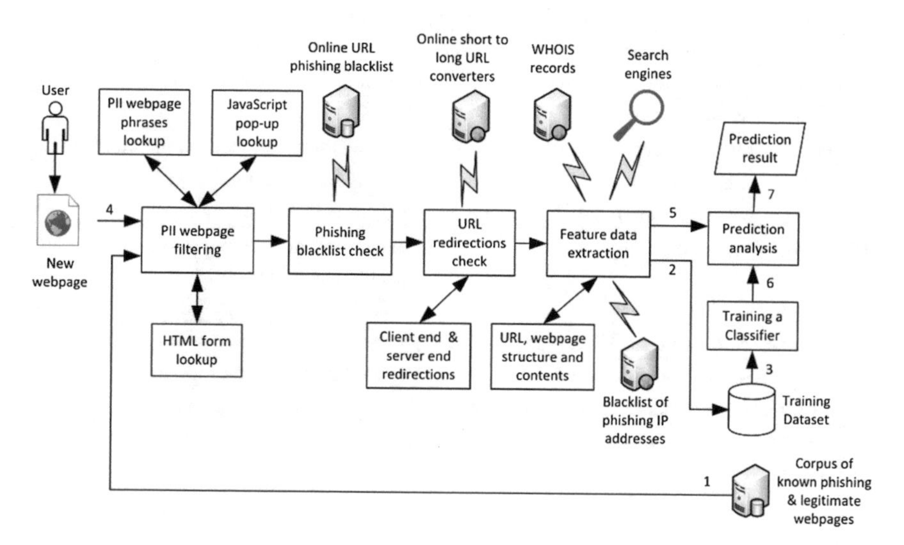

Fig. 1. An architecture of the proposed framework for predicting zero hour phishing websites.

3.2 Phishing Webpage Predictive Features

We have developed 31 webpage features, as listed in Table 1, to model the classifier to predict phishing webpages. The features are categorized into five different types as described below.

Webpage Structure and Contents. The features (F.1–F.7) are related to information contained in a webpage as content or part of its HTML/script structure.

URL Structure. The features (F.8–F.20) define specific decomposition characteristics of a webpage's URL. The features are related to the uses, positions and counts of special characters as well as the uses of third party services to host or form a URL.

Domain Registration Information. The features (F.21–F.24) are related to domain registration information kept by domain registrars. Such information is retrieved from registrars' online WHOIS databases.

SSL Certificate Information. The features (F.25 and F.26) are related to the information contained in a SSL certificate of the webpage.

Webpage Reputation. The features (F.27–F.31) measure reputation of a webpage in both Google and Bing search engines and in a blacklist of IP addresses of PhishTank's phishing websites.

Table 1. The proposed phishing webpage predictive features.

F.1	Domain identity[a] in a webpage	F.17	Number of characters in FQDN
F.2	Domain identity in copyright	F.18	Number of characters in URL path
F.3	Domain name in canonical URL	F.19	Shortened URLs
F.4	Domain name in alternate URL	F.20	Free subdomain services
F.5	Foreign domains in hyperlinks	F.21	Domain name's validity
F.6	Void hyperlinks ratio	F.22	Domain age
F.7	Foreign form handler	F.23	Form handler's domain name validity
F.8	URL path encoding	F.24	Form handler's domain age
F.9	Use of '@' character in a URL	F.25	Type of a SSL certificate
F.10	Domain out positioning	F.26	Domain, certificate and geolocation country matching
F.11	Number of dots in FQDN	F.27	URL search engine ranking
F.12	Number of dots in the URL path	F.28	Domain search engine ranking
F.13	Unconventional port numbers	F.29	FQDN search engine ranking
F.14	Obfuscation characters in FQDN	F.30	FQDN blacklist IP counts
F.15	Obfuscation characters in URL path	F.31	Domain blacklist IP counts
F.16	Number of forward slashes		

[a]Domain identity refers to a second level or third level domain label that represent an identity of the website owner. For instance, for a URL https://accounts.google.com/ServiceLogin, *google* is the domain identity.

We have also designed features related to different URL components (FQDN, domain and path), for instance, F.17, F.18, and F.27–31, to detect phishing websites hosted in either compromised or dedicated phishing domains. This is because phishing websites hosted in compromised domains have similarities with their hosts' legitimate websites in many features, including F.1, F.2, F.21 and F.22. For instance, if F.28 flags 'No' then the website is hosted in a dedicated phishing domain, and if F.28, F.29 and F.27 flag 'Yes', 'Yes' and 'No' respectively, the website is likely to be hosted in a compromised domain.

Of all the features, 18 features were based on the webpage's structure and contents while the other 13 features were based on third party services containing information related to a webpage. Along with the use of five different types of features, such diversity enhances resiliency against current obfuscation techniques deployed by phishers to circumvent detection. Third party services such as WHOIS records can hardly be obfuscated by phishers.

Value of each feature was computed by one of the three approaches; matching of feature's conditions to generate 'Yes', 'No' or 'Unknown' values (example F.2, F.9); identifying the string value answering the feature's question (example F.25); and counting of feature's condition to produce a numeric answer (example F.1, F.11).

As our contributions, we have proposed 12 new features (F.2–F.4, F.18, F.20, F.21, F.23, F.25, F.27–F.30) and modified 14 features from previous works (F.1, F.5–F.7, F.9–F.12, F.14, F.15, F.22, F.24, F.26 and F.31). The other five features (F.8, F.13, F.16, F.17 and F.19) were adopted from previous studies to improve the overall performances.

4 Experiments and Results

Experiments were designed and conducted using eight common ML classification algorithms (classifiers) to determine optimal performances and the best performing classifier for the prediction. Also, they were aimed at evaluating the overall framework's prediction runtime per webpage to determine its overheads. We used Python v3.4 and Scikit-learn v.019 library to build an application for the experiments in a 64x Windows Home environment.

We collected 9,019 phishing URLs from an online repository of PhishTank, a blacklist of phishing URLs. We also collected 1,733 legitimate URLs from Google and Bing search engines by querying the engines using search keywords such as *sign in* and *login*. For each collected phishing and legitimate URL, we confirmed if it was prompting PII by passing it through the PII webpage filtering module. We also filtered each legitimate URL against the PhishTank's blacklist. For each URL, we downloaded its webpage and extracted features' data to create a training dataset.

Missing values in continuous features were replaced with their respective mean values. *One hot encoding* method was used to convert all 17 categorical data into numeric to ensure that linear functions based classifiers are trained smoothly. All features' data was re-scaled to a mean of 0 and standard deviation of 1 to optimize the classifiers. We oversampled legitimate (minority) class by SMOTE technique to a 1:1 balanced dataset to ensure we get accurate predictions.

One ML algorithm from each of the eight common classes of binary classifiers was evaluated for the selection of the final classifier. These are Logistic Regression (LR), k-Nearest Neighbour (k-NN), Decision Tree (DT), Gaussian Naïve Bayes (GNB), Support Vector Classification (SVC), Multilayer Perceptrons (MLPs), Random Forest (RF) and Gradient Boosting (GB). We used accuracy, precision, recall and AUC as performance metrics for evaluation. For model evaluations, we applied stratified k-fold cross validation method with k = 10.

Using feature importance method by RF classifier, prediction influence by weight for each feature was determined and ranked as shown in Fig. 2. With all the features, RF performed better across all the metrics compared to other classifiers with an accuracy, precision, recall and AUC of 98.279%, 0.992, 0.987 and 0.997 respectively. After feature selection, RF achieved the highest accuracy of 98.363% with 20 features (from F.31 to F.8 in Fig. 2), as summarized in Table 2. By performing parameter tuning using Random Search followed by Grid Search methods, the RF achieved an optimum accuracy of 98.45% using the best performing parameters automatically determined by the methods. The classifier achieved false positive (FP), false negative (FN) and classification error of 4.47%, 0.73% and 1.7% respectively.

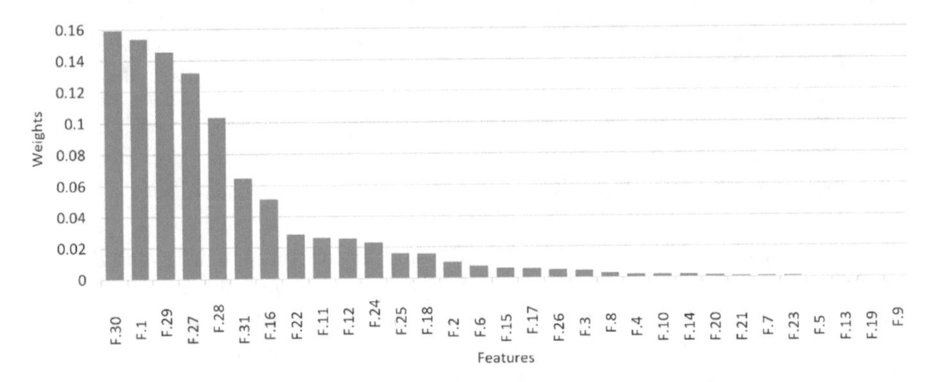

Fig. 2. Distribution of importance weights of the predictive features.

Table 2. Performance scores of all classifiers after applying feature selection.

Classifier	Accuracy (%)	Precision	Recall	AUC
LR	90.234	0.985	0.898	0.962
k-NN	91.239	0.980	0.914	0.947
DT	97.340	0.986	0.983	0.954
GNB	89.137	0.978	0.891	0.938
SVC	91.090	0.984	0.908	0.969
MLP	94.280	0.981	0.951	0.976
RF	98.363	0.994	0.987	0.997
GB	97.638	0.992	0.979	0.995

Each type of features was evaluated individually to determine its performance contributions. Their results in Table 3 shows that domain registration and webpage reputation were the best performers across all metrics whereas SSL certificate was the least performer by far. Similarly, various combinations of types of features were evaluated to analyze their performances (see Table 4). A combination of webpage structure, URL and domain registration features attained the highest performances while that of domain registration and webpage reputation had the lowest performances.

Table 3. RF performance scores of each type of features.

Type of features	Accur. (%)	Prec.	Recall	AUC
Webpage structure and contents	80.265	0.968	0.818	0.748
URL structure	90.485	0.459	0.660	0.888
Domain registration information	99.972	1.000	1.000	1.000
SSL certificate information	34.998	0.211	0.932	0.596
Webpage reputation	97.089	0.898	0.992	0.996

Performance contributions of our new features were also evaluated relative to the adopted features, as summarized in Table 5. Though the new features achieved lesser accuracy than the adopted features, they performed far better in the other metrics and thus contributed significantly to the overall performances of the metrics.

Table 4. RF performance scores of various combinations of types of features.

Combination of types of features	Accur. (%)	Prec.	Recall	AUC
Webpage + URL	95.247	0.743	0.680	0.938
Webpage + URL + Domain	99.851	0.998	1.000	0.998
Webpage + URL + Domain + Reputation	91.137	0.849	0.767	0.943
URL + Domain	99.795	0.998	0.999	0.998
URL + Domain + Reputation	89.806	0.819	0.741	0.931
Domain + Reputation	82.943	0.635	0.682	0.861
Domain + Certificate + Reputation	82.980	0.639	0.672	0.861

Table 5. Performance contributions of new and adopted features on the overall performance.

Subset of features	Accur. (%)	Prec.	Recall	AUC
New + modified	90.867	0.839	0.768	0.938
Adopted	99.591	0.374	0.583	0.876

The runtime of each module was measured to evaluate the framework's prediction time per webpage, thus determining the overall overheads. The prediction runtime, as shown in Table 6, was 7.63 s while training the RF classifier took 16.93 s. We also computed an average downloading time for 1,696 legitimate login webpages used in this study and was found to be 0.843 s.

Table 6. Runtime for each framework's module.

Module	Runtime (s)
PII webpage filtering	0.02030
Phishing blacklist URL	0.27560
URL redirections check	1.69590
Feature data extraction	5.64000
Prediction analysis	0.00012
Total prediction runtime per webpage	**7.63190**
Training a classifier	16.9300

5 Discussions

FN is the most crucial type of an error for this problem, therefore having a relatively small rate is significant in reducing the risk of misdirecting users to phishing webpages. Compared to some of the related works with very high performances (summarized in Table 7), our work compares favorably (in terms of accuracy and FN) against most of them. Other works by [1, 11, 15], with higher accuracy than ours, did not report FNs to compare with. However, our work has used different and more diversified features compared to all works, therefore it is more resilient to detection evasion, in addition to a high performance.

Table 7. Comparison of some of the related works with our work (rates in %).

Study	Acc.	FN	FP	Feature types	Study	Acc.	FN	FP	Feature types
[1]	98.50	–	1.5	4	[13]	99.65	0.34	0.42	4
[8]	97.98	–	–	1	[14]	97.71	–	1.7	4
[9]	–	0.02	1.17	2	[15]	99.55	–	–	4
[10]	97.30	1.61	4.46	2	[16]	92.54	–	0.41	4
[11]	99.09	–	1.25	2	**Ours**	**98.45**	**0.73**	**4.47**	**5**
[12]	92.18	–	–	3					

A slight difference in performances before and after feature selection suggests that all features are collectively effective in the prediction. Although some of the small subsets of features have achieved very high performances, they are still limited with few number of features and diversity of types of features, thus are likely to be vulnerable against detection evasions. A right balance between high prediction performances and resilience to detection evasions should be of high consideration in this problem.

Good performances of the new features suggest that there are many other undiscovered features that are as effective as the previously developed features. This study shows that by combining new features with some of the robust features previously used, new detection models are more likely to yield better performances compared to previous works.

A combined average downloading time for a login webpage and a prediction runtime per webpage, as computed in Sect. 4, is 8.48 s. This time is less than the current average downloading time for all types of webpages, which is 8.66 s [17]. The average downloading time for login webpages is multiple times lesser than that of other types of webpages due to their light weight design, mostly containing few texts only. We therefore argue that our proposed framework brings an insignificant overhead over the current accepted web browsing speed, thus it is potential for real time deployments.

6 Conclusion

We have proposed a framework of new hybrid features to predict zero hour phishing websites using machine learning. A total of 31 features, 26 of them are novel, from five different types of webpage and third party related features were developed to learn the prediction model. Three webpage pre-processing modules were proposed for the framework to improve prediction performance and efficiency. Features' data were extracted and evaluated using eight machine learning classification algorithms, in which Random Forest achieved an optimal accuracy of 98.45% and false negatives of 0.73%. The framework took 7.63 s to predict a new webpage, suggesting that it is promising for real time applications. Further research on new potential features and the use of recent machine learning methodologies such as deep learning and online learning should be pursued to improve prediction performances and efficiency beyond those of the existing works.

Acknowledgement. The research leading to the results presented in the paper was partially funded by the UK Commonwealth Scholarship Commission (CSC).

References

1. Lakshmi, V.S., Vijaya, M.: Efficient prediction of phishing websites using supervised learning algorithms. Procedia Eng. **30**, 798–805 (2012)
2. PhishLabs. https://info.phishlabs.com/2017-phishing-trends-and-intelligence-report-pti. Accessed January 2017
3. Holz, T., Gorecki, C., Rieck, K., Freiling, F.: Measuring and detecting fast-flux service networks. In: Proceedings of 16th Annual Network & Distributed System Security Symposium (NDSS), San Diego, CA (2008)
4. Webroot. https://s3-us-west-1.amazonaws.com/webroot-cms-cdn/8415/0585/3084/Webroot_Quarterly_Threat_Trends_September_2017.pdf. Accessed November 2017
5. Sophos. https://secure2.sophos.com/en-us/medialibrary/Gated-Assets/white-papers/Dont-Take-The-Bait.pdf?la=en. Accessed August 2017
6. Sheng, S., Wardman, B., Warner, G., Cranor, L.F., Hong, J., Zhang, C.: An empirical analysis of phishing blacklists. In: Proceedings of 6th Conference on Email and Anti-Spam, Mountain View, CA (2009)
7. Gupta, B.B., Tewari, A., Jain, A., Agarwal, D.: Fighting against phishing attacks: state of the art and future challenges. Neural Comput. Appl. **28**, 3629–3654 (2017)

8. Sahingoz, O.K., Buber, E., Demir, O., Diri, B.: Machine learning based phishing detection from URLs. Expert Syst. Appl. **117**, 345–357 (2019)
9. Zuhair, H., Selamat, A., Salleh, M.: New hybrid features for phish website prediction. Int. J. Adv. Soft Comput. Appl. **8**, 28–43 (2016)
10. Li, Y., Yang, Z., Chen, X., Yuan, H., Liu, W.: A stacking model using URL and HTML features for phishing webpage detection. Future Gener. Comput. Syst. **94**, 27–39 (2019)
11. Jain, A.K., Gupta, B.B.: Towards detection of phishing websites on client-side using machine learning based approach. Telecommun. Syst. **68**, 687–700 (2018)
12. Mohammad, R.M., Thabtah, F., McCluskey, L.: Predicting phishing websites based on self-structuring neural network. Neural Comput. Appl. **25**, 443–458 (2014)
13. Gowtham, R., Krishnamurthi, I.: A comprehensive and efficacious architecture for detecting phishing webpages. Comput. Secur. **40**, 23–37 (2014)
14. Feng, F., Zhou, Q., Shen, Z., Yang, X., Han, L., Wang, J.: The application of a novel neural network in the detection of phishing websites. J. Ambient Intell. Human. Comput. 1–15 (2018)
15. Rao, R.S., Pais, A.R.: Detection of phishing websites using an efficient feature-based machine learning framework. Neural Comput. Appl. (2018)
16. Xiang, G., Hong, J., Rose, C.P., Cranor, L.: Cantina+ : a feature-rich machine learning framework for detecting phishing web sites. ACM Trans. Inf. Syst. Secur. (TISSEC) **14**, 21 (2011)
17. MachMetrics. https://www.machmetrics.com/speed-blog/average-page-load-times-websites-2018/. Accessed February 2018

Mirkwood: An Online Parallel Crawler

Juan F. García(✉) and Miguel V. Carriegos

RIASC, Instituto CC, Aplicadas a la Ciberseguridad,
Universidad de León, León, Spain
{jfgars,miguel.carriegos}@unileon.es

Abstract. In this research we present Mirkwood, a parallel crawler for fast and online syntactic analysis of websites. Configured by default to behave as a focused crawler, analysing exclusively a limited set of hosts, it includes seed extraction capabilities, which allows it to autonomously obtain high quality sites to crawl. Mirkwood is designed to run in a computer cluster, taking advantage of all the cores of its individual machines (virtual or physical), although it can also run on a single machine. By analysing sites online and not downloading the web content, we achieve crawling speeds several orders of magnitude faster than if we did, while assuring that the content we check is up to date. Our crawler relies on MPI, for the cluster of computers, and threading, for each individual machine of the cluster. Our software has been tested in several platforms, including the Supercomputer Calendula. Mirkwood is entirely written in Java language, making it multi–platform and portable.

Keywords: Crawler · Parallel computation · High performance computing

1 Introduction

Web scraping or harvesting refers to data extraction from websites [16] and can be done manually or automatically using a bot (a web crawler). Crawlers exhaustively visit Internet sites and download (fetch) them into local storage for later retrieval or offline analysis.

In this research we present Mirkwood, a parallel crawler for fast and online syntactic analysis of the content a given set of URL seeds.

Mirkwood strength and novelty, standing out from alternatives in Sect. 5, comes mainly from three features: it can automatically gather its own seeds to crawl, it can perform online (on the fly) analysis of the sites it visits instead of downloading them for offline processing, and it is capable of running in computer clusters, while also optimally working on isolated or single machines.

Our crawler is configured by default to behave as a focused crawler, analysing exclusively a limited set of hosts, but it can work as a general–purpose crawler too.

Regarding availability of our software, we are still evaluating the most suitable distribution solution for it, but not final decision has been taken as of yet.

© Springer Nature Switzerland AG 2020
F. Martínez Álvarez et al. (Eds.): CISIS 2019/ICEUTE 2019, AISC 951, pp. 47–56, 2020.
https://doi.org/10.1007/978-3-030-20005-3_5

The name of the crawler, "Mirkwood", is a homage to J. R. R. Tolkien books: it is the name a fictional forest of Middle-earth (in reality, the name was used for two different forests), whose pockets were dominated by Spiders [12]. Given the architecture of our crawler, based of a forest, and a set of spider nests and spiders (see Sect. 2), we found this to be a suitable name.

The rest of the paper is organized as follows: In Sect. 2, we summarize Mirkwood architecture and implementation. In Sect. 3, we explain seeds extraction process. In Sect. 4, we describe our tests and initial results. In Sect. 5, we sum up related work. Finally, in Sect. 6, we summarize conclusions and envision future work.

2 Architecture

Mirkwood is composed of six modules, the first being autonomous (it is able to work independently from the rest of the application): Seed Extraction, Environment, Forest, Spider, Spider Nest, and Spider Leg.

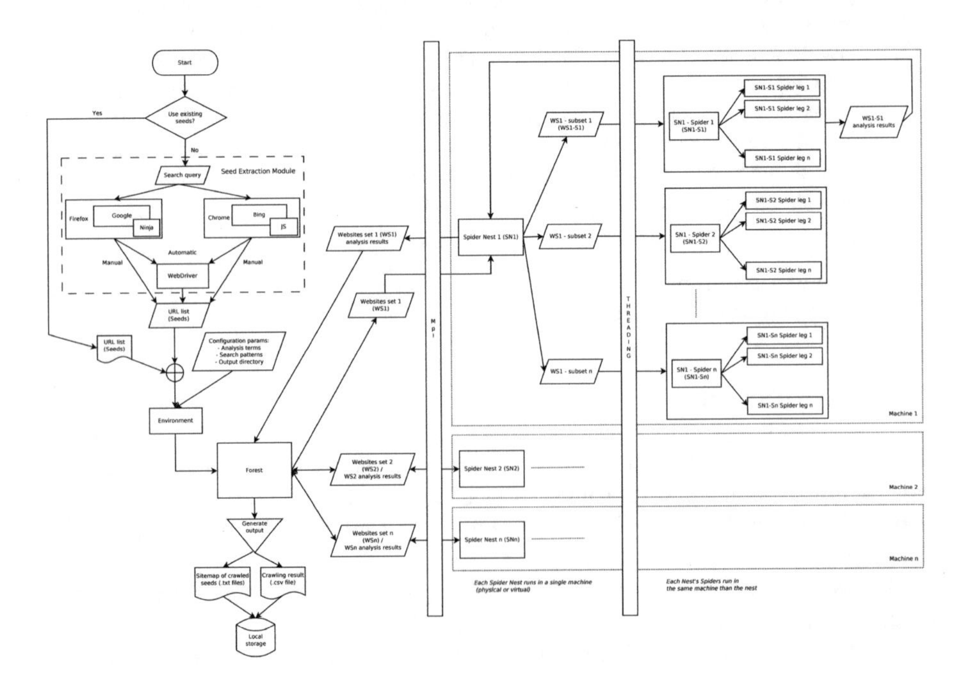

Fig. 1. Mirkwood architecture

Figure 1 shows the architecture of the whole system. Please not that this figure is meant to be properly displayed only in the online version of this paper, since it contains small text that needs zoom in to be read (a full resolution copy of this image is available at [5]).

In the following, we describe all crawler modules in detail, also giving a brief summary of crawling policy followed and technologies used.

2.1 Seed Extraction

This modules feeds the Environment module with the URL of the hosts to crawl. Given its autonomous nature (it can also be run as a standalone application), it is explained separately in Sect. 3.

2.2 Environment

Environment module runs in the main machine, reading all configuration files as well as the list of initial hosts to crawl, which gets from Seed extraction module output. Alternatively, Environment module can get the URL list from an external text file (which can be the result of a previous execution of the Seed extraction module or can provided by a user).

Some of the most relevant parameters set up by this module are the analysis terms (terms to look for at the different domains), the search patterns to perform the searches, and the output directory (or directories) where the crawler will store its results. All this information is sent to Forest module.

2.3 Forest

Forest is the main module of the Mirkwood crawler. It sets up and starts the crawling process according to the information and requirements established by the Environment module.

To start the crawling process, the Forest module creates a set of Spider Nests to which it distributes the target seeds. All interactions and message passing between the Forest module and each Spider Nest is done via Open MPI [7]. Spider Nests do not communicate with each other directly.

A Spider Nest is created for every machine available in the computer system which is running the crawler. Please see Sect. 2.4 for more information regarding how each particular nest operates.

Once the Spider Nests have been created, the Forest module distributes the websites to crawl among them. A lightweight analysis of machine specifications and websites complexity is performed to try to optimize load balance among the servers. In Fig. 1, the set of pages sent to a specific machine are labelled as "*Website set n*" and shortened as "*WSn*".

Forest module periodically receives analysis results from every Nest it has created every time a Spider finishes crawling a full domain. Upon receiving an analysis report, the Forest module updates its (and crawler's) output files, the crawling results file, and the site map of the domains that are being analysed.

2.4 Spider Nest

Spider Nest module receives the list of websites to crawl from Forest module, to whom it will also report progress. There is a maximum of one Spider Nest in every machine (physical or virtual).

Spider Nest functionality encompasses Spiders creation and synchronization, and report–forwarding from Spiders to Forest: every time a report message is received from a Spider, it is immediately sent to Forest module.

The set of pages to crawl received from Forest (labelled as "$Website\ set\ n$" and shortened as "WSn" in Fig. 1) is split into subsets which are distributed between Spiders, one subset for each Spider (these subsets are labelled as "$WSn - subset\ m$" –WSn identifies the package split in Forest and received at the Nest– and are abbreviated as "$WSn - Sm$" in Fig. 1).

Spider Nest communication with Forest is achieved via Open MPI, while information flow to each Spider in the Nest is enabled through Threading (using semaphores to prevent race conditions).

2.5 Spider and Spider Legs

Spider module gets a list of domains to crawl from its Nest. Each Spider goes over the list of hosts it receives, crawling them one by one. By default, Spider module selection policy is set so that crawled URL and content is always internal to the domains in this list.

The crawling process itself is performed by Spider Legs. Spiders have no legs when they are created. For every URL (starting at the domains' home page), the Spider spawns a leg (a Spider Leg), which analyses its content and extracts all links it contains; every time a given URL is fully analysed, the leg which took care of it is de–spawned. Spider legs are not run in the order they are spawned, but in a random manner (this is preferred over fetching them in the order they are parsed out; see [19]). No synchronization is needed since each Spider and its Spider Legs are running in the same machine in a sequential manner.

Our software is capable of performing basic analyses in search of patterns, but human validation may be needed for more complex analysis or proper result interpretation.

The crawling process for each Spider continues until it has no legs left, which means the domain has been fully crawled. As soon as a site is fully crawled, the Spider sends a site final report to its parent Nest.

2.6 Technologies

The application is developed in Java SE 8 following Object–Oriented Programming paradigm (OOP), so each of this modules directly translates into a Java class.

Communication with cluster's machines (happening between Forest and Spider Nest modules) is implemented via Open MPI [7]. Threads used for communication within a machine (happening between a Spider Nest and its Spiders) are an implementation of java.lang.Runnable.

To connect and extract html content from domains, Mirkwood uses jsoup, a Java library for working with real-world HTML. Jsoup is designed to deal with all varieties of HTML found in the wild, creating sensible parse trees [4].

3 Seed Extraction Autonomous Module

Mirkwood's autonomous seed extraction module gathers seeds which are used as input for a crawler (any crawler, not just Mirkwood itself, thus the term "autonomous" used for this module's name). A "seed" is the starting point from where we initiate the crawling process, recursively extracting the links it contains [14].

3.1 Seed Extraction Process

Seeds are obtained in three consecutive steps: Access to the search engine, search, and URL extraction. These steps can be performed either manually or automatically; to achieve automation, the extraction module uses the third party software tool Selenium WebDriver [17].

Our main contribution regarding seed extraction is not just the automation of the process (since the process is already fast enough when done manually), but the quality and quantity of the seeds obtained, as well as the module itself being autonomous, thus usable by other crawlers.

Access to the Search Engine. We use Google or Bing, accessed via the browser Firefox or Google Chrome, respectively. The reason for using one or the other has been to take advantage of other existing software tools which are engine and browser-specific, as well as getting more variety for the seeds gathered.

Search. Search engines group the results by pages, showing a specific set of them (10 by default) in each of them. To reduce the number of queries that the software must perform, we configure the search engines to print the highest number of search results per page (100 in Google and 50 in Bing). By doing this, we also achieve a significant increase in the number of results we obtain.

It is also relevant to point out that search engines usually tell the user that they have found tens or hundreds of thousands, or even millions of coincidences (results), but the user is only given a fraction of them (a maximum of between 100–200 results). If we configure the search so that each page contains more results, we also increase the number of results obtained: by doing so, we can collect several hundred of seeds, even thousands of them for a single search query.

Since the seed extraction module relays in the aforementioned search engines, it can perform both simple searches (i.e. "buy veterinary antibiotics" or "cheap hotel in Madrid", without the quotes) and complex searches (based of patterns and operators; i.e. "%22buy OR purchase%22 %22veterinary OR pet%22 %22antibiotics%22 -%22no antibiotics%22 %22pharmacy OR online store%22").

URL Extraction. Contrary to popular belief, the terms of service of search engines do not prohibit automatic search (using software tools such as crawlers).

What they do prohibit is those searches that are done using means other than those provided by them (example: Graphical interface of the engine) or that abuse the system (automatic searches which make excessive use of the service) [2].

To comply with terms of service, we collect URL directly from the engine interface in the browser (which is known as SERP, Search Engine Results Page). In order to group and collect all the URLs from the end of the SERP, we use additional specific and complementary software for each scenario: In the case of Google, we have equipped the Firefox browser with Greasemonkey and an Internet Marketing Ninjas plug-in [6]. In the case of Bing, we have developed a JavaScript plug-in for Google Chrome with analogous functionality.

The quality of seeds obtained is improved by reducing both human fatigue and propensity for human error, and by automatically discarding ads and fraudulent sites.

4 Experimental Results

In this section, we include some prospective and initial results of our crawler.

4.1 Qualitative Results

We have successfully put Mirkwood to the test for a research project, yet to be published, which dealt with illegal selling of antibiotics for veterinary use through the Internet (Clear Web).

For this study, both Mirkwood and Heritrix (see Sect. 5) crawlers were used in order to compare results obtained from both of them.

Regarding Mirkwood, we configured it to use both simple and complex search strings to automatically gather seeds. The crawler then performed a basic analysis in search for patterns (to conclude if a certain term appears on a web, for example "amoxicillin", "veterinary" or "buy") and human validation was used to evaluate the context in which those terms appear to confirm whether the site really sells these kind of substances or not. The latter was carried out by Pharmacology experts.

Mirkwood managed to crawl and analyse several thousand of potential target (antibiotic–selling) sites, and the research team was overall very satisfied with the crawler performance.

Regarding Heritrix, since it doesn't include seed or URL extraction functionality, URL obtained by Mirkwood Seed Extraction module were used as seeds for it.

After both crawlers finished their jobs, we checked if they have both crawled all the same links (they should, since they used the same seeds): We obtained a 98.68% match up between both tools, with some links only visited by Heritrix and some others only visited y Mirkwood. Upon further inspection, we verified that the difference was due to 11 domains: 6 domains were only crawled by Heritrix, and 4 were only crawled by Mirkwood. There was one domain in which Mirkwood crawled more links that Heritrix.

Upon checking Heritrix log files and Mirkwood result files, we verified that the domains couldn't be visited (by either crawler) because of $404 - -notfound$ or $503 - -serviceunavailable$ errors, which is very likely due to the website being unavailable the day we put that crawler to the test.

From this it follows that our crawler successfully (or at least in a similar fashion as a widely used tool as Heritrix) extracts and follows websites site maps.

4.2 Performance Results

In respect to the testing platform, to validate our implementation from an efficiency point of view (to check if the software does what it is supposed to do as fast as it can), we focused on single machine performance (thus Mirkwood used just a single Spider Nest).

Our software was tested in two platforms, a high performance laptop equipped with an Intel Core CPU i7-6820HK @2.7 GHz of 4 cores each (thus totalling 8 cores) with 16 GB RAM memory, and The Supercomputer Calendula, from SCAYLE (see acknowledgements). For this experiments, we focused on single node computing, and the single best machine (as of May 2018) of SCAYLE calculation cluster was used: an Intel Xeon CPU E5-2670 v2 @ 2.50 GHz of 10 cores each (thus totaling 20 cores) with 128 GB RAM memory.

Performance wise, we didn't stablish a comparison with Heritrix, since both crawlers are different because of the their functionality (Mirkwood is much faster by nature, since it doesn't download content, just analysis it). Instead, we were interested in assessing the quality of Mirkwood parallelization (check if using several Spiders/threads was in fact improving performance over using just one).

In order to put that to the test, we run two configurations of Mirkwood: A single–Nest/single–Spider configuration, and a single–Nest/multi–Spider one.

When running Mirkwood using the single–Nest/multi–Spider configuration instead of the single–Nest/single–Spider one (which is the equivalent of running a parallel implementation of a program instead of the sequential one), we achieved execution times close to six times faster in the laptop and close to eighteen times faster in the Supercomputer in the best scenario (when we had at least 6 domains –18 for the Supercomputer– to crawl). Results in this regards are consistent with our experience in a different asset which also relayed on parallelization [13].

5 Related Work

There is a great variety of web crawlers. Almost all of them use multi-threaded technology to visit URL and harvest their content, as our Spider Nest module does when managing its Spiders.

The biggest advantage of Mirkwood in comparison to other crawlers is that it offers seeds extraction, site crawling and online result analysis in a single tool, unlike other web scrapers, which just allow for crawling and downloading of sites for later processing. Our crawler is also able to optimally run in both isolated machines and computer clusters, while the rest of web scrapers are designed for just one of those scenarios.

In the following, we summarize some of the more relevant focused and general–purpose crawlers published. Crawlers specifically aimed to Dark Web or Deep Web are not considered.

5.1 Focused Crawlers

Focused crawlers harvest content from webs that satisfy a special property. They can crawl pages only from specific domains, or about a specific topic or concept (in which case they are sometimes called "topical crawlers") [8].

As already anticipated in Sect. 1, Mirkwood is configured by default to work as a topical focused crawler. This default behavior can however be turned off, in which case Mirkwood would act as a general–purpose crawler.

The following are some crawlers know to also allow for focused crawling:

ARACHNID, a distributed algorithm for information discovery from early days of the Web, is the precursor of the topical crawling [15].

As an alternative to use external search engines (as we do with Mirkwood), crawlers can also rely on variants of reinforced learning and evolutionary adaptation [9].

Semantic focused crawlers (based of domain ontologies for selection purposes) are enhanced by various semantic web technologies [11]. The major drawback of this approach is that it requires a specific type of web content to work.

Other relevant examples of focused crawlers are new–please, Scrapy, Twitterecho, HAWK, Treasure–Crawler, VRPSOFC, SortSite, or Gnutella.

5.2 General–Purpose Crawlers

As presented in Sect. 1, crawlers exhaustively visit Internet sites and download them for later retrieval or offline analysis. Most general–purpose crawlers can (and have to) be customized to some extent before using. Mirkwood can act as general–purpose upon proper configuration of its modules (see Sects. 2.2 and 3).

Heritrix, Apache Nutch, FAST Crawler, WebRACE, PolyBot, and the crawlers behind search engines are examples of general–purpose crawlers.

Heritrix is a fully configurable and free web crawler designed for web archiving by the Internet Archive [3]. Heritrix recursively and automatically crawls a list of websites (seeds), downloading all their contents to the server it is running on.

Apache Nutch [1] is a highly extensible and scalable (allowing developers to create specific plug–ins) open source web crawler software project. Nutch can run on a single machine, and gains a lot of its strength from running in a Hadoop cluster.

FAST Crawler consists of a cluster of interconnected machines, each of them is assigned a partition of Web space for crawling. These machines exchange information about discovered hyperlinks [18]. WebRACE is a distributed WWW retrieval, annotation, and caching engine [10]. PolyBot consists of a crawl manager, one or more downloaders, and one or more DNS resolvers [19].

6 Conclusions and Future Work

In this research we have presented Mirkwood, a parallel crawler for fast and online syntactic analysis of websites. Mirkwood's main contributions are its Seed extraction capabilities, which allows it to autonomously obtain high quality hosts to crawl from a search query instead of a initial set of URL; its online analysis functionality, which speeds the crawl up by not downloading visited websites; and its dynamically adaptable parallel implementation, which allow it to work on both isolated machines (through threading) and computers clusters (through MPI).

Our proposal is multi–platform, portable, and greatly improves performance of existing crawlers when we are not interested in downloading web content, but just performing specific analysis on it.

Initial results are very promising: We have tested Mirkwood in several platforms, including the Supercomputer Calendula. By not downloading the web content, we achieve crawling speeds several orders of magnitude faster than if we did, while assuring that the content we check is up to date. Even if maximum performance is achieved in computer clusters, the crawler can be run on individual machines too, where it will make use of all logical processors available in the machine running the code.

As future work, we would like to generate seeds files containing links which the crawled tried but couldn't access, since these files can be useful for the user to either start a new crawl to retry scraping them or to manually check what went bad in every case. Also, while many sites can be archived using the Standard crawling technology without issue, some dynamic content proves to still be challenging to archive. Finally, analysis performed by Spider Leg module can be enhanced to enable semantic interaction with crawled content. It would also be interesting to consider using Mirkwood for real environments and critical applications.

Acknowledgements. This work has been partially supported by the Spanish National Cybersecurity Institute (*Instituto Nacional de Ciberseguridad*, INCIBE). This research uses the resources of the *Centro de Supercomputación de Castilla y León* (SCAYLE, www.scayle.es), funded by the "European Regional Development Fund (ERDF)".

References

1. Apache nutch (2018). http://nutch.apache.org/. Accessed 15 Nov 2018
2. Google terms of service (2018). policies.google.com. Accessed 15 Nov 2018
3. Heritrix (2018). https://github.com/internetarchive/heritrix3. Accessed 15 Nov 2018
4. jsoup: Java html parser (2018). https://jsoup.org/. Accessed 15 Nov 2018
5. Mirkwood (2018). http://bit.do/eKLo2. Accessed 15 Nov 2018
6. Ninja plugin (2018). https://www.internetmarketingninjas.com/seo-tools/get-urls-grease. Accessed 15 Nov 2018

7. Open mpi (2018). https://www.open-mpi.org. Accessed 15 Nov 2018
8. Chakrabarti, S.: Focused web crawling. In: Encyclopedia of Database Systems, pp. 1147–1155. Springer (2009)
9. Chakrabarti, S., Punera, K., Subramanyam, M.: Accelerated focused crawling through online relevance feedback. In: Proceedings of the 11th International Conference on World Wide Web, pp. 148–159. ACM (2002)
10. Dikaiakos, M., Zeinalipour-Yiazti, D.: WebRACE: a distributed www retrieval, annotation, and caching engine. In: Proceedings of PADDA01: International Workshop on Performance-Oriented Application Development for Distributed Architectures, April 2001
11. Dong, H., Hussain, F.K.: SOF: a semi-supervised ontology-learning-based focused crawler. Concurrency Comput. Pract. Experience **25**(12), 1755–1770 (2013)
12. Fisher, J.: Dwarves, spiders, and murky woods. In: CS Lewis and the Inklings: Discovering Hidden Truth, pp. 104–115 (2012)
13. García, J.F., Carriegos, M.: On parallel computation of centrality measures of graphs. J. Supercomput., 1–19 (2018)
14. Jamali, M., Sayyadi, H., Hariri, B.B., Abolhassani, H.: A method for focused crawling using combination of link structure and content similarity. In: Proceedings of the 2006 IEEE/WIC/ACM International Conference on Web Intelligence, pp. 753–756. IEEE Computer Society (2006)
15. Menczer, F.: ARACHNID: adaptive retrieval agents choosing heuristic neighborhoods for information discovery. In: Machine Learning-International Workshop Then Conference, pp. 227–235. Morgan Kaufmann Publishers, Inc. (1997)
16. Munzert, S., Rubba, C., Meißner, P., Nyhuis, D.: Automated Data Collection with R. JW & Sons (2014)
17. Ramya, P., Sindhura, V., Sagar, P.V.: Testing using selenium web driver. In: 2017 Second International Conference on Electrical, Computer and Communication Technologies (ICECCT), pp. 1–7. IEEE (2017)
18. Risvik, K.M., Michelsen, R.: Search engines and web dynamics. Comput. Netw. **39**(3), 289–302 (2002)
19. Shkapenyuk, V., Suel, T.: Design and implementation of a high-performance distributed web crawler. In: 2017 Second International Conference on Data Engineering, Proceedings, pp. 357–368. IEEE (2002)

Improving SVM Classification on Imbalanced Datasets for EEG-Based Person Authentication

Nga Tran[1(✉)], Dat Tran[1], Shuangzhe Liu[1], Linh Trinh[2], and Tien Pham[2]

[1] Faculty of Science and Technology, University of Canberra, Canberra, ACT 2601, Australia
nga.tran@canberra.edu.au

[2] Information Security Operations Center, Vietnam Government Information Security Commission, Hanoi, Vietnam

Abstract. Support Vector Machine (SVM) has been widely used in EEG-based person authentication. Current EEG datasets are often imbalanced due to the frequency of genuine clients and impostors, and this issue heavily impacts on the performance of EEG-based person authentication using SVM. In this paper, we propose a new bias method for SVM binary classification to improve the performance of the minority class in imbalanced datasets. Our experiments on EEG datasets and UCI datasets with the proposed method show promising results.

Keywords: SVM · EEG · Authentication · Security · Biometrics

1 Introduction

Recently, using electroencephalogram (EEG) signal as a new type of biometric modality has been established [14] where brain-wave patterns corresponding to particular mental tasks are considered pass-thoughts. Due to the distinctive advantages, such as being impossible to fake or intercept and requiring alive person in EEG recordings, research in EEG-based person authentication has been widely published. There are a variety of classifiers to ensure the best performance, of which Support Vector Machine (SVM) is dominant [1,15].

A typical EEG-based person authentication system consists of two phases: enrollment and verification. In the enrollment phase, a person is required to perform a mental task, such as imagining moving a hand or calculating a simple mathematic expression. The EEG signal corresponding to the task is recorded and pre-processed, and extracted features that form feature vectors are fed into the classifier to train the model for that person.

In the verification phase, a claimed person is required to repeat the task performed in the enrollment phase to login to the security system. The EEG signal related to the task is recorded, pre-processed, and features are extracted similarly to those in the enrollment phase. The obtained features are formed

© Springer Nature Switzerland AG 2020
F. Martínez Álvarez et al. (Eds.): CISIS 2019/ICEUTE 2019, AISC 951, pp. 57–66, 2020.
https://doi.org/10.1007/978-3-030-20005-3_6

feature vectors that are then provided to the classifier to compare with the claimed model to calculate a matching score. Based on this score and a given threshold, the EEG-based person authentication system can verify the claimed user.

The person whose model is trained is a genuine user while other persons are impostors. In EEG datasets, the number of impostors is large and hence the dataset for genuine users is much smaller than the dataset for all impostors, which results in an imbalance. This is a challenge since SVM provides low performance on the imbalanced datasets [3]. Several solutions have been introduced to address this challenge. These solutions can be divided into pre-processing, training, and post-processing strategies.

The pre-processing strategy aims to balance the dataset using a re-sampling technique such as over-sampling or under-sampling. While the over-sampling technique attempts to increase minority class instances, the under-sampling technique attempts to remove a subset of the majority class to reduce its proportion in the training dataset. One of the most popular techniques of this pre-processing strategy is SMOTE [6].

The training strategy aims to modify the standard optimization problem for SVM during learning in order to minimize the cost of error or changing the hyperplane structure. An example of this strategy is weighted-SVM as seen in [7] where the improvement is achieved by optimizing the hyperplane.

The post-processing strategy is applied after the model has been built, aiming to justify weight w or bias b parameters so that SVM can adapt to imbalanced data such as z-SVM [9] and new bias SVM [17]. In z-SVM [9], the improvement focuses on the impact of support vectors on the minority class through a small positive value z for those vectors.

In new bias SVM [17], the standard bias b in Eq. (1) is redefined as seen in Eqs. (2) and (3).

$$b = \frac{1}{N_{SV}} \sum_{i \in SV} \left(y_i - + \sum_{j \in SV-} \alpha_j y_j x_i <x_i, x_j> \right) \tag{1}$$

$$b_p = \frac{N^+\alpha + N^-\beta}{N^+ + N^-} \tag{2}$$

$$b_{p1} = \frac{N_{SV1}\alpha + N_{SV2}\beta}{N_{SV1} + N_{SV2}} \tag{3}$$

where

$$\alpha = \max_{x_k \in SV2} \sum_{i=1}^{N} \alpha_i K(x_i, x_k) \tag{4}$$

$$\beta = \min_{x_k \in SV1} \sum_{i=1}^{N} \alpha_i K(x_i, x_k) \tag{5}$$

Recently, the experiments in [17] on 34 imbalanced datasets from UCI [2] show impressive results compared to z-SVM and SMOTE methods. This reveals that bias justifying could be a good choice for SVM when the number of instances of the negative class is dominant compared to the remaining one. However, bias calculation in Eqs. (2) and (3) depends on both the numbers of feature vectors and support vectors. This means any changes in the dataset can impact on the hyperplane position although that new data could be far from the hyperplane. As a result, a more flexible bias improvement for the imbalanced datasets in EEG-based person authentication system is needed.

This paper's format is structured as follows. In Sect. 2, we propose a new post-processing strategy based on the bias for SVM to deal with the imbalanced datasets in EEG-based person authentication. Experiments and results are presented in Sect. 3. We conclude the paper with a discussion and our future work in Sect. 4.

2 New Bias Proposal for Imbalanced Datasets in SVM

It is worth pointing out that a confusion matrix is usually used to summarize prediction results for a classification problem with the values of true positives (TP), true negatives (TN), false positives (FP), and false negatives (FN). Based on confusion matrices, some popular measures are formulated to evaluate the classification model such as *Precision*, *Recall*, ACC^+ (Accuracy for positive class, also called Sensitivity), ACC^- (Accuracy for negative class, also called Specificity) as follows:

$$Precision = \frac{TP}{TP + FP} \qquad Recall = \frac{TP}{TP + FN} \tag{6}$$

$$ACC^+ = \frac{TP}{TP + FN} \qquad ACC^- = \frac{TN}{TN + FP} \tag{7}$$

However, in imbalanced datasets, there are significant differences between FP and FN values because the numbers of instances for positive and negative classes are unequal. This leads to imbalanced *Precision* and *Recall*. Also, it causes ACC^+ and ACC^- to be differential. Therefore, the geometric mean (g-mean) [10] of ACC^+ and ACC^- and the harmonic mean of *Precision* and *Recall* ($F1$-score) [13] are used for a comprehensive evaluation of a classification model. $F1$-score and g-mean are computed as follows:

$$F1\text{-score} = \frac{2 * Precision * Recall}{Precision + Recall} \tag{8}$$

$$g\text{-mean} = \sqrt{ACC^+ * ACC^-} \tag{9}$$

Similar to other machine learning algorithms, g-mean value is considered as a measure to qualify the trained hyperplane in SVM. The more separation between two classes the hyperplane can achieve, the greater the g-mean on training data is, and that means the hyperplane is more reliable. When a training dataset is imbalanced, standard SVM usually provides a trained hyperplane far from the majority class, so it does not clearly separate two classes or provide a high g-mean value. As a result, increasing g-mean value is the goal of justifying the hyperplane to an optimized position between two classes. This can be seen in the previous imbalanced data improvement approaches for SVM. For example, in weighted-SVM methods, g-mean was an objective function to embed weight in the model. For z-SVM methods in [9], z-value was defined to increase the weights of the positive support vectors in order to maximize the g-mean value. Similarly, a higher obtained g-mean value after redefining bias b is considered to be a useful measurement to compare other improvement proposals in [17].

As aforementioned, recent publications on improving SVM for imbalanced datasets in [17] proposed a new bias as defined in Eqs. 2 and 3 with impressive g-mean values on 34 imbalanced UCI datasets compared to the standard SVM, weighted-SVM, and z-SVM. However, new bias formulas in [17] seem not to be suitable with almost all imbalanced datasets. For example, let A equal an imbalanced dataset where formulas in [17] work well. When some instances x are added to A, and x are far from the hyperplane, the proposal bias is modified and that leads to a change in the position of the hyperplane while the old hyperplane is actually better than the updated one.

As illustrated in Fig. 1, the circles present data in the negative class, and the triangles present the positive class in which the circles are the majority. Assume that the standard hyperplane is L. After applying new bias in [17], the imbalanced dataset provides hyperplane L2 that is in a good position to separate two classes. If some data or support vectors of negative class far from the current hyperplane are added to this dataset, the new bias in [17] forces L2 to L1. It can be seen that L1 actually is not superior to L2 because the added data or support vectors do not impact on the hyperplane. This means the new bias in [17] still has its own limitations when it depends on both numbers of data instances and support vectors while not always increasing the quality of the hyperplane.

We propose a method to enhance determining bias value. Assume that b_0 is the obtained bias after training the model. The proposed new bias is defined as follows:

$$b^* = b_0 * \epsilon \tag{10}$$

where ϵ is defined by maximizing g-mean on the training dataset.

The decision function is redefined as follows:

$$f(x, \epsilon) = \sum_{x_i \in SV} \alpha_i y_i K(x, x_i) + b_0 * \epsilon \tag{11}$$

The objective function is maximizing g-mean value, and the optimization problem is:

$$\max_{\epsilon} J(\epsilon) = \sum_{x_i, y_i \in T; y_i > 0} I(y_i f(x_i, \epsilon)) * \sum_{x_i, y_i \in T; y_i < 0} I(y_i f(x_i, \epsilon)) \tag{12}$$

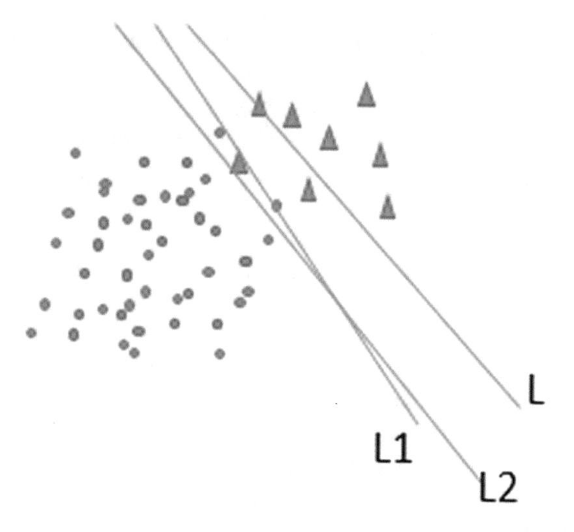

Fig. 1. Illustration of hyperplane changing position caused by added data

where

$$I(u) = \begin{cases} 1 & u \geq 0 \\ 0 & u < 0 \end{cases} \tag{13}$$

When ϵ changes in a range, it changes the value of g-mean that can reach a maximum value before reducing because of the change of ϵ. We apply Golden Section Search algorithm [8] to find ϵ. The optimized value ϵ^* is the one that makes g-mean reach the maximum value on the training set. Some ranges will be investigated to observe and discover the best one for searching the optimized ϵ^*. The performance of the proposed method will be evaluated on both $F1$-score and g-mean values.

3 Experiments and Results

We conducted experiments following two steps. First, the proposed method was tested on binary datasets from the UCI repository [2] to prove itself in terms of improving the imbalanced data in SVM. Next, it was applied to an EEG-based person authentication system where the number of feature vectors of genuine users in the positive class in training datasets are minor compared to the instance number of impostors in the negative class.

The performance of the new bias proposal was experimented on 21 UCI binary datasets which are in different levels of imbalanced positive and negative classes as shown in Table 1. The LibSVM [5] was used with a separate training set and testing set. The feature vectors related to each class were divided into 70% for training and 30% for testing as it is a popular proportion in research community. The LibSVM [5] was used with the RBF kernel function $K(x_i, x_j) = e^{-\gamma||(x_i - x_j||^2}$.

Regarding classifier parameters, the 5-fold cross validation was applied to the training set and then the best found parameters were used to train the models. The parameters for training were γ and C in which γ was searched in $\{2k : k = -4, -3, ..., 1\}$ while C was searched in $\{2k : k = -2, -1, ..., 3\}$. The found best C and γ can be seen in Table 1.

As aforementioned, the performance of the new bias determining method was measured by both the $F1$-score and g-mean which are represented in Table 1. These results reveal some interesting findings. Firstly, some UCI binary datasets had high accuracy whereas the values of g-mean and $F1$-score were low. That is an indicator of serious imbalance between the accuracy of positive and negative classes. This result is consistent with previous studies such as [17], and it is confirmed that the $F1$-score and g-mean are correct metrics to evaluate classifier performance in imbalanced datasets. Secondly, different datasets provided different g-mean improvement although all of them achieved greater g-mean values. For example, Bupa, Ecoli1, Heart, and Pima datasets did not have considerable improvement compared to the remaining datasets. This could be caused by different levels of imbalances of positive and negative classes in those datasets.

Compared with previous studies in the literature, new bias in [17] and z-SVM [9] methods were used. As aforementioned, the bias calculation strategy in [17] depends on both the number of feature vectors and support vectors. Therefore, modified bias could change the position of the hyperplane while the old hyperplane is actually better than the updated one when new data, that is far from the old hyperplane, is added. The proposed method can avoid that limitation and might be more flexible for imbalanced datasets. Compared to z-SVM [9] where five datasets were tested and only Yeast dataset is in the binary UCI datasets, which are used in our study, and in this case our method shows a superior result with g-mean = 0.8231 compared to their result with g-mean = 0.7281.

Having good experimental results from chosen UCI binary datasets, we applied the proposed new bias to improve the EEG-based person authentication where users use their brain-wave patterns corresponding to mental tasks as credentials to login to the security system. The Graz BCI2008A [4] and BCI2008B [12] datasets are selected for conducting experiments as they are well-known for BCI (Brain-Computer Interface) systems. Both datasets contain EEG signals of 9 people. In the BCI2008A dataset, subjects performed four covert tasks including the motor imagery of left hand, right hand, foot, and tongue while BCI2008B participants performed two motor imagery of left hand and right hand. EEG signals were acquired at a sampling frequency of 250 Hz.

For signal pre-processing, a bandpass filter between 0.5 Hz and 100 Hz was used, and then the 50 Hz notch filter was applied to eliminate the power line noise. The EEG signals were investigated in the length of 6 s and 7.5 s segments for BCI2008A and BCI2008B datasets, respectively, based on expert knowledge in previous studies [16].

Table 1. Datasets with parameters, and $F1$-score and g-mean values for each dataset

Name of dataset	% of positive labels	bestC	bestG	Original $F1$-score	Original g-mean	Improved $F1$-score	Improved g-mean
Banana	44.83	32	0.500	0.4873	0.5704	0.6745	0.7025
Appendicitis	20.48	64	0.062	0.6000	0.6934	0.7692	0.8771
Btissue	34.51	128	1,000	0.5882	0.6580	0.6400	0.7207
Bupa	42.04	16	2,000	0.5000	0.5774	0.5169	0.5743
Ecoli1	22.98	128	0.016	0.8182	0.8675	0.7170	0.8424
Ecoli3	10.51	64	0.125	0.7368	0.8273	0.6923	0.9110
Glass0	32.86	128	0.031	0.2759	0.4156	0.6071	0.6860
Haberman	26.49	64	1,000	0.2857	0.4355	0.5424	0.6911
Heart	44.48	128	1,000	0.6944	0.7244	0.6849	0.7136
Page-blocks0	10.22	64	0.062	0.7305	0.7801	0.6500	0.9112
Pima.mat	34.90	128	2,000	0.5730	0.6652	0.6108	0.6781
Pima-5-1tra	34.87	32	0.500	0.4694	0.5713	0.6667	0.7310
Pima-5-2tra	34.87	128	4,000	0.5968	0.6836	0.6165	0.7008
Pima-5-3tra	34.87	8	4,000	0.6061	0.6922	0.6447	0.7189
Pima-5-4tra	34.97	2	4,000	0.6387	0.7139	0.6939	0.7644
Pima-5-5tra	34.97	0.5	1,000	0.6239	0.6947	0.6974	0.7659
Vehicle1	25.70	128	0.125	0.5893	0.6856	0.7108	0.8402
Vehicle3	25.11	32	1,000	0.5138	0.6312	0.6260	0.7436
Wisconsin	35.01	64	0.250	0.9517	0.9641	0.9353	0.9430
Yeast	20.49	128	0.500	0.7290	0.8054	0.6550	0.8231
Robot	21.15	64	0.250	0.8717	0.9147	0.8377	0.9359

For feature extraction, the signals from all three channels C3, C4, and Cz were selected to extract the autoregressive (AR) model parameters applying Burg's lattice-based method with the order 21st as suggested by authors in [16].

In addition to AR model parameters, power spectral density (PSD) was also extracted as features using Welch's averaged modified periodogram method [18] in the band 8–30 Hz resulting in 12 power components. Combining AR and PSD parameters, the feature vectors consist of $3 * (12 + 21) = 99$ features. They were divided into 60% for training and 40% for testing as suggested by EEG processing research community such as [21] and [11]. Similar to the above UCI datasets experiments, LibSVM with RBF kernel was applied with both original bias and the proposed one. Tables 2 and 3 present performances of the EEG-based person authentication system when users elicited EEG signals by covert tasks in BCI2008A and BCI2008B datasets, respectively. These measurements are with decision threshold = 0.

An EEG-based person authentication system is evaluated based on two types of errors: False Acceptance (FA) and False Rejection (FR). FA error occurs when the system accepts an impostor and it is measured by False Acceptance Rate (FAR) while FR error happens when the system rejects a true client and it is weighed by False Rejection Rate (FRR) as follows:

$$FAR = \frac{FP}{FP + TN} \qquad FRR = \frac{FN}{FN + TP} \tag{14}$$

Table 2. Average performance of the EEG-based person authentication system with BCI2008A dataset when users perform 4 tasks (left hand, right hand, foot, and tongue motor imagery).

Subject ID	Original FAR	Original FRR	Improved FAR	Improved FRR	Original HTER	Improved HTER
A01	0.0012	0.3362	0.0271	0.0954	0.1687	0.0613
A02	0.0000	0.8223	0.0373	0.2425	0.4112	0.1399
A03	0.0000	0.6845	0.0486	0.0840	0.3422	0.0663
A04	0.0025	0.6963	0.1118	0.0577	0.3494	0.0847
A05	0.0013	0.7180	0.0793	0.1565	0.3598	0.1022
A06	0.0000	0.9413	0.0886	0.2045	0.4707	0.1466
A07	0.0000	0.3008	0.0099	0.1882	0.1504	0.0991
A08	0.0000	0.9310	0.0431	0.1685	0.4654	0.1059
A09	0.0000	0.9360	0.0255	0.0342	0.4680	0.0299

Table 3. Average performance of the EEG-based person authentication system with BCI2008B dataset when users perform 2 tasks (left hand and right hand motor imagery).

Subject ID	Original FAR	Original FRR	Improved FAR	Improved FRR	Original HTER	Improved HTER
B01	0.0000	0.6750	0.0435	0.1480	0.3370	0.0960
B02	0.0085	0.5085	0.0920	0.0000	0.2580	0.0460
B03	0.000	0.8495	0.0010	0.0075	0.4250	0.0040
B04	0.0000	0.8505	0.0990	0.1700	0.4210	0.1340
B05	0.0000	0.9715	0.0230	0.3200	0.4855	0.1715
B06	0.0000	0.7320	0.0145	0.1160	0.3660	0.0650
B07	0.0000	0.7380	0.0140	0.3210	0.3690	0.1495
B08	0.0000	0.5040	0.0215	0.1395	0.2520	0.0805
B09	0.0000	0.8420	0.0415	0.0665	0.4210	0.0540

At first glance, the FAR values of the original SVM in Tables 2 and 3 show an interesting person authentication system in terms of not opening to imposters compared to the FAR values of improved method. However, the original FRR values reveal that the unimproved system has unacceptable false rejection rates particularly with subjects A02, A06, A08, and A09 in dataset BCI2008A and subjects B03, B04, B05, and B09 in dataset BCI2008B. Consequently, this system does not allow not only impostors but also almost all genuine clients to access it, and that is not practical in real life application. On the other hand, the improved system gained significantly reduced FRR values. Although the proposed method increased FAR values, this variation is not considerable. Moreover, compared to previous studies in EEG-based person authentication, the improved system still achieved impressive false acceptance rate as seen in Table 4.

Table 4. A comparison to previous studies in the literature in terms of FAR values

Study	Number of subject	False Acceptance Rate (FAR)
Marcel et al. [14]	9	0.144
Zúquete et al. [21]	70	0.062
Yeom et al. [20]	10	0.145
Wu et al. [19]	15	0.067
Our work (BCI2008B dataset)	9	0.039
Our work (BCI2008A dataset)	9	0.052

Half total error rate (HTER = (FAR + FRR)/2) is also seen in some studies to present the performance of an EEG-based person authentication system. The smaller HTER is considered the better system. Tables 2 and 3 confirm that errors of an EEG-based person authentication system are significantly reduced with the proposed new bias. That means the performance of SVM for imbalanced classes problem in EEG datasets is enhanced.

4 Conclusion and Future Work

Using EEG signals regarded as individualized passwords for person authentication has the merits of both password-based and biometric-based authentication approaches, yet without their weaknesses. However, EEG datasets for person authentication purpose are imbalanced due to the nature of the numbers of genuine and impostor instances, and this heavily impacts on the performance of the system when using machine learning algorithms, particularly SVM. In this paper, a new bias method for binary classification has been proposed, and it showed that the performance of SVM for the imbalanced classes problem in EEG-based person authentication datasets was significantly improved.

In the future, we will conduct experiments with our proposed method on larger datasets. A more comprehensive comparison between the new bias method and other published proposals in imbalanced datasets problem will also be conducted.

References

1. Armstrong, B.C., Ruiz-Blondet, M.V., Khalifian, N., Kurtz, K.J., Jin, Z., Laszlo, S.: Brainprint: assessing the uniqueness, collectability, and permanence of a novel method for ERP biometrics. Neurocomputing **166**, 59–67 (2015)
2. Asuncion, A., Newman, D.: UCI machine learning repository. https://archive.ics.uci.edu/ml/index.php (2007)

3. Batuwita, R., Palade, V.: Class imbalance learning methods for support vector machines. In: Imbalanced Learning: Foundations, Algorithms, and Applications, p. 83 (2013)
4. Brunner, C., Leeb, R., Müller-Putz, G., Schlögl, A., Pfurtscheller, G.: BCI competition 2008–Graz data set A, pp. 136–142. Institute for Knowledge Discovery (Laboratory of Brain-Computer Interfaces), Graz University of Technology (2008)
5. Chang, C.C., Lin, C.J.: LIBSVM: a library for support vector machines. ACM Trans. Intell. Syst. Technol. (TIST) **2**(3), 27 (2011)
6. Chawla, N.V., Bowyer, K.W., Hall, L.O., Kegelmeyer, W.P.: SMOTE: synthetic minority over-sampling technique. J. Artif. Intell. Res. **16**, 321–357 (2002)
7. Cortes, C., Vapnik, V.: Support vector machine. Mach. Learn. **20**(3), 273–297 (1995)
8. Gill, P.E., Murray, W., Wright, M.H.: Practical Optimization. Academic Press, Cambridge (1981)
9. Imam, T., Ting, K.M., Kamruzzaman, J.: z-SVM: an SVM for improved classification of imbalanced data. In: Australasian Joint Conference on Artificial Intelligence, pp. 264–273. Springer (2006)
10. Kubat, M., Matwin, S., et al.: Addressing the curse of imbalanced training sets: one-sided selection. In: ICML, vol. 97, pp. 179–186, Nashville, USA (1997)
11. Kumar, T.S., Kanhangad, V., Pachori, R.B.: Classification of seizure and seizure-free EEG signals using local binary patterns. Biomed. Signal Process. Control **15**, 33–40 (2015)
12. Leeb, R., Brunner, C., Müller-Putz, G., Schlögl, A., Pfurtscheller, G.: BCI competition 2008-Graz data set B. Graz University of Technology, Austria (2008)
13. Maratea, A., Petrosino, A., Manzo, M.: Adjusted F-measure and kernel scaling for imbalanced data learning. Inf. Sci. **257**, 331–341 (2014)
14. Marcel, S., Millán, J.d.R: Person authentication using brainwaves (EEG) and maximum a posteriori model adaptation. IEEE Trans. Pattern Anal. Mach. Intell. **29**(4), 743–752 (2007)
15. Nakamura, T., Goverdovsky, V., Mandic, D.P.: In-ear EEG biometrics for feasible and readily collectable real-world person authentication. IEEE Trans. Inf. Forensics Secur. **13**(3), 648–661 (2018)
16. Nguyen, P., Tran, D., Huang, X., Ma, W.: Motor imagery EEG-based person verification. In: International Work-Conference on Artificial Neural Networks, pp. 430–438. Springer (2013)
17. Núñez, H., Gonzalez-Abril, L., Angulo, C.: Improving SVM classification on imbalanced datasets by introducing a new bias. J. Classif. **34**(3), 427–443 (2017)
18. Welch, P.: The use of fast Fourier transform for the estimation of power spectra: a method based on time averaging over short, modified periodograms. IEEE Trans. Audio Electroacoust. **15**, 70–73 (1967)
19. Wu, Q., Zeng, Y., Zhang, C., Tong, L., Yan, B.: An EEG-based person authentication system with open-set capability combining eye blinking signals. Sensors **18**(2), 335 (2018)
20. Yeom, S.K., Suk, H.I., Lee, S.W.: EEG-based person authentication using face stimuli. In: 2013 International Winter Workshop on Brain-Computer Interface (BCI), pp. 58–61. IEEE (2013)
21. Zúquete, A., Quintela, B., da Silva Cunha, J.P.: Biometric authentication using brain responses to visual stimuli. In: Biosignals, pp. 103–112 (2010)

Solving the Test Case Prioritization Problem with Secure Features Using Ant Colony System

Andreea Vescan[1], Camelia-M. Pintea[2(✉)], and Petrica C. Pop[2]

[1] Computer Science Department, Babes-Bolyai University, Cluj-Napoca, Romania
[2] Department of Mathematics and Computer Science, Technical University Cluj-Napoca, North University Center at Baia Mare, Baia-Mare, Romania
avescan@cs.ubbcluj.ro, dr.camelia.pintea@ieee.org, petrica.pop@cunbm.utcluj.ro

Abstract. Nowadays, the correctness of a program is a must and this comes with a plus when is about security. It is a major benefit to have confident results when the software is tested. In particular, when a regression test is made, it ensures that when a program system is modified, the existing and good functionality is not affected. NP-hard problems, including complex optimization problems necessitate high quality and intensively tested software. It is described an optimized test case prioritization method inspired by ant colony optimization, called ***T****est* ***C****ase* ***P****rioritization* ***ANT*** and denoted by *TCP-ANT*. The current approach uses the *Average Percentage of Fault Detected (APFD)* metric as selection criterion, and tries to uncover maximum fault and to reduce the regression testing time. Furthermore, there are considered some metrics to better encapsulate the *TCP-ANT* execution cost and a criterion for a proper number of test cases, hopefully covering all possible faults. The main aim of the paper is to illustrate the *Test Case Prioritization ANT* from security perspective. This approach includes a severity factor of an identified fault, when using *APFD* metric and compares the *TCP-ANT* beneficent results with random, reverse and no prioritization techniques.

Keywords: Regression testing · Security features · Faults severity · Prioritization · Ant algorithms · Optimization

1 Introduction

Software feasibility is an increasing demand. That is why testing should provide the quality levels of the developed software. Testing should be a well conducted process especially when are involved secure environments and security-related faults features. Security testing identifies if the specified or intended security features are correctly implemented. Security functional testing validates if the specified security requirements are implemented in a proper way, both in terms of

© Springer Nature Switzerland AG 2020
F. Martínez Álvarez et al. (Eds.): CISIS 2019/ICEUTE 2019, AISC 951, pp. 67–76, 2020.
https://doi.org/10.1007/978-3-030-20005-3_7

security properties and as well of security functionalities. Security vulnerability testing addresses the identification of unintended system vulnerabilities.

Severity by definition describes the gravity of an undesirable occurrence. Thus, in testing severity defines the extent to which a particular defect could create an impact on the application or system.

Testing [1] could include processes to find functional or coding errors. Complex optimizations problems are mainly solved today using heuristics approaches, for more information please refer to [2–4]. Testing software comes with a cost. In order to reduce the testing costs, the test selection, minimization, and prioritization are used. The regression test [5] is a test that ensures the stability and correctness of a program when modifications are made. In [5], it is also specified that regression testing the non-functional security requirement is difficult and further specific regression investigation is needed.

The Test case prioritization problem [6] deals with optimizing the order of test cases for regression testing in order to amplify a given criterion. Possible criterion include the fault detection rate criterion [7], rate of detection of high-risk faults, regression errors, and confidence of the reliability criterion [8].

Bio-inspired algorithms are one of the best heuristics used to solve efficiently complex optimization problems. Ant colony optimization [9] is used here as the basis for the *Test Case Prioritization-ANT (TCP-ANT)*. *TCP-ANT* is an optimized test case prioritization algorithm implemented in C++ that uncovers maximum fault, each fault reviewing a "severity" value and reduce the regression testing time. The *Average Percentage of Fault Detected (APFD)* metric [7] is used. The algorithm also tries to encapsulate better the execution cost feature and there are involved a many test cases to cover all possible faults.

The aim of this paper is to investigate the Test Case Prioritization Problem, proposing an ant-based optimization algorithm with various criterion: number of covered faults, cost of execution and severity of faults. The rest of the paper is organized as follows: Sect. 2 presents the main definitions and the state-of-art related to the regression testing and the Test Case Prioritization Problem. Section 3 illustrates the *Test Case Prioritization-ANT* technique and Sect. 4 describes the security perspective of the TCP problem, followed by a discussion of the achieved results in comparison to the existing prioritization algorithms. Some further work ideas and the conclusions reiterates the *TCP-ANT* promising results and considering other security faults features.

2 Regression Testing and the Test Case Prioritization Problem: Definitions and Algorithms

The definition of the **Regression test** [10] follows.

Definition 1. *Let P be a procedure or program; let P' be a modified version of P; and let T be a test suite for P. A typical* **regression test** *proceeds as follows:*

1. *Select $T' \subseteq T$, a set of test cases to execute on P'.*
2. *Test P' with T', establishing P''s correctness with respect to T'.*

3. *If necessary, create T'', a set of new functional or structural test cases for P'.*
4. *Test P' with T'', establishing P''s correctness with respect to T''.*
5. *Create T''', a new test suite and test execution profile for P', from T, T', and T''.*

The steps of the previous definition [10] rises some issues. A regression testing method could retest the entire method, safely executing all test cases. The time and costs efficiency of Regression testing could be reached while using: *test case selection techniques* to reduce the number of test cases (by a given criterion) and *test case prioritization techniques* to order the test cases (by the rate of early fault-detection).

The **Test case prioritization problem** [7,10] is defined as follows:

Definition 2. *Given a test suite (T), the set of permutations of T (PT) and a function from PT to real numbers (f), the problem is to find $T' \in PT$ such that $(\forall T'')(T'' \in PT)(T'' \neq T)[f(T') \geq f(T'')]$.*

In the previous definition, the function f illustrates the unknown rate of fault detection. The *Average Percentage of Fault Detected (APFD)* metric [7] is used for $f : PT \rightarrow \Re$ in order to early detecting the faults. In order to solve the *Test Case Prioritization (TCP)* problem, diverse techniques were involved. There were also considered investigations to increase the rate of fault detection, total statement coverage and function coverage [7,11,12].

In [13], there were developed techniques to automatically select test cases for security policy evolution [14]. For the regression-test selection the authors select T$'$ from an existing test suite T; T illustrates different policy behaviors for P and P$'$ policies.

Potter and McGraw [15] operationalized the main security approaches and distinguished testing security mechanisms in order to ensure that their functionality is properly implemented, and performing risk based security testing motivated by understanding and simulating the attackers approach.

In order to prioritize test cases, Zhang et al. [16] changed the priorities and test case costs. A new metric to evaluate rate of fault dependency and a new way to test case prioritization: observing fast dependency among faults was described by Graves et al. [17].

From the heuristics used to solve the *Test Case Prioritization* problem the genetic algorithm [18] and the ant colony optimization algorithm [19–21]. The genetic algorithm [18] had better results when compared with some greedy versions and hill climbing to prioritize test cases. A review of Test Case Prioritization approached may be found in the review type paper [22].

In the ant-based algorithm [21] it is a time constraint environment. In the new ant-based approach, the *Test Case Prioritization-ANT* algorithm it is also considered *the remaining set of uncovered faults* and obtain multiple solutions in a running process. The presented papers considered the cost, the faults and fault dependencies. There is also used the APFD metric and make comparisons with un-prioritized approach. The current work uses also a reasonable number of test cases covering all possible faults as a criterion.

In what it follows is described the proposed *Test Case Prioritization ANT* algorithm.

3 Test Case Prioritization ANT (TCP-ANT) Algorithm

The current section describes the *Test case prioritization-ANT (TCP-ANT)* algorithm. At first the introduced factors during the prioritization process are defined based on the *Average Percentage Faults Detected (APFD)* [11] (the weighted average of the percentage of faults while executing a test suite), the *Rate of Fault Detection (RTF)* (the average number of defects found in a minute by a test case) and the *Percentage of Fault Detected (PFD)* [23]. A higher APFD value, $APFD \in [0..100]$, implies a higher fault detection rate. In *TCP-ANT*, the APFD factor define the best solution, and for the solution cost are considered just the test cases that entirely covers the set of faults.

3.1 Definitions

Next follows the definitions: the *Rate of Remaining Fault Detection (RRTF)* based on the RTF and the *Percentage of Remaining Fault Detected (PRFD)* based on PFD considering the *remaining* set of faults to be detected.

Definition 3. *The Rate of Remaining Fault Detection (RRTF) is the average number of defects found per minute by a test case, from the remaining set of uncovered faults.*

For the test case T_j, $RRTF_j$, Eq. (1), use the number of defects found by T_j from the remaining set of uncovered faults and the time used by T_j to detect the same defects.

RN_j is the number of remaining faults detected by T_j test case and $time_j$ the time used by T_j test case.

$$RRTF_j = \frac{RN_j}{time_j} \times 100, \tag{1}$$

Definition 4. *The Percentage of Remaining Fault Detected (PRFD) for test case T_j is computed with the number of faults found by test case T_j from the set of remaining faults to be covered and total number of faults.*

The PRFD is expressed as in Eq. (2), where RN_j is the number of remaining faults detected by T_j test case and N the total number of faults.

$$PRFD_j = \frac{RN_j}{N} \times 100, \tag{2}$$

```
Initialize
Loop /* at this level each loop is called an iteration */
  Each ant is positioned on a starting node
  Loop /* at this level each loop is called a step */
    Each ant applies a state transition rule to incrementally build a solution
    and a local pheromone updating rule
  Until all ants have built a complete solution
  A global pheromone updating rule is applied
Until End_condition
```

Fig. 1. The Ant Colony System (ACS) algorithm [9].

3.2 Description of the TCP-ANT Algorithm

The *Test case prioritization-ANT* algorithm is based on the *Ant Colony System (ACS)* [9]. The used factors are: *APFD, RRTF, PRFD* and the cost of the test cases; during the selection of the next test case, there are considered the execution cost and the uncovered faults of test case, and also the severity of faults. Figure 1 illustrates the Ant Colony System (ACS) algorithm [9].

For the *TCP-ANT*, at first ants are placed within the nodes, in particular test cases; there are made the initialization of the parameters including the pheromone trails.

The main phase is the construction solution phase; it includes a repetitive probabilistically selection of next tour node (test case) from available test cases from neighborhood, based on the number of faults of the test case (from uncovered yet faults) and based on the pheromone trails. The factors *RRTF* and *PRFD* also influence the choice for the next test case.

A *TCP-ANT* constraint during optimization is to consider just **the test cases covering all the faults** to reduce the cost of executing a prioritization.

After an iteration a solution (a tour of test cases) with a maximum fault detection rate at minimum cost and maximal/highest severity faults is found. As in ACS, after each step, a local update rule follows.

The last phase includes the global pheromone update rule and finding the best global solution. The algorithm stops after a given number of iterations. A solution includes the sum of APFD and the solution cost, for test cases covering all the faults. As $APFD \in [0, 100]$ the same interval goes for the sum of test cases costs. The sum is optimized while considering the cost for all covered faults, $\frac{1}{costAllCoveredFaults}$. Therefore, for the TCP problem, criterion includes covered faults, solution cost and severity of faults.

The *Test Case Prioritization ANT* algorithm reach goals similar to other related methods when considering covered faults and cost. The *TCP-ANT* differs by the fact that the solution of the prioritization is verified against duplicate coverage of faults by test cases. The prioritization solution ends with test cases that do not discover new faults.

The algorithm runs for a given maximum number of iterations. For each iteration an ant is first "assigned" to a test case; furthermore is considered the next test case to "visit" based on the number of faults not yet covered by the neighbor test case; the ant selects the test case with the maximum number of faults not yet covered.

4 Security Perspective of TCP. Numerical Results and Discussions

Security testing shows if a security feature is correctly implemented. When it comes to bugs, the severity of a bug would indicate the effect it has on the system in terms of its impact. Testing security mechanisms can be performed by standard test organizations with classical functional test techniques, whereas risk-based security testing requires specific expertise and sophisticated analysis. The current paper considers the *severity* of an identified fault, each fault reviewing a "severity" value.

The *TCP-ANT* algorithm was implemented[1] in C/C++ and it is based on a version of the algorithm developed by Perez-Uribe [24] in order to solve the classical *Traveling Salesman Problem (TSP)*. The conducted experiments considers various criterion combinations (faults, costs, number of test cases, severity) on the considered case study.

Table 1 shows the data set used in the experiments [25]. It consists of the test suite of test cases, the exposed faults of each test case, the cost of test cases, and the fault severities of the faults detected by the test cases.

The *TCP-ANT* parameters for all tests are: five ants, ten iterations, $q_0 = 0.5$; a parameter to control the influence of pheromone, $\alpha = 1$; a parameter to control the influence of the desirability of state transition $\beta = 2$; the pheromone evaporation coefficient, $\rho = 0.0001$, and the initial amount of pheromone deposited on the graph's edges $\tau_0 = 0.01$.

In order to build confidence for the ant-based technique were used one hundred runs for each test suite. Each run yielded the path found whether or not optimal.

Table 1. Experimental instance: test cases, faults identified and cost and severity of faults based on [25].

Test case/Faults	$F1$	$F2$	$F3$	$F4$	$F5$	*cost*
T1	1	0	0	1	0	4
T2	0	1	0	0	1	2
T3	1	1	1	0	0	1
T4	0	1	0	0	1	3
T5	1	0	1	0	1	2
Severity	2	1	4	5	2	

[1] http://www.cs.ubbcluj.ro/~avescan/?q=node/220.

Table 2. Order of test cases for various prioritization approaches.

	No. order	Reverse order	Random order	TCP-ANT	Optimal order
	T1	T5	T4	T4	T2
	T2	T4	T3	T2	T4
	T3	T3	T2	T1	T3
	T4	T2	T5	T3	T5
	T5	T1	T1	T5	T1
APFD	0.54	0.50	0.46	0.42	

An initial investigation on the *TCP-ANT* parameters was made; it was considered various initial parameters values including the number of ants and iterations. The investigation concludes to use the stated parameters.

Experiments with No Severity of Test Cases. The first experimental instance does not considered the severity of test cases. There were used five test cases with five faults; each test case was evaluated with a cost execution, i.e. time execution of the test case. Table 1 includes the relation between the test cases, the identified faults, and the cost for each test case.

The $APFD$ value for the given set of test cases is: $APFD_{init} = 0.54$. Figure 2 illustrates the plot for the initial $APFD$, the *No order* option *Reverse order and TCP-ANT order*, and the optimal order. The results shows that 100% fault coverage was obtain after 3 test cases from the test suite.

Numerical Experiments When Considering Only Faults. Several comparison were made with respect to the following ordering: *No order, Reverse order, Random order*, and *TCP-ANT* of the test cases. The orderings with respect to these approaches are shown in Table 2, the last line of the table containing the AFPD values. Figure 2 shows that the best solution for *TCP-ANT*-order has a lower APFD as the solution for No-order and Reverse-order and Random-order. Table 2 shows for the *TCP-ANT*-order $4-2-1-3-5$ the $APFD = 0.42$, where for Random-order $4-3-2-5-1$ the $APFD = 0.46$.

Experiments with Severity of Test Cases. The second experiments considered also the severity of test cases provided in Table 1. Five different solutions were obtained by executing considering only the number of faults that are not yet covered at each step, i.e. having thus five different APFD: $0.38, 0.42, 0.46, 0.5$, and 0.54. The Table 2 values are compared by calculating the $APFD$ on the same example; the results are illustrated on Fig. 2. For example, $4-2-1-3-5$ with $APFD = 0.42$ is computed as follows: $AFPD = 1-(3+1+4+3+1)/(5*5)-1/(2*5)$, i.e. the first position of the test cases that reviled the first fault is 3 (the test cases that discover the first bug are T1, T3 and T5 and the first one in the order of test cases is T1 that is on the third position in the vector).

The strategy for incorporating also the severity when a test case is selected to be added in the prioritization order considers both the number of faults not covered yet and the sum of severity of faults not covered yet, with the same weights, i.e. 0.5. Five different solution were obtained when use both the number of faults not yet covered at each step and the severity of faults, i.e. having thus five different APFD: 0.38, 0.42, 0.46, 0.50, and 0.54.

Figure 3 depicts the number of obtained solutions for each APFD; it is visible that the number of solution with a better APFD is decreasing for the *withSeverity*

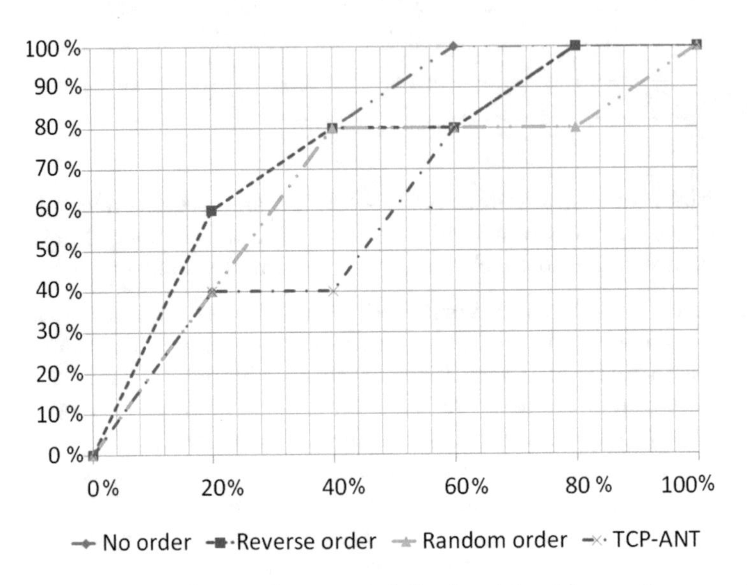

Fig. 2. The percentage of test suite executed (horizontal axis) over the percentage of faults (vertical axis): the APFD values for prioritization approaches from Table 2.

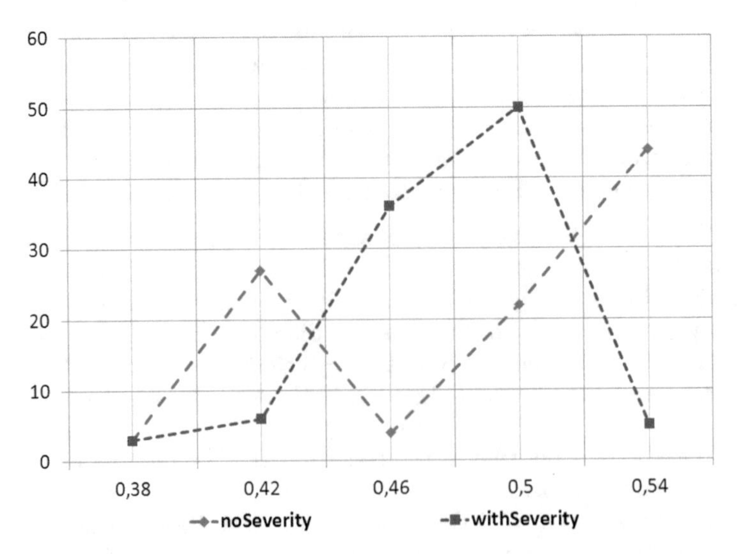

Fig. 3. The number of solution for each APFD from experiments: comparison between without (*noSeverity*) and with severity feature (*withSeverity*).

strategy for several solutions. This is due to the fact that the severity of a considered test case when selecting the next prioritization test case was also considered with the same weights as number of faults not covered yet.

A future work could use Pareto instead of weights when selecting next test case using both number of faults not covered yet and the security feature related to the severity of a fault.

5 Conclusions and Further Work

For complex problems involving security issues, making a confident software it is a necessity. There are many testing software involved in this process. The paper describes the *Test Case Prioritization-ANT (TCP-ANT)* an ant-based model in order to find an optimal ordering solution. *TCP-ANT* outperforms the existing no ordering, random prioritization and reverse prioritization techniques from the literature. Severity of faults, a security feature, is also used as criterion when selecting next test case. In order to reduce the cost of executing the found prioritization, a new criterion referring to the number of test cases to cover the faults is introduced. In the future, we will try to improve the *TCP-ANT* by changing the local update rule [26] and choosing next test case and/or using beneficent metrics (related to the testing type) for large test cases suites as for sensor networks security [27], secure transportation [28] or secure and green supply chain networks [29]. In the investigation, from the security perspective, the *severity* of an identified fault, each fault reviewing a "severity" value was considered. Another future approach would be to consider Safety-Critical, Non-Critical, and Unclassified faults.

References

1. Pezzand, M., Young, M.: Software Testing and Analysis: Process, Principles and Techniques. Wiley, New York (2008)
2. Battiti, R.: Reactive search: toward self-tuning heuristics. In: Rayward-Smith, V.J., et al. (eds.) Modern Heuristic Search Methods, Chap. 4, pp. 61–83. Wiley, New York (1996)
3. Cook, W., Cunningham, W., Pulleyblank, W., Schrijver, A.: Combinatorial Optimization. Wiley, New York (1998)
4. Pop, P.: Generalized Network Design Problems. Modeling and Optimization. De Gruyter, Berlin (2012)
5. Yoo, S., Harman, M.: Regression testing minimization, selection and prioritization: a survey. Softw. Test. Verif. Reliab. **22**(2), 67–120 (2012)
6. Panigrahi, C., Mall, R.: An approach to prioritize the regression test cases of object-oriented programs. CSI Trans. ICT **1**(2), 159–173 (2013)
7. Rothermel, G., et al.: Test case prioritization: an empirical study. In: Conference on Software Maintenance, ICSM, Oxford, UK, pp. 179–188 (1999)
8. Khalilian, A., Azgomi, M., Fazlalizadeh, Y.: An improved method for test case prioritization by incorporating historical test case data. Sci. Comput. Program. **78**(1), 93–116 (2012)

9. Dorigo, M., Gambardella, L.: Ant colony system: a cooperative learning approach to the traveling salesman problem. IEEE Trans. Evol. Comput. **1**(1), 53–66 (1997)
10. Graves, T., et al.: An empirical study of regression test selection techniques. In: Proceedings of the International Conference on Software Engineering, ICSE, Kyoto, Japan, pp. 188–197 (1998)
11. Elbaum, S., Malishevsky, A., Rothermel, G.: Prioritizing test cases for regression testing. In: Proceedings of the International Symposium on Software Testing and Analysis, ISSTA Portland, USA, pp. 102–112 (2000)
12. Malishevsky, A., et al.: Cost-cognizant test case prioritization. Technical report TR-UNL-CSE-2006-004, University of Nebraska (2006)
13. Hwang, J., et al.: Selection of regression system tests for security policy evolution. In: IEEE/ACM International Conference on Automated Software Engineering, ASE, Essen, Germany, pp. 266–269 (2012)
14. OASIS: extensible access control markup language (XACML) (2005)
15. McGraw, G., Potter, B.: Software security testing. IEEE Secur. Priv. **2**(5), 81–85 (2004)
16. Zhang, X., et al.: Test case prioritization based on varying testing requirement priorities and test case costs. In: Conference on Quality Software (QSIC 2007), Portland, Oregon, USA, pp. 15–24 (2007)
17. Kayes, M.: Test case prioritization for regression testing based on fault dependency. In: 2011 3rd International Conference on Electronics Computer Technology, vol. 5, pp. 48–52 (2011)
18. Li, Z., Harman, M., Hierons, R.: Search algorithms for regression test case prioritization. IEEE Trans. Softw. Eng. **33**(4), 225–237 (2007)
19. Agrawal, A., Kaur, A.: A comprehensive comparison of ant colony and hybrid particle swarm optimization algorithms through test case selection. In: Data Engineering and Intelligent Computing, pp. 397–405. Springer, Singapore (2018)
20. Ahmad, S., Singh, D., Suman, P.: Prioritization for regression testing using ant colony optimization based on test factors. In: Intelligent Communication, Control and Devices, pp. 1353–1360. Springer, Singapore (2018)
21. Singh, Y., Kaur, A., Suri, B.: Test case prioritization using ant colony optimization. ACM SIGSOFT Softw. Eng. Notes **35**(4), 1–7 (2010)
22. Saraswat, P., Singhal, A., Bansal, A.: A review of test case prioritization and optimization techniques. Advances in Intelligent Systems and Computing, pp. 507–516. Springer, Singapore (2018)
23. Kavitha, N.: Test case prioritization for regression testing based on severity of fault. Int. J. Comput. Sci. Eng. **2**(5), 1462–1466 (2010)
24. Perez-Uribe, A.: Ant colony system algorithm in C/C++ (2002)
25. Huang, Y.C., Peng, K.L., Huang, C.Y.: A history-based cost-cognizant test case prioritization technique in regression testing. J. Syst. Softw. **85**(3), 626–637 (2012)
26. Pintea, C., Dumitrescu, D., Pop, P.: Combining heuristics and modifying local information to guide ant-based search. Carpath. J. Math. **24**(1), 94–103 (2008)
27. Pintea, C.M., Pop, P.: Sensor networks security based on sensitive robots agents: a conceptual model. Advances in Intelligent Systems and Computing, vol. 189, pp. 47–56. Springer, Heidelberg (2013)
28. Pintea, C.M., Crişan, G., Pop, P.: Towards secure transportation based on intelligent transport systems. Novel approach and concepts. Advances in Intelligent Systems and Computing, pp. 469–477. Springer, Cham (2018)
29. Pintea, C.M., Calinescu, A., Pop Sitar, C., Pop, C.: Towards secure & green two-stage supply chain networks. Logic J. IGPL (jzy028) (2018)

Toward an Privacy Protection Based on Access Control Model in Hybrid Cloud for Healthcare Systems

Ha Xuan Son[1(✉)], Minh Hoang Nguyen[2], Hong Khanh Vo[1], and The Phuc Nguyen[3]

[1] FPT University, Cantho, Vietnam
hxson@ctuet.edu.vn, {sonhx4,khanhvh}@fe.edu.vn
[2] Hanoi University of Science and Technology, Ha Noi, Viet Nam
hoang.nguyenminh@hust.edu.vn
[3] University of Trento, Trento, Italy
thephuc.nguyen@studenti.unitn.it

Abstract. The Electronic Patient Health Recorded systems have been designed and implemented to provide the ability of collaboration among healthcare providers. Such collaborations raise the need for an authorization system to implement network security and protect sensitive data of each patient. Without a proper security protocol, this confidential information could be exploited by malicious attackers. This paper presents a novel access control system called "Access Control Model in Hybrid Cloud for Healthcare Systems" will help us to deal with any security vulnerabilities within the system. The proposed system is based upon two levels of access control: (1) protecting shared data containing basic information of patients among hospitals in the systems and (2) protecting private's patient data that can be accessed by only the treating doctor. The model is implemented in a real-case application to demonstrate its effectiveness in managing the different level of security and privacy concerns.

Keywords: ABAC · Privacy · Large-scale data access · Access control in healthcare systems · Hybrid cloud

1 Introduction

Due to the ever-growing development of sensor networks, distributed systems, and the Internet of Things, the health care methods have been gradually transformed from traditional service to online healthcare services. In fact, healthcare organizations are moving toward the Electronic Patient Health Records (EPHR) for storing their patients' health information in cloud systems. In a speech at the annual conference of the American Medical Association 2009, US President Barack Obama raised some of the vital issues related to medical data [19]. He mentioned that all the medical information should be stored securely in a private medical record so that patients' information can be shared between specialists,

© Springer Nature Switzerland AG 2020
F. Martínez Álvarez et al. (Eds.): CISIS 2019/ICEUTE 2019, AISC 951, pp. 77–86, 2020.
https://doi.org/10.1007/978-3-030-20005-3_8

doctors, and hospitals - thus reducing treatment cost and providing prompt service in an emergency. Even though billions of dollars have been poured into a push to make EPHR become the universal standard, the process still has encountered many difficulties.

The healthcare systems must support interactions among parties including patients, their relatives, doctors, nurses, insurance companies and pharmacies. For sensitive data, these systems should fulfill a tradeoff between two requirements: (i) availability to ensure the best healthcare for patients; and (ii) confidentiality for users' sensitive data (phone number, address detail, health history, examination reports). Therefore, the priority concern of the healthcare systems is sensitive data protection from unauthorized users. To address this problem, a simple approach is to use Access Control (AC) mechanism. However, traditional AC models are inflexible and difficult to apply in such a large system as healthcare systems in which "nothing interferes with the delivery of care". It is therefore necessary to construct an electronic medical records management system that can share patient information while protecting their confidentiality.

The trade-off between availability and security in hybrid cloud for healthcare system has been discussed in several studies [1,2,6]. Among them, ABAC model (Attribute-based Access Control) are the most widely used due to its ability to protect patient data at fine-grained levels [5,10–15]. Our approach is to provide a model that integrates both types of policies including access policy for public cloud and private policy for private cloud. Particularly, access policy protects shared data with other hospitals in the system (public data). They provide basic security definitions based on user roles, access objects, access rights, and access environments. Whereas, privacy policy protects private data containing sensitive information that only doctors who directly treat the patient know. Here, the private policy provides security at a fine-grained resolution: retrieving protected data requires satisfaction of all requirements. Figure 1 explains in detail the concept of our access control model.

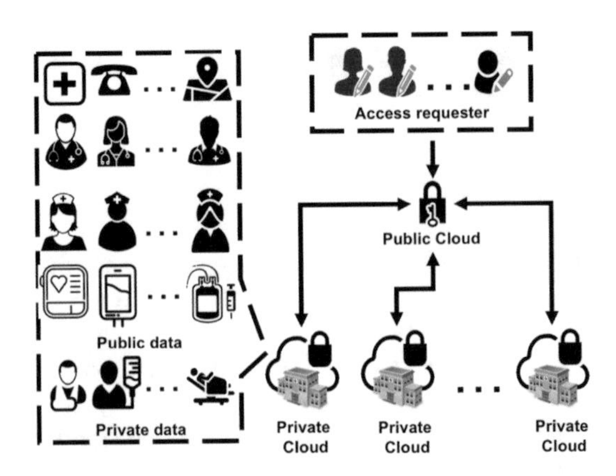

Fig. 1. Protecting electronic patient health records in hybrid cloud

In addition, a healthcare system must have an efficient mechanism to detect and handle redundancy and conflicts between policies, especially when multiple administrators can create security policies for the system [1,3,5]. The aim of this paper is to present a model specifically developed to resolve the problem of privacy, conflict policies, and availability of shared data. To do so, we used a combining algorithm and implemented priority function to resolve the conflict between privacy policies.

According to us the main contribution of this work to the current literature is the development of an access control model for healthcare system and a mechanism to detect and handle conflict and redundancy policies.

This paper is organized as follow. In the second section, we briefly review related works. The following section illustrates policy structure, our proposed architecture concept in detail, and the main algorithms. The fourth section describes our experimental designs and discusses the results. Finally, Sect. 5 presents our conclusions and future works.

2 Related Works

Zhang et al. [18] proposed a data sharing solution between hospitals belonging to a system. The system uses MAC model to protect shared data areas as privacy data is encrypted using a shared key and permissions are granted to different user groups. However, as MAC model is inflexibility, it is not suitable with larger systems supporting different users with different policy structures. Other studies (Ray et al. [7]; Yang et al. [16]) used ABAC model since it provided an efficient approach to define policies for healthcare systems; however, they did not focused on protecting privacy of data.

Yu et al. [17] proposed a Fine-grained Distributed Data Access Control (FDAC) model based on Attribute-Based Encryption. They provided a distributed data access control to support access control on sensor data. A network controller containing access structures operates as a distribution center to assign keys to its users. Only user possesses the right key with a proper structure can access the requested data. The access structure will be different for each user depending on their level of access. A drawback of this approach is that since the model is controlled by a centralized architecture, in the case of vulnerabilities, no security provision is provided in the network. To avoid a single point of failure, Ruj et al. [9] proposed an encryption-based access control model based on Multi-Authority Attribute-Based Encryption (MABE). Data access control granted completely using some Distribution Centers (DCs). All access structures from each DC must satisfy attributes from the sensor nodes, while being ANDed together to get full access for single users. As this approach has no detailed explanation on how to combine access structures, users have to store all structures which enlarged the complexity of the storage system. Maw et al. [4] proposed an Adaptive Access Control (A2C) model based on priority level and behavioral monitoring to provide access control for medical data in Wireless Sensor Networks. Those studies encrypted privacy data for storage and access control, but systems still need to pre-define attributes, roles and policies before deploying. This is a difficult task as real-world applications are often unpredictable.

Our model covered two issues which described above, including (i) protect the privacy area by using privacy policy, (ii) manage the system as decentralized architecture by supporting multiple administrators. Our proposed model does not base on cryptography as cryptography-based access control is designed for an unreliable environment in which there is a lack of global knowledge and control mechanism.

3 Our Approach

3.1 Policy Structure

Our model propose two data (policies) filtering layers before returning results to the user, including access policy (public cloud) and privacy policy (private cloud). An access policy defines a condition expression modeled in the function tree structure. The access policy returns the value specified in the element `Effect` if the condition is `true`. Each access policy is defined as following:

- **policy_id:** identifier of policy
- **data_name:** name of collection or table containing resource data
- **rule_combining:** responsible for solving the conflict of rules
- **is_attribute_resource_required:** a derived field used to determine whether the policy needs attribute resource to evaluate conditions of target or rules.
- **target:** conditional expression specifies when the policy should be applied.
- **rule:** an array field which contains `id` field, `effect` field (value of this field can either `Permit` or `Deny`) and condition.

On the other hand, a privacy policy is an array field which contains `id` field, `field_effect` field, and condition. Beside, the `field_effect` field has an array type that describes the list of data closure level for each field of data contained in privacy policy's rules. Each element in `field_effects` has two components: (i) component `name` storing the path to the field; and (ii) component `effect_function` containing `DomainPrivacy` and `Function` value where `DomainPrivacy` denotes the privacy domain and `Function` denotes the name of privacy functions on that domain including `Hide` or `Show`.

For hospital data security, firstly, policies are converted to XML format and classified to access policy (for public data) and private policy (for private data). Then they are synchronized with the other existing policies in the system. Depending on the hospital and the requirement of each patient, the classification of access policy and private policy is different.

3.2 Privacy Protection Based on Access Control Model in Hybrid Cloud for Healthcare Systems

Figure 2 depicts the proposed model. The architecture contains the following main components.

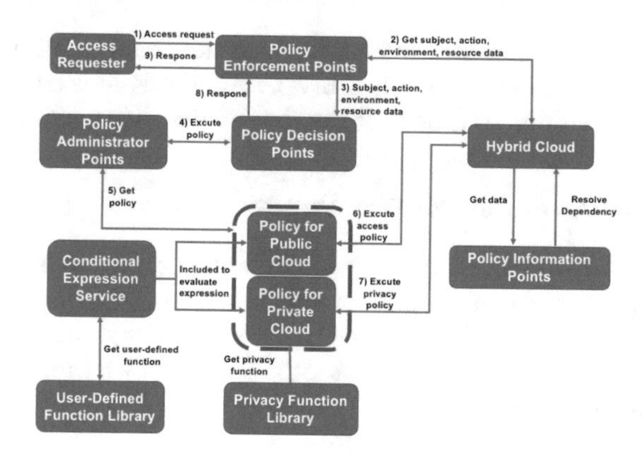

Fig. 2. Access control model in hybrid cloud for healthcare systems

- **Policy Enforcement Points (PEP):** responsible for receiving requests from users. Moreover, it performs access control by making decision requests and enforcing authorization decisions.
- **Hybrid Cloud:** store the public and private data. Other components can send request to **Hybrid Cloud** whenever they need more information or data.
- **Policy Information Points (PIP):** serves as the source of attribute values, or the data required for policy evaluation.
- **Policy Decision Points (PDP):** responsible for receiving and examining requests. It retrieves and evaluates applicable policies. After the evaluation process, it returns the authorization decision to PEP. It is the core component of the model.
- **Policy Administrator Points (PAP):** responsible for creating access policies, privacy policies and storing them in the repository.

Access control and privacy protection works as follows. Firstly, when a user sends an access request to the PEP module, PEP generates a request to the *Hybrid Cloud*, asking for user information (step 2). This information contains user's data, what action he could perform with the requested data and the resource environment data. After getting the data from *Hybrid Cloud*, PEP sends the request and user's information to PDP (step 3). The PDP gets the applicable policies from the PAP and retrieves required attribute for evaluation. PAP contains two difference policies, there are access policy and privacy policy. Firstly, the request is checked with the list of access policy. Apparently, if any condition is not satisfied, the returned is `Deny` and user's request is not granted access. If all conditions are satisfied, the access policy is executed and the requested data is loaded from the MongoDB through the *Hybrid Cloud* (step 6). Meanwhile, privacy policy checking procedure is enabled (step 7) and privacy function is loaded from the *Privacy Function Library.* The request is then evaluated through each privacy policy. The more condition it satisfies, the more data can be revealed to the user. In this way, if a user possesses an administration permission to the requested data, he can perform his action (execute, read, modify or delete) on the whole data. On the other hand, if his permission is limited to a certain, he can only perform actions to a part of the data while the remaining is hided.

3.3 The Resolution for Conflict Policy

Access Policy Conflict: to avoid conflicts among access policies or rules, our approaches applied a set of combining algorithm to the policy and policy set. Those solution are inherited from XACML v3.0 [8]. In general, the combining algorithm is represented by a structure called "PolicyCombining" described by two components as below:

- **policies_id:** An array of policy identifiers
- **combining_algorithm:** The name of algorithm is used to solve conflict when multiple policies are contained in `policies_id` field.

Privacy Policy Conflict: a conflict can be created as multiple rules from the same policy simultaneously satisfied a condition. Additionally, the conflict can occur when several privacy functions can be applied to a particular field of the object (`PrivacyDomain`). To handle this situation, we added a structure called `hierarchy`. It contains two elements including:

- **name:** The name of function (Show, Anonymity or Hide).
- **priority:** The value of priority of function.

The privacy function will be chosen by the following rule:

$$\text{“}Optional\text{”} < \text{“}DefaultArea.Show\text{”} < ... < \text{“}DefaultArea.Hide\text{”} \quad (1)$$

3.4 Algorithm

Algorithm 1 describes the evaluation between the list of policy and the request. The **Input** of this algorithm is the list of the policy stored in PAP and the request sent from an access requester. We assume that *listPolicy* is a list storing the policies including *access_policy* in PAP and *request* is a variable storing the value of subject, action, environment, resource. First, we find the best *access_policy* which allows the request to access the data resource. If the subject value, resource value, and the action value between *access_policy* and *request* does not equal, the system will consider the next policy (line 2). Next, the value of *is_sub_policy* is checked. If returns **true**, we check the value of *Overlap_Area*. In this case **true**, the conflict occurs in the evaluation process, and the function with lower priority is executed (lines 3–4). Otherwise, the value of *response* is the value of the function of `PolicyCombining`(*sub_policy*). If the value of *is_sub_policy* is **false** compared to the value of *request* to the `Target` element. If the *request* can fulfill all target constraints, the `Access Policy` is evaluated. Otherwise, we move to the next policy (line 5). *listAccess* and *listPrivacy* are the variable storing the `Rule` of `Access Policy` and `Privacy Policy` respectively (lines 6–13). Apparently, if a single condition is not satisfied, the returned value is **false** and user's request is not granted access (from line 13 to 16). We only execute the `Privacy Policy` if and only if the access request in *access_policy* returns `Permit` (line 14). According to the `name` of `field_effect`, the `effect_function` is executed. Finally, the algorithm continues with the remaining policies (line 25) and returns the value of *response* (line 28).

Algorithm 1. Evaluating policy and request

Input: *listPolicy*, *request* **Output:** *response*

1: **for** *policy* **in** *listPolicy* **do**
2:   **if** GetSubject (*access_policy*, *request*) **and** GetCollection (*access_policy*, *request*) **and** GetAction (*access_policy*, *request*) **then**
3:     **if** *is_sub_policy* **and** *Overlap_Area* **then**
4:       //Execute the function with lower priority
5:     **else if** Target(*access_policy*, *request*) **then**
6:       *listAccess* = policy.GetAccessRule(*access_policy*)
7:       *flag* = **true**
8:       **while** *accRule* **in** *listAccess* **and** *flag* **do**
9:         **if** !Condition(*accRule*,*request*) **then**
10:           *flag* = **false**
11:         **end if**
12:       **end while**
13:       *response* = RuleCombining(*accRule*.GetEffect())
14:       **if** *flag* **then**
15:         *listPrivacy* = policy.GetPrivacyRule($privacy_policy$)
16:         **while** *priRule* **in** *listPrivacy* **do**
17:           **if** Condition(*priRule*,*request*) **then**
18:             //Choose `field_effects` by `name`
19:             //Execute `effect_function`
20:           **end if**
21:         **end while**
22:       **end if**
23:     **end if**
24:   **else**
25:     //Continue with next *policy* in *listPolicy*
26:   **end if**
27: **end for**
28: **return** *response*

4 Evaluation

To validate the effectiveness of the proposed model, the system was implemented for three cases: (i) with single dataset (the data of one hospital); (ii) with two datasets; and (iii) with three datasets representing three hospitals in the same system. It is worth to note that as the system contains both security policy and privacy policy, the structure will be more complex, i.e. the process takes longer. This can be a problem as in real-case application, we have to balance the trade-off between processing time and system's benefits. In this experiment, we varied the number of hospitals from one to three and observe the difference between the basic system and our proposed system with privacy policy. To do that, we used `mockaroo tool` to generate sample dataset. For the first scenario, the sample database describes a hospital which contain doctor collection (1000 records), patient collection (from 1000 records to 15000 recodes), department collection (10 recodes), and pharmacy collection (100 recodes). On the other hand, the second one contains three datasets which includes difference to data structure, number of recodes, and collection structure. Each record has an array of embedded documents field containing at least five elements inside. In general, all experiments are included in total five policies. Moreover, to observe the difference between the performances of with and without privacy policy, the processing time of each case is recorded. The system configuration for the experiments is a 64-bit machine with 8GB of RAM and 2.8 GHz Intel Core i5 CPU running macOS High Sierra. The prototype is implemented by C#, .NET Core and MongoDB v4.0 for storing policies and database[1].

Table 1 compare the performances of the model on two cases including one dataset, two dataset and three different datasets. It can be observed that as the number of records increases, the gap difference between the processing time of both cases expands. Considering the case of one dataset, when the number of records is 5000, this gap is nearly 3 s; it increases slightly to 3.3 s as the number of records reaches 15000. A similar situation happens in the case of three different datasets as this difference rises from 2.7 s to 4.6 s when the number of recode increases between 10000 records and 30000 records. It is worth to note that as the number of hospitals in the same system increases, the time needed to process record increases. The time difference between with and without privacy policy is spent on conflict resolution. Furthermore, it can be seen that the average processing time for each record increases as the number of datasets increases. While one hospital spent 0.6 ms to process one record, two hospitals and three hospitals need 1.0 ms and 1.1 ms, respectively. However, with the development of computer system nowadays, this difference is acceptable.

Regarding the time difference between the basic system and our model, we showed that the propose approach is quite effective for hospital systems. The gap between two models is large in the case of only one hospital presented (0.32 ms and 0.67 ms). However, as the number of hospitals increases, this gap is reduced. When three datasets are presented, the time different is not significant as they are 0.91 and 1.10 ms, respectively. Thus, as the hospitals system expands, the proposed model does take the similar amount of time with the basic model; ensuring the system latency at acceptable levels.

[1] https://github.com/xuansonha17031991/privacy-aware-access-control-model-for-healthcare-system.

Table 1. Processing time (measure in millisecond) for the model with and without privacy policy on different datasets

One dataset			Two different dataset			Three different datasets		
Num. records	No privacy policy	Privacy policy	Num. records	No privacy policy	Privacy policy	Num. records	No privacy policy	Privacy policy
5000	2529	5528	8000	6649	9877	10000	8942	11730
7000	2853	5814	10000	7630	11572	15000	13287	16972
9000	2930	6305	12000	10928	14938	20000	17829	22012
11000	3230	6804	20000	14638	18430	25000	23410	26834
15000	3804	7096	25000	18942	21238	30000	28341	32902
Aver. each record	**0.327**	**0.671**	**Aver. each record**	**0.784**	**1.014**	**Aver. each record**	**0.918**	**1.105**

5 Conclusion

In this work, we introduced an access control model focusing on privacy protection for the healthcare systems. As data will be protected by two sets of policies including access policy and privacy policy, hospitals in the system can easily share the patient records for emergency cases while protecting privacy information. On the other hand, we also gave the structure for both types of policies based on XML format, including 4 components: subject, action, environment and resource. Comparing to traditional AC models in which the returns are either Permit or Deny, the access control model for fine-grained privacy protection provide a smooth resolution mechanism that enabling access to a portion of the protected data, depend on the level of its policy satisfaction. In addition, policy conflict detection and handling are also described in this work. Specifically, for the access policy, conflict handling is done by combining algorithms while for the privacy policy, it is done by analyzing the priority of function. The system assessment was performed from one to three datasets representing from one to three hospitals of the same system and the number of records is gradually increased during the evaluation process. In addition, to investigate the efficiency of the system with a complexity database, the policy structure is generated with different level of complexity. As a future work we plan to expand our objectives to (i) evaluate the system in the real data environment; and (ii) detects and provides solutions for the "break the glass" problem when an emergency occurs.

References

1. Abomhara, M., Yang, H., Køien, G.M., Lazreg, M.B.: Work-based access control model for cooperative healthcare environments: formal specification and verification. J. Healthc. Inf. Res. **1**(1), 19–51 (2017)
2. Haas, S., Wohlgemuth, S., Echizen, I., Sonehara, N., Müller, G.: Aspects of privacy for electronic health records. Int. J. Med. Informatics **80**(2), e26–e31 (2011)

3. Karimi, L., Joshi, J.: Multi-owner multi-stakeholder access control model for a healthcare environment. In: 2017 IEEE 3rd International Conference on Collaboration and Internet Computing (CIC), pp. 359–368. IEEE (2017)
4. Maw, H.A., Xiao, H., Christianson, B.: An adaptive access control model for medical data in wireless sensor networks. In: 15th International Conference on e-Health Networking, Applications & Services (Healthcom), pp. 303–309. IEEE (2013)
5. Nguyen, M.H., Son, H.X.: A dynamic solution for fine-grained policy conflict resolution. In: International Conference on Cryptography, Security and Privacy. ACM (2019)
6. Ni, J., Zhang, K., Lin, X., Shen, X.S.: Securing fog computing for internet of things applications: challenges and solutions. IEEE Commun. Surv. Tutorials **20**(1), 601–628 (2017)
7. Ray, I., Alangot, B., Nair, S., Achuthan, K.: Using attribute-based access control for remote healthcare monitoring. In: 2017 Fourth International Conference on Software Defined Systems (SDS), pp. 137–142. IEEE (2017)
8. Rissanen, E.: Extensible access control markup language (XACML) version 3.0 (2013)
9. Ruj, S., Nayak, A., Stojmenovic, I.: Distributed fine-grained access control in wireless sensor networks. In: Parallel & Distributed Processing Symposium (IPDPS), pp. 352–362. IEEE (2011)
10. Son, H.X., Chen, E.: Towards a fine-grained access control mechanism for privacy protection and policy conflict resolution. Int. J. Adv. Comput. Sci. Appl. **10**(2) (2019)
11. Son, H.X., Dang, T.K., Massacci, F.: REW-SMT: a new approach for rewriting XACML request with dynamic big data security policies. In: International Conference on Security, Privacy and Anonymity in Computation, Communication and Storage, pp. 501–515. Springer (2017)
12. Son, H.X., Dang, T.K., Tran, L.K.: Xacs–dypol: towards an XACML–based access control model for dynamic security policy
13. Son, H.X., Nguyen, M.H.: A novel attribute-based access control system for fine-grained privacy protection. In: International Conference on Cryptography, Security and Privacy. ACM (2019)
14. Son, H.X., Tran, L.K., Dang, T.K., Pham, Y.N.: REW-XAC: an approach to rewriting request for elastic ABAC enforcement with dynamic policies. In: 2016 International Conference on Advanced Computing and Applications (ACOMP), pp. 25–31. IEEE (2016)
15. Thi, Q.N.T., et al.: Using JSON to specify privacy preserving-enabled attribute-based access control policies. In: International Conference on Security, Privacy and Anonymity in Computation, Communication and Storage, pp. 561–570. Springer (2017)
16. Yang, Y., Li, X., Qamar, N., Liu, P., Ke, W., Shen, B., Liu, Z.: Medshare: a novel hybrid cloud for medical resource sharing among autonomous healthcare providers. IEEE Access **6**, 46949–46961 (2018)
17. Yu, S., Ren, K., Lou, W.: FDAC: toward fine-grained distributed data access control in wireless sensor networks. IEEE Trans. Parallel Distrib. Syst. **22**(4), 673–686 (2011)
18. Zhang, A., Bacchus, A., Lin, X.: Consent-based access control for secure and privacy-preserving health information exchange. Secur. Commun. Netw. **9**(16), 3496–3508 (2016)
19. Zhang, R., Liu, L., Xue, R.: Role-based and time-bound access and management of EHR data. Secur. Commun. Netw. **7**(6), 994–1015 (2014)

Video Anomaly Detection and Localization in Crowded Scenes

Mariem Gnouma(✉), Ridha Ejbali, and Mourad Zaied

Research Team in Intelligent Machines (RTIM),
National Engineering School of Gabes (ENIG), University of Gabes, Gabés, Tunisia
mariem1gnouma@gmail.com, ridha_Ejbali@ieee.org, mourad.zaied@ieee.org

Abstract. Nowadays, the analysis of abnormal events becomes more and more exhausting due to the divine use of surveillance cameras. This paper proposes a novel approach to predict and localize anomaly events. In this paper, a new framework for motion extraction called BQM is proposed. Then, the regions of interest are extracted and a filtering process is applied to eliminate the non-significant ones. However, for more precision, the HFG descriptor is applied for each region already divided into non-overlapping cells, Finally, we have evaluated our method using UCSD and Avenues datasets. The Sparse Auto-encoder, an instance of a deep learning strategy is presented for efficient abnormal activity detection and the Softmax for the classification.

Keywords: Anomaly detection and localization · Binary Quantization Map · Feature extraction · Sparse Auto-encoder

1 Introduction

Safety and security have always been a major problem for shopping centers, banks, government institutions, homes, kindergartens and businesses, etc.

Nowadays, almost everyone is looking for a way to keep their belongings safe. With the recent abstain in the cost of video surveillance, a large number of places can be checked everlastingly. Though, monitoring spaces are senseless if the situation can not be identified and appropriate responses are made immediately.

A significant number of surveillance cameras have been deployed in public spaces as a tool to reduce crime and risk management. These conventional video surveillance systems rely heavily on human operators to monitor activities and determine what action to take in case incident, e.g. track suspicious target from one camera to another or alert agencies affected areas of concern. Unfortunately, many incidents that can lead to abnormal actions are simply erroneously detected in such a manual system due to the inherent limitations of deploying humans monitoring multiple video surveillance.

In congested scenes, we need a method that can achieve high performance in a congested environment, since the performance of tracking algorithms decreases due to the temporary loss of an object related to the blocking of the vision camera [1].

© Springer Nature Switzerland AG 2020
F. Martínez Álvarez et al. (Eds.): CISIS 2019/ICEUTE 2019, AISC 951, pp. 87–96, 2020.
https://doi.org/10.1007/978-3-030-20005-3_9

In low-level feature extraction approaches, moving objects are not recognized as independent entities. Features such as color, motion, optical flow, and texture are extracted for each cell. These methods have the advantage of being robust anti occlusions, which affects the accuracy of monitoring. this approach can work in over-crowded scenes because no object is extracted from the image [2–4]. In most of these low-level methods, the input video is divided into a set of non-overlapping 2D or 3D cells.

Subsequently, these cells are represented in the form of appropriate descriptors such as the Oriented Gradient Histogram (HOG) [5], the Optical Flow Histogram (HOF) [6]. Finally, cells that contain abnormal events are detected using a classifier.

In recent years, divers' approaches have been developed to detect abnormal events without the required of human tracking. To detect global and local anomaly Kim et al. [7] design the local optical flow using mixtures of probabilistic principal component analyzers. They employ a spatio-temporal Markov random field model to detect global and local abnormal activities. As well, their model can update to adapt to environmental changes. In a similar work, Mehran et al. [3] presents the abstraction of the social force in which the interaction forces are considered using optical flow. To overcome the same, recently Xu et al. [8] models crowd scenes for anomaly detection in congested and complicated areas. their approach identifies anomalies based on a discovery of hierarchical activity models, taking into account local and global spatio-temporal contexts based on the deep neural network [9,10]. Biswas et al. [11] propose an approach in H.264/AVC plan to identify anomalies in crowded videos using the magnitude and orientation of motion vectors. Fang et al. [12] in their recent study, based on in depth learning. They draw the saliency information from the histogram of the optical flow and the spatial domain in the time domain. Subsequently, they combine them in spatial-temporal characteristics to extract high-level information for anomaly detection.

Even if some of these approaches may have achieved a proper accuracy in terms of abnormality detection, but they don't give good accuracy in abnormality localization. In real-world monitoring systems, it is necessary to discover the location of the anomaly in addition to the detection of abnormal frames. Moreover, the technical and temporal complexity of the technique should be such that it can operate in the real-time environment, but the main problem of most existing approaches is in the complex algorithms that have avoided their implementation in the real world.

In this article, we propose a new approach for detecting and locating anomalies in overcrowded scenes. Key contributions include: (1) using a new binary quantization map (BQM) for identifying regions of interest in an image. (2) It is no longer necessary to proceed by trial and error in estimating the appropriate size for each image. The new proposal was to divide the regions obtained into overlapping cells and analyze the anomalies of each cell to obtain more precise locations.

2 Feature Extraction

As shown in Fig. 1, Initially, the optical flow is estimated via a consecutive set of frames. In the next step, a new binary quantization map (BQM) is constructed of each two successive frames. Thereafter, the connected components are extracted from the BQM image using a proprietary region map. A connected component in a binary image of contiguous pixels. The feature vectors are subsequently constructed as a histogram from the connected components of the images considered. The classification of the events is done by means of deep learning on a subset of features derived from the learning phase.

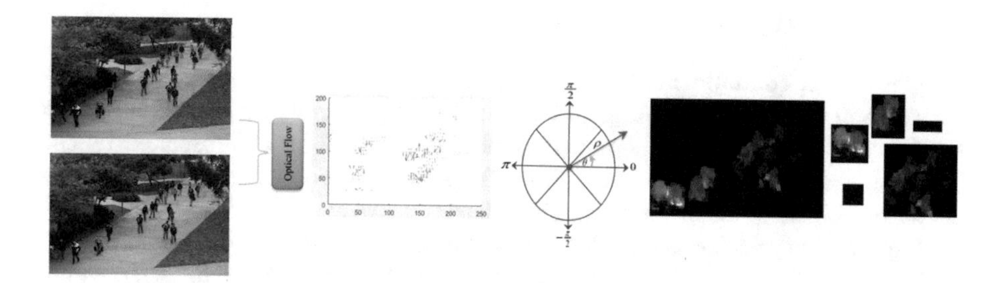

Fig. 1. Overview of the proposed system.

2.1 Optical Flow Estimation

The proposed approach is based on estimating the optical flux to generate a feature vector from a sequence of consecutive frames. The approach eases the process of object detection since, there is no need for a primordial phase for the detection of each object in motion this contributed to a saving of computing time. However, the ability to update our BQM model is influenced by important factors, including weather, clutter and other environmental conditions.

Unlike our ancient works [13,14] which, we used the method of Lukas Kanade and Horn Schunk for motion estimation. In this work a new optical flow technique is generated. This new method is based on polynomial expansion of a frame. According to this method, the neighborhood of each pixel can be approximated by a given quadratic polynomial as:

$$I1(x) = x^T M_1 x + b_1^T x + c_1 \tag{1}$$

where M_1 is a matrix, b_1 is a vector, and c_1 is a scalar. This makes it possible to present a presentation of the local signal. Applying the weighted least squares approach can estimate the coefficients by matching neighborhood signal values. After a global displacement d, the new signal obtained is:

$$I2(x) = x^T M_2 x + b_2^T x + c_2 \tag{2}$$

To reach the value of d, the coefficients of the quadratic polynomials (1) and (2) must be equivalent:

$$d = -\frac{1}{2} M_1^{-1} (b_2 - b_1) \tag{3}$$

The Farneback algorithm, associate the polynomial approximation with multi scale resolution to invent optical flow results for each pixel of the image.

Given the existence of an abnormal movement in a small area between two images, t and t + 1, the displaced patch must be readily located in the neighboring area at the level of the new image.

2.2 Binary Quantization Map

To generate a common model for image motion; it is necessary to group the entity components. The coding of the binary mappings is performed based on the results obtained from the calculation of the optical flow. Subsequently, the ideal solution is the use of a map. This allows the representation of the intensity of the movement of each object in the scene.

Each velocity vector is encoded by the angle it indicates by placing its origin in the center of the map. Depending on the angle the point takes a different value, its intensity varies from black color to complete white color according to the standard of the vector. The maximum speed is represented by white dots. This mapping ultimately returns to construct a binary quantization that associates a point in space with the velocity vector. This type of representation makes it possible to give a color for each pixel and to quickly visualize the coherence of the results as shown in Fig. 1. Based on the magnitude and the angle values for the optical vector, the gray Map BQM is considered as the discretized color for the flow vector as presented in Algorithm 1.

Algorithm 1. Encoding binary maps

Input: u, v, ϱ
Output: BQM
Initialize current angles and color step to:
anglestep: $pi/3$, **colorstep**: $1/anglestep$, **Colorpart**:[0..255]
BQM: zeros(size(u,1), size(u,2),3)
for i=1, i<=size(u,1)
for j=1, j<=size(v,2)
angle=atan2(u(i,j),v(i,j))
angle = angle+2*pi
part=angle/anglestep
BQM(i,j,k)=colorstep*mod(part,1)*length/ ϱ
End
End

2.3 Partition of the Region

Due to the extraction of connected components of the BQM frames, it is not necessary to divide the entire frame into non-overlapping cells (see Fig. 2). Indeed, the computing load is considerably reduced and the amount of data required. To determine the size of each extracted region, it is necessary to recognize the connected components in the BQM image. A connected component is a set of contiguous pixels belonging to one or more objects. A unique random number will be assigned to each component connected as a region code to differentiate the different images obtained and to facilitate the task of recognition and location of the abnormal event.

A filtering process is applied to eliminate small areas due to sudden lighting problems, tree leaves etc. It consists of rejecting the different small regions whose size are less than half of the average of different regions obtained. After analyzing the BQM image and getting the different regions associated with each BQM. The process of decoding each region on the same cells begins. Subsequently a new matrix "M = size (each connect component (BQM))" is constructed which contains the number of regions of the same size. After analyzing all BQM images in the training data, a value with the maximum number is considered the fixed cell size (FCS).

$$FCS = max(M) \tag{4}$$

The BQM is then partitioned into non-overlapping cells. Thereafter, we eliminate all the cells which do not contain the white color since as presented beforehand the white color presents the maximum of speed. Thereafter HFG is generated. As shown in Fig. 2 after calculating the optical flux using Farneback, HFG is calculated for each cell. It should be noted that cells that do not contain important information are rejected to avoid further calculation.

As a modified version of HOG is called Histogram of Flow Gradient. The HFG is very similar to that of HOG, but it operates on a local optical flux field, as shown in Fig. 1. In addition, the implementation of HFG is faster in terms of calculation than that of HOG and HOF.

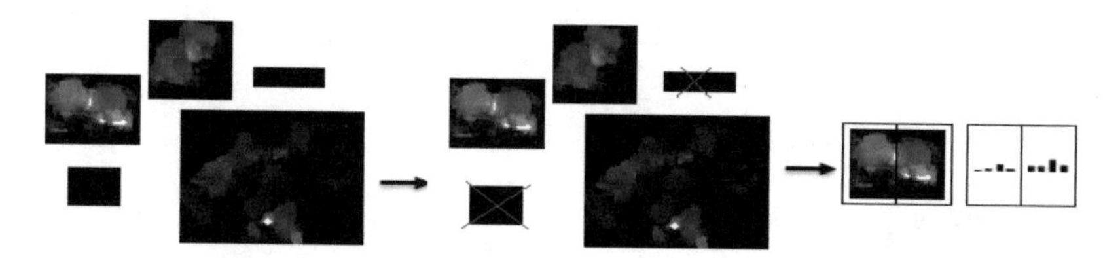

Fig. 2. Division each region into non-overlapping cells and HFG generation.

2.4 Anomaly Detection

Deep learning is widely used for abnormal event detection [2,7,9,12,15,16]. In this work we use a Sparse Auto-Encoder, which is one approach to automatically learn features from unlabeled data. SAE is a type of AE with a sparsity applicator that instructs a single layer network to learn a codebook that minimizes reconstruction errors while limiting reconstruction. In fact, the classification task becomes a kind of algorithm that reduces and minimizes the prediction error.

The SAE is introduced with a single hidden layer, which is connected to the input x with a weighting matrix. The hidden layer is then sent back to his reconstruction $\widetilde{x}$.

To lead to our Deep learning, we use series of the SAE. So, the hidden layer in the auto-encoder is considered as a hidden layer in our SAE and the input layer for the second training to obtain another layer for the SAE. Two hidden layer permits the creation of our SAE. The steps of an example of the creation with two hidden layers and a Softmax classifier are: (Fig. 3).

- We train an SAE on the raw inputs x to learn primary features on the raw input.
- We then use these primary features as the "raw input" to another SAE to learn secondary features on these primary features.
- Finally, we combine all two layers together to form our Deep SAE with 2 hidden layers and we add a linear classifier layer capable of classifying the images as desired.

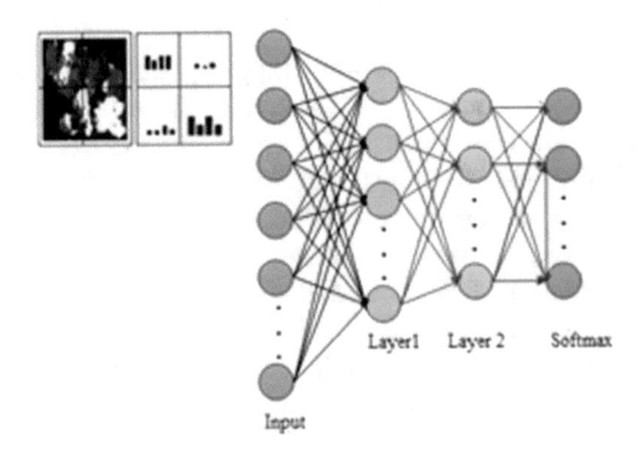

Fig. 3. A deep neural network with two hidden layers and a linear classifier layer.

3 Experimental Results

3.1 Datasets

In order to evaluate our approach for the use of the different characteristics according to the movement of the optical flow using the proposed processing, two different datasets are taken for the evaluation process and its performance is compared with other state of the art approaches in literature.

- UCSD dataset: The data [17] was split into two subsets; each scene was separated into different clips of around 200 images. The first one Contains 34 training video samples and 36, and Ped 2 includes 16 training videos and 12 for the test.

- Avenue dataset: The dataset [18] contains 15 sequences. Each sequence is about 2 min long. The total number of frames is 35,240; 16 training and 21 testing clips. The videos are captured with 15328 training and 15324 for the test. It's a complex streamlined database containing various types of abnormal events.

In experiments, the calculated FCS value is $35 * 17$ for Ped 1, $41 * 18$ for Ped 2, $39 * 18$ for Avenue.

3.2 Comparative Analysis

Comparative analysis for the proposed method compared to existing approaches which are applied recently for abnormal event detection on the UCSD and Avenue datasets are shown in Fig. 4.

Fig. 4. Examples of abnormal event detection results on the UCSD and Avenue datasets.

As well, Table 1 presents the results of our proposed approach where the defined model BQM + HFG + SAE was used for the detection of motion anomalies.

However, EER allows detection at the frame level. It checks whether the detection coincides with the actual location of the anomaly.

Table 1. Quantitative comparison with popular approaches of state of the art

Method	Frame EER Ped 1	Frame EER Ped 2	AUC
SF [19]	0.31	0.42	55.6
MDT [20]	0.25	0.25	82.9
150FPS [18]	0.15	-	-
STC [21]	0.13	-	-
SHD [22]	012	-	-
AMDN [16]	0.16	0.17	90.8
MAD [4]	0.279	0.21	80
GMFCA [23]	0.11	0.12	92.2
Proposed method	0.107	0.117	97.8

As shown in the first table, the EER of our approach is drastically reduced using the BQM mechanism with FCS for motion anomaly detection. Table 1 indicate that the results does not check if the abnormal event really matches the location of the anomaly. For this, to clarify and test the actual accuracy of the location of abnormal events, the receiver operating characteristic (ROC) of the method is shown in Fig. 5.

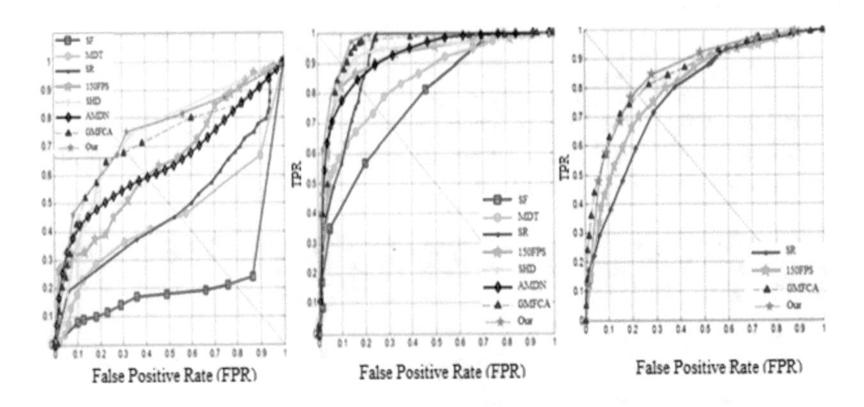

Fig. 5. ROC curves for the UCSD Ped 1, Ped 2, Avenue datasets.

The ROC illustrates that our approach is comparable to several methods for the UCSD and Avenue database. Our approaches achieved 97.8% AUC for UCSD and 96.2% for Avenue dataset, this surpasses all methods used for comparison.

Only a few approaches have been tested on the Avenue dataset, which is publicly released recently. However, the anomalies are labeled by rectangular regions, but are not really rectangular, the truth of ground contains pixels of foreground and background. For this reason, we dismiss the measurement at the pixel level and use only the measurement of the optical flow presented previously BQM. Compared to advanced approaches, our method shows the effectiveness of our proposed system for detecting abnormal events and this is shown in Table 2.

Table 2. Comparisons of detection accuracy on the Avenue dataset

Method	AUC Avenue
150FPS [18]	80.9
DF [24]	78.3
GMFCA [23]	83.4
Proposed method	96.2

In our experiments we perceive that in most of the condition the proposed approaches hit the abnormality correctly.

To detect abnormal events in the proposed approach, due to the extraction of connected components from the original frames, there is no need for dividing of the full frame into the non-overlapping cells. Therefore, the amount of required data and thus the computational load is significantly reduced, as can be seen in Tables 1 and 2.

The comparison of these methods proves that our approach can accurately detect anomalies using a little number of feature vectors.

4 Conclusion and Future Works

Abnormal event detection and localization presented in the last years a challenge for most researchers. In this paper, we present a new method for detecting anomalous events in crowded scenes.

One of the innovations of the proposed method is a pre-treatment applied to the training frameworks by extracting their important regions using a new BQM descriptor. After analyzing the BQM image and getting the different regions. The appropriate cell size FCS is defined for each region. Then, the anomaly is detected using the HFG descriptions of the extracted regions which presents the entries to our SAE for the training and the Softmax for the classification.

The experimental results show the lead of the proposed approach over state of the art methods on the UCSD and Avenue databases. In the follow up of this research, we aim to explore the possibilities of detecting an abnormal event and, at the same time, categorize them according to the severity or the need for an immediate response.

Acknowledgments. The authors would like to acknowledge the financial support of this work by grants from General Direction of Scientific Research (DGRST), Tunisia, under the ARUB program.

References

1. Tang, S., Andriluka, M., Schiele, B.: Detection and tracking of occluded people. Int. J. Comput. Vis. **110**(1), 58–69 (2014)
2. Gnouma, M., Ejbali, R., Zaied, M.: Human fall detection based on block matching and silhouette area. In: Ninth International Conference on Machine Vision (ICMV 2016), vol. 10341, p. 1034105. International Society for Optics and Photonics (2017)
3. Mehran,R., Oyama, A., Shah, M.: Abnormal crowd behavior detection using social force model. In: CVPR, pp. 935–942 (2009)
4. Amraee, S., Vafaei, A., Jamshidi, K., Adibi, P.: Abnormal event detection inf crowded scenes using one-class SVM. Signal Image Video Process. **12**(6), 1–9 (2018)
5. Dalal, N., Triggs, B.: Histograms of oriented gradients for human detection. In: CVPR, pp. 886–893 (2005)
6. Dalal, N., Triggs, B., Schmid, C.: Human detection using oriented histograms of flow and appearance. In: Proceedings of European Conference on Computer Vision, pp. 428–441 (2006)

7. Kim, J., Grauman, K.: Observe locally, infer globally: a space time MRF for detecting abnormal activities with incremental updates. In: CVPR, pp. 2921–2928 (2009)
8. Xu, D., Song, R., Wu, X., Li, N., Feng, W., Qian, H.: Video anomaly detection based on a hierarchical activity discovery within spatiotemporal contexts. Neurocomputing **143**(1), 144–152 (2014)
9. Khatrouch., M., Gnouma, M., Ejbali, R., Zaied, M.: Deep learning architecture for recognition of abnormal activities. In: The 10th International Conference on Machine Vision, p. 106960F. International Society for Optics and Photonics, Vienna, Austria (2018)
10. ElAdel, A., Ejbali, R., Zaied, M., Amar, C. B.: Dyadic multi-resolution analysis-based deep learning for Arabic handwritten character classification. In: ICTAI, pp. 807–812 (2015)
11. Biswas, S., Babu, R.V.: Anomaly detection in compressed H.264/AVC video. Multimed. Tools Appl. **74**(24), 11099–11115 (2015)
12. Fang, Z., Fei, F., Fang, Y., Lee, C., Xiong, N., Shu, L., Chen, S.: Abnormal event detection in crowded scenes based on deep learning. Multimed. Tools Appl. **75**(22), 14617 (2016)
13. Gnouma, M., Ladjailia, A., Ejbali, R., Zaied, M.: Stacked sparse autoencoder and history of binary motion image for human activity recognition. Multimed. Tools Appl. **78**, 1–23 (2018)
14. Gnouma, M., Ejbali, R., Zaied, M.: Abnormal events' detection in crowded scenes. Multimed. Tools Appl., 1–22 (2018) https://doi.org/10.1007/s11042-018-5701-6
15. Ejbali, R., Zaied, M., Amar, C. B.: Face recognition based on Beta 2D elastic bunch graph matching. In: HIS pp. 88–92 (2013)
16. Xu, D., Yan, Y., Ricci, E., Sebe, N.: Detecting anomalous events in videos by learning deep representations of appearance and motion. Comput. Vis. Image Underst. **156**(C), 117–127 (2017)
17. UCSD Anomaly Detection Dataset. http://www.svcl.ucsd.edu/projects/anomaly/dataset
18. Lu, C., Shi, J., Jia, J.: Abnormal event detection at 150 fps in Matlab. In: Proceedings of the IEEE International Conference on Computer Vision, pp. 2720–2727 (2013)
19. Mehran, R., Oyama, A., Shah, M.: Abnormal crowd behavior detection using social force model. In: CVPR, pp. 935–942 (2009)
20. Mahadevan, V., Li, W., Bhalodia, V., Vasconcelos, N.: Anomaly detection in crowded scenes. In: CVPR, pp. 1975–1981 (2010)
21. Roshtkhari, M.J., Levine, M.D.: An on-line, real-time learning method for detecting anomalies in videos using spatio-temporal compositions. Comput. Vis. Image Underst. **117**(10), 1436–1452 (2013)
22. Yuan, Y., Feng, Y., Lu, X.: Statistical hypothesis detector for anomalous event detection in crowded scenes. IEEE Trans. Cybern. **99**, 1–12 (2016)
23. Fan, Y., Wen, G., Li, D., Qiu, S., Levine, M.D.: Video anomaly detection and localization via Gaussian mixture fully convolutional variational autoencoder. arXiv preprint arXiv:1805.11223 (2018)
24. Del Giorno, A., Bagnell, J.A., Hebert, M.: A discriminative framework for anomaly detection in large videos. In: European Conference on Computer Vision, pp. 334–349. Springer, Cham (2016)

Beta Chaotic Map Based Image Steganography

Zahmoul Rim(✉), Abbes Afef, Ejbali Ridha, and Zaied Mourad

Research Team in Intelligent Machines, National Engineering School of Gabes, 6072 Gabes, Tunisia
rima.zahmoul@gmail.com

Abstract. Steganography is one of the growing security fields. It is the science of embedding sensitive data in images using different techniques. In this paper, we proposed a steganographic scheme based on Beta chaotic map in order to embed secret images bits into cover images. Thus a secure transmission of data could be established. We applied the PSNR, MSE and histogram tests on the obtained images and comparisons were made with different existing methods; prove that our proposed scheme has a good steganographic quality.

Keywords: Beta chaotic map · Steganography · Chaos · Beta function

1 Introduction

Since the beginning of civilization, the need to hide information has been growing. Cryptography provides a way to protect secret data by making it unintelligible to unauthorized persons [1]. Yet communication with encrypted messages attracts attention. This can be problematic in the case of a third-party monitored communication channel that can destroy communication between the two parties. Today, most of the applications require secure data transiting between users, via a vector of information such as telecommunications networks. To maintain the secrecy of transmitted information, steganography is an effective solution. It is defined as the science of hiding data. It is characterized by three main proprieties: capacity, robustness, and invisibility. The capacity presents the amount of information that can be presented without appreciable lapse. The robustness is that the hidden data cannot be destroyed without degrading the image [2]. Invisibility ensures that the resulting message is not modified by the secret information inserted. Steganography in images has many different methods [3–5]. One of the most secure techniques is the use of chaos theory. The latter has witnessed a remarkable development in recent decades. Due to inherent attacks, communication between people should be secure, the use of chaos prove its high confidentiality.

Our Paper presents an efficient method in hiding images using beta chaotic maps which are characterized by the variety of parameters and its high sensibility to any small changes in the initial condition. The paper consists of three main sections; section two presents some related work, while in section three we developed the proposed approach and the fourth section contains the analysis results.

© Springer Nature Switzerland AG 2020
F. Martínez Álvarez et al. (Eds.): CISIS 2019/ICEUTE 2019, AISC 951, pp. 97–104, 2020.
https://doi.org/10.1007/978-3-030-20005-3_10

2 Related Works

Many proposed steganography techniques using chaos theory were proposed all over the years. Sharif et al. [6] presented a new steganography scheme for digital images which is based on a new created chaotic map. The technique consists of the generation of three chaotic sequences from the created maps. The embedding process depends on the LSB (Least Significant Bit) and MSB (Most Significant Bit). Result proves the efficiency of the proposed approach. Authors in [7] introduce a hiding process of secret data into a cover image. It was based on LSB or PVD (Pixel Value Differencing) techniques. The logistic chaotic map was used in the embedding system. Simulation results assure high-security performances and good stego-image quality. In [8] researchers suggested an efficient method of RGB color image steganography established on overlapping block based PVD. The proposed approach has been tested using different metrics. It shows an acceptable quality of stego-image. Recently, Dash et al. [9] proposed an improved hiding image method based on chaotic encryption. The cryptography process was implemented with the chaotic neural network system (CNN) in order to protect hiding data. After the encryption phase, the encoded data was scrambled and embedded into the cover image's edge region. Analysis results show the strengths of the scheme. In our scheme, we have used the Beta chaotic map [10,11] to generate the chaotic sequence, which is employed to select, in a random manner, the values of pixels for hiding in the cover image.

3 Proposed Approach

3.1 Beta Chaotic Map

The use of chaotic maps in recent years was increased due to its efficiency in maintaining high security for digital steganography. As a result of its specific proprieties like the ergodicity, random behavior and control parameters, chaotic maps were qualified as a good choice in steganography and encryption [12,13]. One of the recent chaotic maps was applied in our hiding process. Beta map has important features in the security domain [10,11]. It is characterized by a large range of bifurcation parameter, a strong chaotic behavior and a high amount of parameters. It is based on the beta function [14–16,21–23]. This function was used in different domains [24–27]. It is written as indicated in Eqs. (1) and (2).

$$Beta(x; j, h, x_1, x_2) = \begin{cases} (\frac{(x-x_1)}{(x_c-x_1)})^j(\frac{(x_2-x)}{(x_2-x_c)})^h & \text{if} x \in]x_1, x_2[\\ 0 & \text{else} \end{cases} \tag{1}$$

Where j, h, x_1 *and* $x_2 \in R$, $x_1 < x_2$ and x_c:

$$x_c = \frac{(jx_2 + hx_1)}{(j+h)} \tag{2}$$

Figure 1 shows the map, that we used in our scheme, which have the following parameters:
$a = 0.1$, $x_1 = -1$, $x_2 = 1$, $k = 0.85$, $b_1 = 5$, $c_1 = 1$, $b_2 = 3$ and $c_2 = -1$

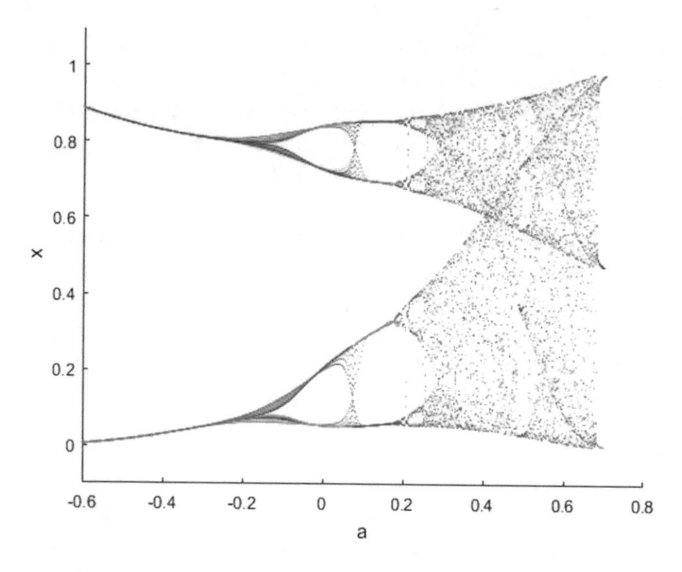

Fig. 1. a = [−0.8:0.7], $x_1 = -1$, $x_2 = 1$, $k = 0.85$, $b_1 = 5$, $c_1 = 1$, $b_2 = 3$ and $c_2 = -1$.

3.2 Least Significant Bit Technique

Historically, the technique of substitution of low-order bits (LSB Substitution) is the first method of steganography in the literature. It is one of the most famous steganography tools. Based on the binary form of the secret image, it hides the data in the binary cover image. This method consists of embedding the secret image bits into the least significant bits of the cover image pixels. In other words, to insert a message $m = (m_1, ..., m_m)$, the last bit of low weight $b(i, 0)$ of each pixel is changed by a bit of the message to be hidden. The direction of browsing the pixels is usually chosen by a pseudo-random path [17]. To do this, the transmitter and the receiver must first exchange a key k, used as the seed of a pseudo-random number generator.

3.3 Image Hiding Process

The process shows different steps in order to hide the secret image into a cover image and guarantee a secure transmission. First, we choose two images a secret one S and a cover one C. Then, a random chaotic sequence Y, defined by $Y = Y_1, Y_2, ..., Y_m$ with m is the size of rows or columns of the cover image C, was generated from the Beta map. The generated chaotic sequence from the beta map shows a variety of parameters $a = 0.1$, $x_1 = -1$, $x_2 = 1$, $k = 0.85$, $b_1 = 5$, $c_1 = 1$, $b_2 = 3$ et $c_2 = -1$.

We applied a sorting on the obtained list and we extract the index of the sorted values in a vector idx as follows:

$$[Y, idx] = sort(Y1) \tag{3}$$

Where *idx* represent the index vector of the sorted elements of the chaotic map and $Y1$ is the sorted sequence. Then, we converted the secret image and the cover image into a binary form to prepare our process for the embedding technique. Using the LSB technique, we integrated the secret image bits into the coverage image according to the values of the obtained index *idx*. Thus we got the new Stego-Image *SG*.

The extraction process is the reverse of the embedding. After generating the chaotic sequence using Beta map, a sorting was applied and a new index vector was obtained. The extraction of bits happened in a random way and depended on the index vector. The Stego-image was converted into binary form. Thus, the extraction of the secret image using LSB technique was applied.

4 Analysis Results

The evaluation of the proposed method was done by the use of different metrics. Figure 2 shows the used cover image *a*, the secret image *b* and the obtained stego-image *c*. The tests were applied on different USC-SIPI Image Database. Comparison was done with several methods [18–20]. Our tests were implemented on different images (Elaine, Baboon and Airplane), they were presented in Fig. 3.

Fig. 2. (a) Cover image; (b) Secret image; (c) Stego-image

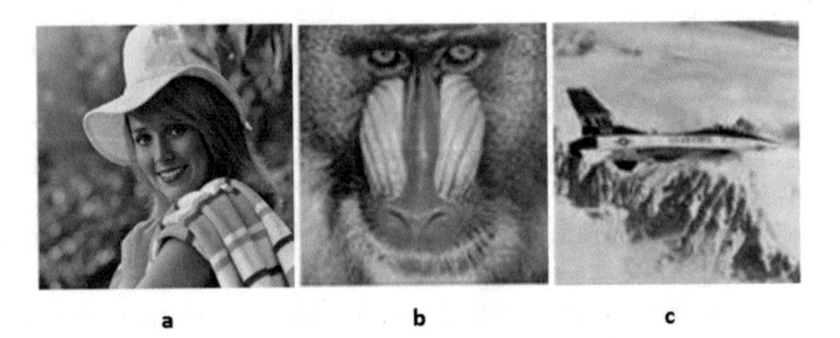

Fig. 3. (a) Elaine image; (b) Baboon image; (c) Airplane image

4.1 Histogram Analysis

This kind of tests studies the similarity between the stego-image pixels and the cover image pixels. After testing our obtained images, results display a tiny deformity in the obtained stego-image after applying the hidden process. Figure 4 presents the histograms of the cover image *a* and of the resulting image after embedding the secret image into it *b*. We conclude that the introduced method is robust against visual attacks.

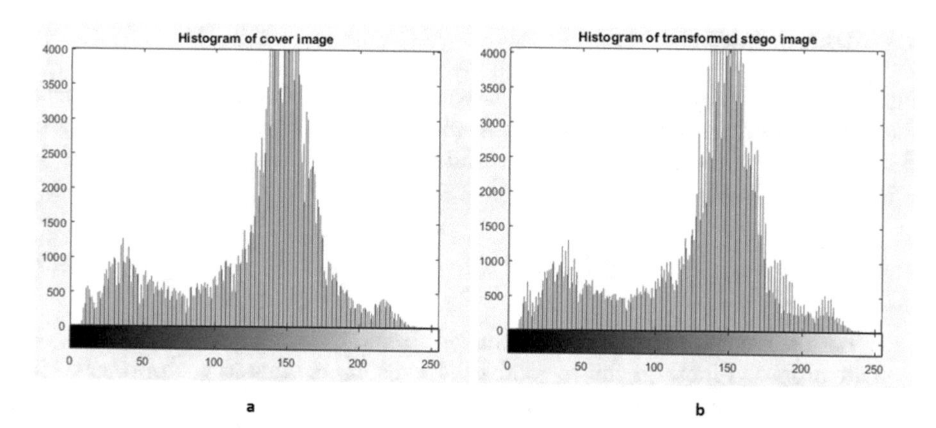

Fig. 4. (a) Histogram of the cover image; (b) Histogram of the transformed stego-image

4.2 MSE Test

Mean Squared Error (MSE) is standard technique to calculate the difference between two digital images. It is written as indicated in Eq. (4).

$$MSE = \frac{1}{\mathrm{MN}} \sum_{i=1}^{N} \sum_{j=1}^{M} [IM1\,(i,j) - IM2(i,j)] \tag{4}$$

Where IM_1 is the cover or the original image, and IM_2 is the obtained stego-image. This test indicates that the more the difference is low the more the steganography result is good and vice versa. Our approach proves, in Table 1, that the difference between the cover and stego images is small compared with methods in [18–20], thus the embedding quality is good.

Table 1. MSE values of different methods using Boat image as secret image

Cover image	Basic LSB	[18]	[19]	[20]	Our approach
Elaine	17.55	16.09	9.78	2.28	0.49
Airplane	17.32	15.47	10.13	2.34	0.46
Baboon	17.51	15.89	9.68	2.28	0.47
Average	17.46	15.81	9.86	2.30	0.47

4.3 PSNR Test

Pick Signal to Noise Ratio (PSNR) is similar to the MSE test. It introduced the ration between the signal strength and the noise strength. High values of PSNR indicate good steganography results, and poor ones gives a bad sign. It is introduced by Eq. (5).

$$\text{PSNR} = 10 \times log_{10}\left[\frac{R^2}{MSE}\right] \tag{5}$$

Where R consider the highest of image pixels.

Our proposed method shows good PSNR results compared to approaches in [18–20].The obtained values were drawn in Table 2.

Table 2. PSNR values of different methods using Boat image as secret image

Cover image	Basic LSB	[18]	[19]	[20]	Our approach
Elaine	35.55	36.02	39.36	44.53	57.21
Airplane	35.70	36.08	39.19	44.42	57.50
Baboon	35.61	36.14	39.29	44.54	56.79
Average	35.62	36.08	39.28	44.49	57.16

5 Conclusion

In this paper, an image masking approach based on a chaotic map has been proposed using Beta chaotic map. Our algorithm embeds the secret image bits in a random position of the cover image according to the sorted result of the chaotic sequence. The used map shows better chaotic behavior, thus generated sequences are better in term of randomness. Experimental analysis, based on PSNR, MSE and histogram tests, and comparisons with other existing approaches which use other chaotic maps like the logistic map and other embedding systems like the genetic algorithm prove the resistant of our method against some steganalytic attacks. Thus, the use of Beta map in image steganography was a good choice in order to perform an improved embedding quality.

Acknowledgment. The authors would like to acknowledge the financial support of this work by grants from General Direction of Scientific Research (DGRST), Tunisia, under the ARUB program.

References

1. Borda, M.: Fundamentals in Information Theory and Coding. Springer, Heidelberg (2011). https://doi.org/10.1007/978-3-642-20347-3
2. Katzenbeisser, S., Petitcolas, F.: Information Hiding Techniques for Steganography and Digital Watermarking. Artech house, Norwood (2000)
3. Tang, W., Li, B., Tan, S., Barni, M., Huang, J.: CNN-based adversarial embedding for image steganography. IEEE Trans. Inf. Forensics Secur. **PP**, 1 (2019)
4. Pal, A.K., Naik, K., Agarwal, R.: A steganography scheme on JPEG compressed cover image with high embedding capacity. Int. Arab J. Inf. Technol. **16**(1), 116–124 (2019)
5. Duan, X., Jia, K., Li, B., Guo, D., Zhang, E., Qin, C.: Reversible image steganography scheme based on a U-Net structure. IEEE Access **7**, 9314–9323 (2019)
6. Sharif, A., Mollaeefar, M., Nazari, M.: Multimed. Tools Appl. **76**, 7849 (2017). https://doi.org/10.1007/s11042-016-3398-y
7. Prasad, S., Pal, A.K.: Logistic map-based image steganography scheme using combined LSB and PVD for security enhancement. In: Abraham, A., Dutta, P., Mandal, J., Bhattacharya, A., Dutta, S. (eds.) Emerging Technologies in Data Mining and Information Security. Advances in Intelligent Systems and Computing, vol. 814. Springer, Singapore (2019)
8. Prasad, S., Pal, A.K.: An RGB colour image steganography scheme using overlapping block-based pixel-value differencing. R. Soc. Open Sci. **4**(4), 161066 (2017)
9. Dash, S., Das, M.N., Das, M.: Secured image transmission through region-based steganography using chaotic encryption. In: Behera, H., Nayak, J., Naik, B., Abraham, A. (eds.) Computational Intelligence in Data Mining. Advances in Intelligent Systems and Computing, vol. 711. Springer, Singapore (2019)
10. Zahmoul, R., Ejbali, R., Zaied, M.: Image encryption based on new Beta chaotic maps. Opt. Lasers Eng. **96**, 39–49 (2017)
11. Zahmoul, R., Zaied, M.: Toward new family beta maps for chaotic image encryption. In: 2016 IEEE International Conference on Systems, Man, and Cybernetics (SMC), pp. 004052–004057. IEEE, October 2016
12. Gashim, L.L., Hussein, K.Q.: A new algorithm of encryption and decryption of image using combine chaotic mapping. Iraqi J. Inf. Technol. **9**, 1–16 (2018)
13. Jain, M.: Medical image steganography using dynamic decision tree, piecewise linear chaotic map, and hybrid cryptosystem. Int. J. Appl. Eng. Res. **13**(15), 12353–12363 (2018)
14. Amar, C.B., Zaied, M., Alimi, A.: Beta wavelets. synthesis and application to lossy image compression. Adv. Eng. Softw. **36**(7), 459–474 (2005)
15. ElAdel, A., Ejbali, R., Zaied, M., Amar, C.B.: A hybrid approach for content-based image retrieval based on fast beta wavelet network and fuzzy decision support system. Mach. Vis. Appl. **27**(6), 781–799 (2016)
16. Ejbali, R., Zaied, M., Amar, C.B.: Face recognition based on Beta 2D elastic bunch graph matching. In: HIS, pp. 88–92, December 2013
17. Chanu, Y.J., Singh, K.M., Tuithung, T.: Image steganography and steganalysis: a survey. Int. J. Comput. Appl. **52**(2), 1–11 (2012)

18. Khodaei, M., Faez, K.: Image hiding by using genetic algorithm and LSB substitution. In: International Conference on Image and Signal Processing, pp. 404–411. Springer, Heidelberg, June 2010
19. Bedi, P., Bansal, R., Sehgal, P.: Using PSO in image hiding scheme based on LSB substitution. In: International Conference on Advances in Computing and Communications, pp. 259–268. Springer, Heidelberg, July 2011
20. Rajendran, S., Doraipandian, M.: Chaotic map based random image steganography using LSB technique. IJ Netw. Secur. **19**(4), 593–598 (2017)
21. El Adel, A., Zaied, M., Amar, C.B.: Learning wavelet networks based on multiresolution analysis: application to images copy detection. In 2011 International Conference on Communications, Computing and Control Applications (CCCA), pp. 1–6. IEEE, March 2011
22. Ejbali, R., Zaied, M., Amar, C.B.: Multi-input multi-output beta wavelet network: modeling of acoustic units for speech recognition. arXiv preprint arXiv:1211.2007 (2012)
23. Zaied, M., Mohamed, R., Amar, C.B.: A power tool for Content-based image retrieval using multiresolution wavelet network modeling and dynamic histograms. Int. Rev. Comput. Softw. (IRECOS) **7**(4), 1435–1444 (2012)
24. Jemai, O., Ejbali, R., Zaied, M., Amar, C.B.: A speech recognition system based on hybrid wavelet network including a fuzzy decision support system. In: Seventh International Conference on Machine Vision (ICMV 2014), vol. 9445, p. 944503. International Society for Optics and Photonics, February 2015
25. Teyeb, I., Jemai, O., Zaied, M., Amar, C.B.: A drowsy driver detection system based on a new method of head posture estimation. In: International Conference on Intelligent Data Engineering and Automated Learning, pp. 362–369. Springer, Cham, September 2014
26. Ejbali, R., Zaied, M., Amar, C.B.: Intelligent approach to train wavelet networks for recognition system of arabic words. In: KDIR, pp. 518–522, October 2010
27. Hassairi, S., Ejbali, R., Zaied, M.: Supervised image classification using deep convolutional wavelets network. In: 2015 IEEE 27th International Conference on Tools with Artificial Intelligence (ICTAI), pp. 265-271. IEEE, November 2015

Deep Wavelet Extreme Learning Machine for Data Classification

Siwar Yahia[1,2](✉), Salwa Said[1,2], and Mourad Zaied[1,2]

[1] National Engineering School of Gabes, Gabès, Tunisia
yahiasiwar@yahoo.fr, salwa.said@isimg.tn, mourad.zaied@ieee.org
[2] RTIM: Research Team in Intelligent Machines, Gabès, Tunisia

Abstract. Recently, Extreme Learning Machine (ELM) has drawn an increasing attention, Due to its fast and good generalization ability. This paper proposes a new learning method for Extreme Learning Machine based wavelet and deep architecture. We have applied a composite wavelet activation function at the hidden nodes of ELM and the learning is done by a Deep Extreme Learning Machine. To evaluate the performance of our approach we have used a standard benchmark dataset for multi-class image classification (MNIST). Results show that our approach offers a significantly better performance relative to others approaches.

Keywords: Extreme Learning Machine · Wavelet Neural Networks · Deep learning · ELM auto-encoder

1 Introduction

In the last few years, there has been a growing interest in the domain of machine learning. In fact, it was well studied by numerous research projects such as feature identification [1] and object recognition [2,3]. Consequently, several learning methods have evolved.

These methods include regression algorithm (Ordinary Least Squares Regression, Linear Regression, Logistic Regression ...), Instance-based Algorithms (k-Nearest Neighbour (kNN), Self-Organizing Map (SOM), Learning Vector Quantization (LVQ) ...), Decision Tree (Classification and Regression Tree (CART), M5, Conditional Decision Trees ...) and Artificial Neural Network (Back-Propagation, Radial Basis Function Network (RBFN)) [4].

Quite recently, considerable attention has been paid to ANN. So, it has emerged in several theories. Indeed, the combination of wavelet theory and artificial neural networks has led to the introduction of wavelet neural networks [5].

It is proposed by Zhang and Benveniste in 1992 [5]. This technique is able to produce a universal approximation. It performs well in classification applications. Previous studies indicate that WNN guarantees a better generalization performance in nonlinear system problems due to the properties of wavelet decomposition for localization in both time and frequency domain [1,6,7].

© Springer Nature Switzerland AG 2020
F. Martínez Álvarez et al. (Eds.): CISIS 2019/ICEUTE 2019, AISC 951, pp. 105–113, 2020.
https://doi.org/10.1007/978-3-030-20005-3_11

Besides, Extreme Learning Machine is a learning algorithm for generalized Feed-Forward Neural Networks. It was introduced by [8]. It is based on the random initialization of the input weights and the analytical determination of the outputs. So, once the input parameters are initialized, they remain unchangeable during the learning phase. It has a lot of advantages such as the speed of learning, the better generalization ability, and minimal human intervention [9,10].

For several years great effort has been devoted to the study of ELM. Therefore, it much emerged in theories. Indeed, various approaches have been proposed as ELM with the kernel learning [11], ELM for regression [12] ELM using the multilayer architecture [11] and The Online Sequence Extreme Learning Machine that makes learning data by blocks [13]. In addition, ELM has been extended for both semi-supervised and unsupervised learning [14]. It has also been transformed in a stacked (S-ELMs) [15] and Hierarchical [16] version.

This paper introduced a new learning algorithm (Deep wavelet extreme learning machine auto-encoder) for Extreme learning machine. The learning is based on a composite wavelet function used as an activation function of ELM Auto Encoder (ELM-AE). Our approach used also a Deep network as an architecture.

2 Related Works

2.1 Extreme Learning Machine

Extreme Learning Machine Presentation

Extreme Learning Machine [8] is a learning algorithm proposed for training SLFNs (Fig. 1). The concept of ELM consists of the randomness of input hidden layer parameters (weights and bias) and analytical resolution of output weights. So it is a resolution of a linear system. In fact, the hidden node parameters are randomly assigned and the output weights will be derived analytically by a simple generalized inverse of the hidden layer output matrix. So, there is no need to tune all the network parameters such the traditional gradient-based algorithms and we can overcome many problems of gradient-based algorithms such as local minima, learning rate, and learning epochs. Elm with direct learning produces a good generalization performance and fast learning speed.

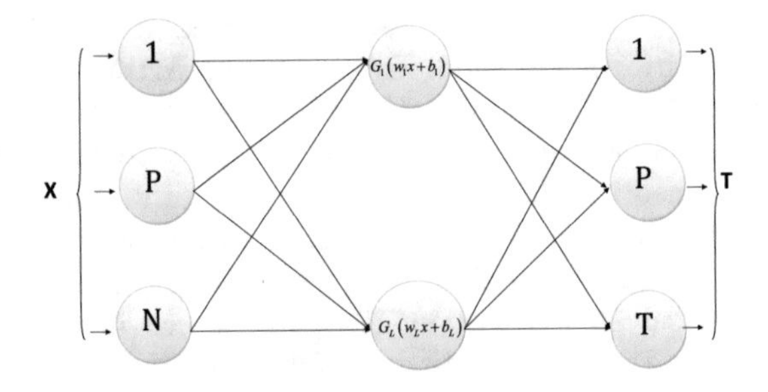

Fig. 1. Extreme Learning Machine

ELM Algorithm

The ELM algorithm pass through three main steps. At the first, we choose randomly the parameters of input nodes (w_i: the weight vector connecting the ith hidden node and the input nodes, and $bias_i$: the bias of the ith hidden node). Then, it computes the output matrix of hidden node H.

$$\mathbf{H} = \begin{bmatrix} G(\mathbf{w}_1, bias_1, \mathbf{x}_1) & \cdots & G(\mathbf{w}_m, bias_m, \mathbf{x}_1) \\ \vdots & \vdots & \vdots \\ G(\mathbf{w}_1, bias_1, \mathbf{x}_N) & \cdots & G(\mathbf{w}_m, bias_m, \mathbf{x}_N) \end{bmatrix} \quad (1)$$

where:
N: is the number of training samples.
m: is the number of hidden nodes.
H: the output matrix of hidden layer.

Let $\mathbf{T}$ is the training data target matrix:

$$\mathbf{T} = \begin{bmatrix} \mathbf{t}_1 \\ \vdots \\ \mathbf{t}_N \end{bmatrix} \quad (2)$$

The single layer feed forward (SLFN) neuron networks with m hidden nodes and activation function $g(x)$ are presented as:

$$g(\mathbf{x}) = \sum_{i=1}^{m} \boldsymbol{\beta}_i h_i(\mathbf{x}) \quad (3)$$

SLFN can approximate N samples

$$g(\mathbf{x}) = \sum_{i=1}^{m} \boldsymbol{\beta}_i h_i(\mathbf{x}) = T \quad (4)$$

The above equation can be written as:

$$H\boldsymbol{\beta} = T \quad (5)$$

Calculate the output weight by the application of Moore-Penrose generalized inverse

$$\boldsymbol{\beta} = H^{\dagger} * T \quad (6)$$

ELM Auto-encoder

The main objective of ELM-AE is to represent the input features meaningfully in compressed representations.

The Extreme learning machine as auto-encoder is presented by Fig. 2 where x the input data is the same as output (inputs=outputs=x). The parameters (weights and biases) of the hidden nodes are chosen randomly.

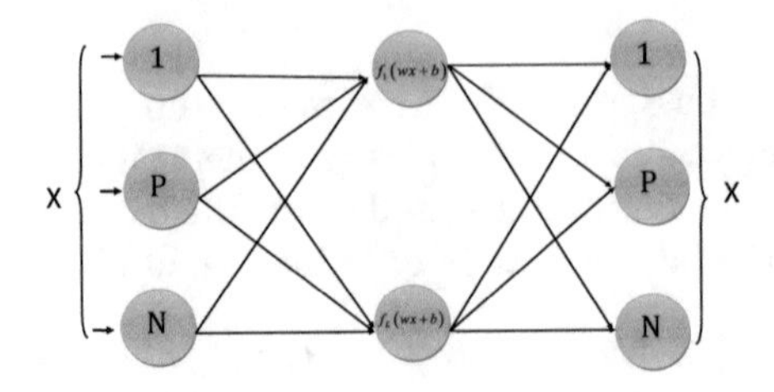

Fig. 2. Extreme Learning Machine auto-encoder

2.2 Wavelet Theory

Description of Wavelet Theory

The concept of wavelet transform (WT) is based on small basic wave function called wavelet. The object is to generate a signal by some translated and dilated wavelet [17]. Therefore, the wavelet transform gives a frequency and temporal analysis of the signal. It spreads when low-frequency behavior is observed and shrinks when analyzing high frequency using the translation and dilatation parameters. The continuous wavelet transform (CWT) is a type of wavelet analysis presented by the following equation:

$$CWT(a,b) = \frac{1}{\sqrt{a}} \int_{-\infty}^{+\infty} f(t)\psi\left(\frac{t-b}{a}\right) dt \tag{7}$$

Where ψ is the wavelet function, a the variation of the frequency parameter and b is the variation of time parameter.

The DWT is another type of wavelet analysis which is used to compute the digital signals. The objective is to generate a set of wavelets using sampled values of (a) and (b).

Let (j = 1..m) is the scale parameter, a_0^j is the parameter that presents the number of points contained in the signal. Thus, we use only the following family wavelets:

$$\psi_{j,n} = \psi(a_0^{-j}x - nb_0) \tag{8}$$

where $n = 1..a_0^{m-j}$.
n represent the position parameter and m the number of scales. The Discrete Wavelet Transform (DWT) is given by the following formula:

$$W(j,n) =< f(x), \psi_{j,n} > \tag{9}$$

Wavelet Neural Networks

The Wavelet Neural Networks were introduced by Zhang and Benveniste [5]. It is a combination of wavelet theory and artificial neural networks [18,19]. The objective consists in the approximation of signals using wavelets [3,20–22]. Three layers (input layer, hidden layers, and output layer) composed the architecture of WNN [23–25]. The input layer represents the value of the input signal. The hidden node is activated by a wavelet function. The output layer represents the summations of the outputs of the hidden neurons multiplied by the connection weights.

The output of WNN is given by the following equation:

$$\tilde{f} = \sum_{i=1}^{n} \omega_i \psi_i \tag{10}$$

3 Proposed Approach

As shown in Fig. 3 the deep wavelet extreme learning machine auto-encoder presented by input layer that presented the input data X. This input date will be learned by an ELM-Auto-encoder where the output layer t = x. The parameters (w, bias) are generated randomly. A composite wavelet activation function is applied. Deep wavelet ELM-AE projected features to a different or equal dimension space using the following equation:

$$h = w * g(\psi(x)) + bias \tag{11}$$

The output matrix H is defined like that:

$$H = \begin{bmatrix} g(\psi_{a_1 b_1}(x_1))...g(\psi_{a_L b_L}(x_1)) \\ g(\psi_{a_1 b_1}(x_N))...g(\psi_{a_L b_L}(x_N)) \end{bmatrix} \tag{12}$$

where: a, b are the translation and dilation parameters.
ψ: is a wavelet function.
x: is the input data.
N: number of inputs nodes.
L: number of hidden nodes.
As shown in Fig. 4 the β presents the output weights of ELM-autoencoder. It is calculated by the flowing equation:

$$\beta = H^{-1} X \tag{13}$$

$$\beta \beta^T = I \tag{14}$$

let L^K the number of nodes used in the deep extreme learning machine architecture. The number of nodes in each hidden layer weights are initialized using the Extreme Learning Machine Auto-encoder.

The projection matrix of Kth hidden layer is calculated by this equation

$$H^K = g((\beta^K)^T H^{K-1}) \tag{15}$$

The output of the connections between the last hidden layer and the output node t is analytically calculated using the Moore-Penrose generalized inverse.

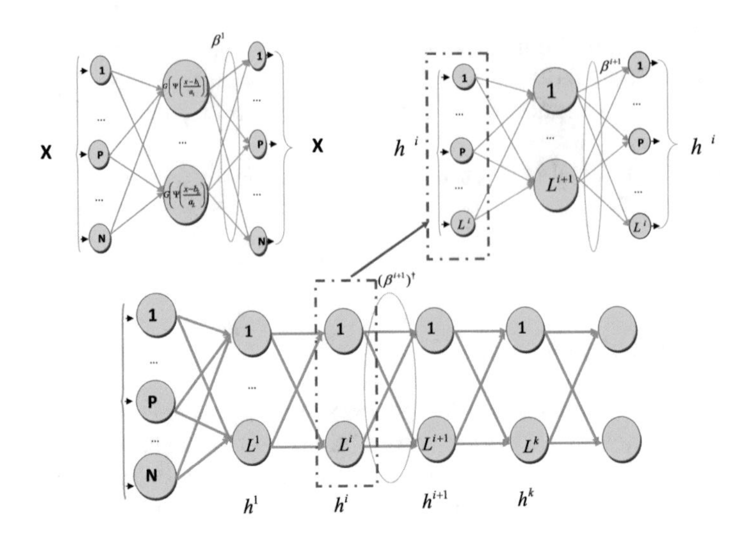

Fig. 3. Deep wavelet extreme learning machine auto-encoder

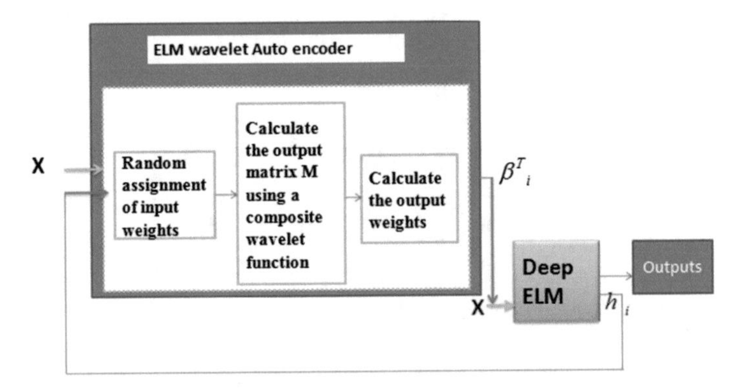

Fig. 4. Approach description

To better explain our approach we presented the steps of the implementation in Fig. 4. As shown, In each hidden layer of deep architecture, we used input weights that will be computed by the Extreme Learning Machine autoencoder where the number of input neurons is equal to the number of output neurons. The ELM auto-encode pass through three steps. In the first step, we project the inputs randomly using random weights. After (step 2), we calculate the output matrix of the hidden layer. In phase 3 we used the Moore-Penrose generalized inverse to calculate the output weights. The ELM autoencoder output weights will be used as the weight for the hidden layer of deep ELM. We repeat these steps for each hidden layer of deep ELM. Finally, The output weights are analytically calculated using the Moore-Penrose generalized inverse.

4 Results

4.1 Database Description

As shown in Table 1, the MNIST database contains 70000 handwritten digits image between 0 and 9 (Fig. 5). It contains 10 classes which are approximately equally distributed in the training and testing sets. Each image is composed by size-normalized of a 28 * 28 pixel from 256 level gray-scale images.

Table 1. Description of MNIST database

Number of examples	70000
Number of classes	10
Number of training images	60000
Number of testing images	10000

4.2 Results and Discussion

To evaluate our approach, we are compared our approach and 2 others algorithms. The first algorithm used in comparison is the MLP ELM. It is an extreme learning machine based on a Multilayer Neural Network where the hidden layer weights will be initialized by an unsupervised training. The whole MLP ELM is fine-tuned with Back propagation algorithm. The second technique used for the comparison is the deep ELM-auto encoder. It is an ELM with a deep architecture where the hidden layer weights are initialized with ELM-AE. In this paper, we have used the same protocol to evaluate our results. For each technique, there are 60000 of images are used for training and 10000 images are used for the test. Table 2 presents all methods introduced for the comparison with the number of the hidden layer used in each method. The percentage of Cases Correctly Classified (CCR) is considered to compute the performance of each method. It is

Fig. 5. Mnist example

the most obvious accuracy measure. It represents the percentage of the correct classification rate. It is obtained by the subdivision of the number of correct classes of the test divided by the number of all tested data. As shown the best rate is given by our approach with 95.96%. Results prove that our approach is better than the deep ELM-auto-encoder and MLP ELM.

Table 2. Comparison between our approach and other techniques

Techniques names	Number of hidden nodes in the first layer	Number of hidden nodes in the second layer	Rates
MLP ELM	784	1000	94.95
Deep ELM-auto-encoder	2000	1000	93.64
Deep wavelet ELM auto-encoder	2000	1000	95.96

5 Conclusion

This paper presents a new learning approach (Deep wavelet ELM-AE) for Extreme Learning Machine algorithm. We have used a deep architecture with a composite wavelet activation function. We have evaluated our approach using the MNIST Database. Results prove that our approach is able to produce a better rate in comparison with other algorithms well used in the field of machine learning.

References

1. Zaied, M., Said, S., Jemai, O., Amar, C.B.: A novel approach for face recognition based on fast learning algorithm and wavelet network theory. Int. J. Wavelets Multiresolut. Inf. Process. **9**(06), 923–945 (2011)
2. Siwar, Y., Ridha, E., Olfa, J., Mourad, Z.: Improving the classification of emotions by studying facial feature. Int. J. Comput. Theor. Eng. **8**(5), 419 (2016)
3. Said, S., Jemai, O., Zaied, M., Amar, C.B.: Wavelet networks for facial emotion recognition. In: 2015 15th International Conference on Intelligent Systems Design and Applications (ISDA), pp. 295–300. IEEE (2015)
4. Dengsheng, L., Weng, Q.: A survey of image classification methods and techniques for improving classification performance. Int. J. Remote Sens. **28**(5), 823–870 (2007)
5. Zhang, Q., Benveniste, A.: Wavelet networks. IEEE Trans. Neural Netw. **3**(6), 889–898 (1992)
6. Jemai, O., Zaied, M., Amar, C.B., Alimi, M.A.: Fast learning algorithm of wavelet network based on fast wavelet transform. Int. J. Pattern Recogn. Artif. Intell. **25**(08), 1297–1319 (2011)
7. Pourtaghi, A., Lotfollahi-Yaghin, M.A.: Wavenet ability assessment in comparison to ANN for predicting the maximum surface settlement caused by tunneling. Tunn. Undergr. Space Technol. **28**, 257–271 (2012)

8. Huang, G.-B., Zhu, Q.-Y., Siew, C.-K.: Extreme learning machine: a new learning scheme of feedforward neural networks. In: Proceedings of the 2004 IEEE International Joint Conference on Neural Networks, vol. 2, pp. 985–990. IEEE (2004)
9. Ding, S., Zhao, H., Zhang, Y., Xinzheng, X., Nie, R.: Extreme learning machine: algorithm, theory and applications. Artif. Intell. Rev. **44**(1), 103–115 (2015)
10. Rajesh, R., Siva Prakash, J.: Extreme learning machines a review and state of the art. Int. J. Wisdom Based Comput. **1**(1), 35–49 (2011)
11. Frénay, B., Verleysen, M.: Parameter-insensitive kernel in extreme learning for non-linear support vector regression. Neurocomputing **74**(16), 2526–2531 (2011)
12. Huang, G.-B., Zhou, H., Ding, X., Zhang, R.: Extreme learning machine for regression and multiclass classification. IEEE Trans. Syst. Man Cybern. Part B (Cybern.) **42**(2), 513–529 (2012)
13. Liang, N.Y., Huang, G.B., Saratchandran, P., Sundararajan, N.: A fast and accurate online sequential learning algorithm for feedforward networks. IEEE Trans. Neural Netw. **17**(6), 1411–1423 (2006)
14. Huang, G., Song, S., Gupta, J.N.D., Wu, C.: Semi supervised and unsupervised extreme learning machines. IEEE Trans. Cybern. **44**(12), 2405–2417 (2014)
15. Zhou, H., Huang, G.-B., Lin, Z., Wang, H., Soh, Y.C.: Stacked extreme learning machines. IEEE Trans. Cybern. **45**(9), 2013–2025 (2015)
16. Tang, J., Deng, C., Huang, G.-B.: Extreme learning machine for multilayer perceptron. IEEE Trans. Neural Netw. Learn. Syst. **27**(4), 809–821 (2016)
17. Ejbali, R., Zaied, M., Amar, C.B.: Multi-input multi-output beta wavelet network: modeling of acoustic units for speech recognition. arXiv preprint arXiv:1211.2007 (2012)
18. Ejbali, R., Zaied, M., Amar, C.B.: Intelligent approach to train wavelet networks for recognition system of Arabic words. In: KDIR, pp. 518–522 (2010)
19. Hassairi, S., Ejbali, R., Zaied, M.: Supervised image classification using deep convolutional wavelets network. In: 2015 IEEE 27th International Conference on Tools with Artificial Intelligence (ICTAI), pp. 265–271. IEEE (2015)
20. Yahia, S., Said, S., Jemai, O., Zaied, M., Amar, C.B.: Comparison between extreme learning machine and wavelet neural networks in data classification. In: Ninth International Conference on Machine Vision (ICMV 2016) (2016)
21. Said, S., Jemai, O., Zaied, M., Amar, C.B.: 3D fast wavelet network model-assisted 3D face recognition. In: Eighth International Conference on Machine Vision (ICMV 2015), vol. 9875, p. 98750E. International Society for Optics and Photonics (2015)
22. Said, S., Jemai, O., Hassairi, S., Ejbali, R., Zaied, M., Amar, C.B.: Deep wavelet network for image classification. In: 2016 IEEE International Conference on Systems, Man, and Cybernetics (SMC), pp. 000922–000927. IEEE (2016)
23. Zaied, M., Mohamed, R., Amar, C.B.:. A power tool for content-based image retrieval using multiresolution wavelet network modeling and dynamic histograms. Int. REv. Comput. Softw. (IRECOS) **7**(4) 2012
24. Zaied, M., Abdennour, I.B., Amar, C.B.: Decision support system including fuzzy logic and multiresolution wavelet network modeling for content-based image retrieval. Wulfenia J. **19**(10), 200–218 (2012)
25. Teyeb, I., Jemai, O., Zaied, M., Amar, C.B.: A novel approach for drowsy driver detection using head posture estimation and eyes recognition system based on wavelet network. In: The 5th International Conference on Information, Intelligence, Systems and Applications, IISA 2014, pp. 379–384. IEEE (2014)

Palm Vein Age and Gender Estimation Using Center Symmetric-Local Binary Pattern

Wafa Damak(✉), Randa Boukhris Trabelsi, Alima Damak Masmoudi, and Dorra Sellami

Computers Imaging Electronics and Systems Group (CIELS) from Advanced Control and Energy Management Laboratory (CEM-Lab), University of Sfax, Sfax Engineering School, Sfax, Tunisia
damakwafa@gmail.com, trabelsiboukhrisranda@live.fr, damak_alima@yahoo.fr, dorra.sellami@enis.tn

Abstract. Many investigations on hand veins modality have been done in the literature for identification and recognition systems. However, researches on age and gender estimation by hand veins are very limited and very preliminary. Our contribution in this paper is to propose a system able to estimate the age and the gender of a person from its hand veins. Accordingly, we are interested in studying the discriminating features for the prediction of a person's age and gender. In fact, hand vein images are very rich in orientation and contour characteristics and they are faced with poor quality and illumination variation. Hence, we investigate texture analysis invariant to illumination as well as venous pattern gradient information determination by Center Symmetric-Local Binary Patterns (CSLBP) descriptor. Since Region Of Interest (ROI) extraction is important in a biometric system, we aim to cover the whole informative region of hand veins by our dynamic ROI extraction method. Our experimental study is based on palm vein VERA database. As considered database has a class imbalance problem, we remediate this problem by using Weighted K-Nearest Neighbor (WKNN). The obtained performance metrics demonstrate the effectiveness of our proposed system for gender classification and age estimation respectively: 95.8% and 94.2% for F-measure, 95.9% and 94.4% for G-mean.

Keywords: Palm vein · CSLBP · Age · Gender · Classification

1 Introduction

Biometrics, as defined by [1], is a technique of recognizing individuals based on their physiological characteristics (iris, face, fingerprint, hand vein, etc.) or behavioral ones (voice, signature, gait, etc.). Nevertheless, these biometric characteristics can achieve much more than the identity recognition. Indeed, demographic attributes such as age, gender and race can be extracted from biometric

© Springer Nature Switzerland AG 2020
F. Martínez Álvarez et al. (Eds.): CISIS 2019/ICEUTE 2019, AISC 951, pp. 114–123, 2020.
https://doi.org/10.1007/978-3-030-20005-3_12

characteristics [2]. Many popular applications are based on demographic analysis. Sun et al. [2] classify them into various categories: human-computer interaction, security control and surveillance, multimedia retrieval, biometrics, and targeted advertising. Several modalities have been explored in the literature for age and gender estimation such as face, speech, gait, iris, fingerprint, skin and hand veins. Hand veins are hidden biometric features since they are under the skin. This guarantees their robustness against forgery. In addition, they are independent of the body appearance (hands in gloves, dusty hands, etc.). Moreover, hand veins sensors are contactless, which ensures user comfort. In recent decades, hand vein recognition has attracted special attention of researchers in biometrics, industrials in the security domain and the products provide good results in the test of the International Biometric Group (IBG). Considering its many advantages, we choose this modality for an estimation of the gender and the age. The remaining of this paper is organized as follows: in Sect. 2, we present recent work on hand veins age and gender detection. Then, in Sect. 3, we describe our proposed system for estimating the age and classifying the gender of individuals by their palm veins. Thereafter, experimental results are presented in Sect. 4. Finally, a conclusion is drawn and some perspectives are discussed in Sect. 5.

2 Related Works

In [3], authors deal the discrimination of young and old by dorsal hand veins. They extract gray level histogram and gray level mean value as features. Furthermore, they use k-means for classification and Euclidean distance for similarity measure. Accordingly, the obtained aging recognition rate for elderly is 82% by gray level histogram and 76% by gray level mean value, whereas they get 5% for young by histogram and 2% by gray level mean value. In [4], authors propose an age and gender recognition system by finger vein patterns. An important step in the system is image enhancement where Guided Filter based Singe Scale Retinex (GFSSR) method is applied. In order to characterize venous texture from finger veins, feature extraction step is based on Local Binary Pattern (LBP) descriptor. The validation database is MMCBNU_6000 finger vein [5]. In order to evaluate the system performance, data are divided into two sets: training set (70% of data) and test set (30% of data). The effect of the hand type (left or right) and the effect of the finger type (middle, ring, fore) are tested. Experimental results prove that extracted features from finger vein can define the gender and the age class. The middle finger of right hand provides the best age recognition rate (99.67%, 99.78% and 97.33%) for different class number (2, 3 and 4 respectively). The best gender recognition rate achieved is 98%. It is obtained by the middle finger of left hand. In [6], author deduce that the remarkable difference between young and old dorsal veins is characterized by both blood flow and skin state. Thus, he analyzes venous patterns and skin areas separately for an old and a young based on their gray level histograms. Then, he studies histogram statistical parameters: minimum, maximum, median, mean, mode and standard deviation. For the classification of old and young by dorsal hand vein, Linear

Discriminant Analysis (LDA) and K-Nearest Neighbors (KNN) are used. It is illustrated that skin features lead to better prediction rates: 92.6% by KNN and 91.8% by LDA, whereas venous features reach 89.6% by LDA and 90.4% by KNN. Moreover, Wang et al. [7] prove that dorsal hand veins can classify the gender of individuals. Accordingly, they use four feature extraction methods: Mean Curvature, Two-Dimensional Principal Component Analysis (2DPCA), LBP, and Scale-Invariant Feature Transform (SIFT). In addition, they adopt Support Vector Machine (SVM) for gender classification. They get as performances low accuracy rates: 32.5% with mean curvature, 57.3% with 2PDCA, 25.7% with LBP and 19.2% with SIFT. In the same work, and in order to improve the classification result, an unsupervised learning model of sparse features is used, leading to a classification accuracy (which is variable with features number): a high rate of 98.2% with 450 features and a low rate of 89.2% with 50 features.

Recent researches on age and gender estimation show that works on hand veins in this area are very limited. A preliminary analysis in [6] addresses the discrimination of young and old only. This analysis has shown that age subgroups such as children and adults can't be separated easily. Thus, other features must be taken into account.

In this paper, we are interested in studying the discriminating features for the prediction of a person's age and gender. We aim at implementing age and gender estimation system based on hand vein. Accordingly, our goal is the discrimination between a man and a woman and the distinction of the following age groups: child (<20 years), young (20–40 years), adult (41–65 years).

3 Proposed Palm Vein Based Age and Gender Estimation System

Our proposed system for palm vein age and gender estimation is shown in Fig. 1. It is composed by three main steps: ROI detection, feature extraction and classification. The ROI detection identifies the informative region from the palm vein image. We use our dynamic ROI extraction method for hand vein images proposed in [8]. The challenge of our approach is that it does not exceed the hand borders and can be adjusted dynamically. Feature extraction aims at characterizing the palm vein ROI by extracting a feature vector based on CSLBP descriptor. The purpose of classification step is to give a decision about the age and the gender of individuals using WKNN.

3.1 Feature Extraction

Feature extraction is an essential step in age and gender estimation system. It aims to represent each image by a discriminant feature vector. Several methods have been proposed for feature extraction from the hand vein [9–15]. Indeed, age and gender estimation system performance depends on extracted feature relevance. Thus, our goal is to identify an effective method for describing the

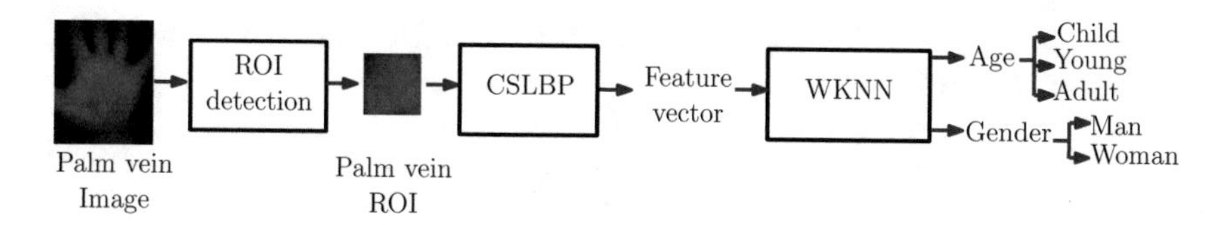

Fig. 1. Proposed palm vein based age and gender estimation system.

age and the gender of hand veins. In fact, hand vein images are very rich in orientation and contour characteristics. Moreover, these images are faced with poor quality and illumination variation. This led us to opt for the determination of the venous pattern gradient information, as well as vein texture analysis invariant to illumination. Hence, we investigate to extract local texture features by CSLBP descriptor. This descriptor is robust to illumination changes and captures gradient information. It is proposed by [16] as a modified version of Local Binary Patterns (LBP) for texture analysis. It allows capturing the gradient information better than LBP. Moreover, CSLBP produces more compact binary patterns than LBP by using only the center-symmetric pairs of pixels. Thus, for 8 neighbors and a radius equal to 1, the number of features is reduced to 16. It is defined as follows:

$$CSLBP_{P,R,T} = \sum_{k=0}^{(P/2)-1} 2^k s(g_k - g_{k+(P/2)}), s(x) = \begin{cases} 1, \texttt{if } x > T \\ 0, \texttt{else} \end{cases} \quad (1)$$

where P is the number of neighbors, R is the neighborhood radius and T is the threshold value.

It should be noted that histogram of an image reconstructed by CSLBP descriptor is considered as a feature vector. However, this vector does not contain any location information. In order to integrate this missing information, the palm vein ROI is decomposed into 4×4 non-overlapped cells. For each cell, the CSLBP histogram extracts 16 different features. The corresponding histograms are concatenated thereafter to form a vector of $(4 \times 4 \times 16)$ 256 features. In order to ensure data consistency, we use the min-max normalization technique.

3.2 Classification

For sake of simplicity, we adopt K-Nearest Neighbors (KNN) in age estimation as well as gender identification. It consists on identifying the closest neighbors of an instance X_i based on a distance between X_i and the learning set. The class of the instance X_i is defined according to the majority class among its K nearest neighbors. However, KNN performances decrease in case of unbalanced classes, since it depends on the K value choice and the number of samples. As a proposed solution for class imbalance: the Weighted K-Nearest Neighbors (WKNN) [17] in which we assign different weights for each class with respect to its number of samples. It attributes a high weight to the neighbors of the minority classes and a low weight for the neighbors of the majority classes. The weight of class C_m is defined as follows:

$$Weight(C_m) = \frac{1}{(\frac{Num(C_m)}{min(Num(C_m)|m=1..N_c)})^{\frac{1}{p}}} \quad (2)$$

where p is an exponent greater than 1.

4 Experimental Results

In this section, we introduce the database considered to validate our system. We then present the performance metrics adopted in our evaluation. Thereafter, we analyze the factors that can influence the performance of the proposed system. Thus, we prove the effectiveness of the extracted features for both age and gender classification by palm veins.

4.1 Database

To validate our proposed system, we consider VERA palm vein database [18]. It consists of 2200 palm vein images, containing 110 individuals: 40 women and 70 men, aged from 18 to 60 years. For each individual, five images of palm veins of left and right hand in two different sessions are collected. In order to make a balanced use of the database, we use 5-fold cross-validation. Thus, we divide the dataset into five parts. We use the four parts for learning (80% of the data) and the fifth part for the test (20% of the data). Then, we iterate the learning and the test procedures five times and we calculate the mean of the different obtained performance metrics as a final result. As a consequence, each instance is used exactly four times for learning and one time for testing. The different classes are represented as follows: the child by 60 images, the young by 1520 images and the adult by 620 images. The corpus used for gender classification contains 800 images for women and 1400 images for men. As can be noticed, the database is unbalanced (a class imbalance problem occurs when one class contains more samples than the other classes). The young class represents the majority class, while other classes (child and adult) represent minority classes. Even for gender classification, we note a majority of men and a minority of women.

4.2 Performance Metrics

Accuracy is a widely used evaluation metric to examine the classifier performance. However, it is not appropriate when dealing with unbalanced data. F-measure and G-mean are among the popular metrics for evaluating the classification of unbalanced data. These metrics are defined based on a confusion matrix.

$$F - measure = 2 * \frac{Precision * Recall}{Precision + Recall} \quad (3)$$

where:

$$Precision = \frac{TP}{TP + FP} \tag{4}$$

$$Recall = \frac{TP}{TP + FN} \tag{5}$$

$$G - mean = \sqrt{\frac{TP}{TP + FN} * \frac{TN}{TN + FP}} \tag{6}$$

where:

TP: True Positives: The positive predictions that are really positive.
FP: False Positives: The positive predictions that are in fact negative.
TN: True Negatives: The negative predictions that are really negative.
FN: False Negatives: The negative predictions that are in fact positive.

4.3 Evaluation of Feature Extraction Settings

In this section, we undertake a fitting of the feature extraction settings for getting best performances. Indeed, the characterization region of an image by CSLBP can be divided into different grids: 1×1 (1 cell), 2×2 (4 cells), 3×3 (9 cells), 4×4 (16 cells), as shown in Fig. 2. In Fig. 3, we illustrate the obtained F-measure mean values of age estimation and gender identification for different cell size of the CSLBP descriptor. This Figure shows that the division into 4×4 (16 cells) yields the best results.

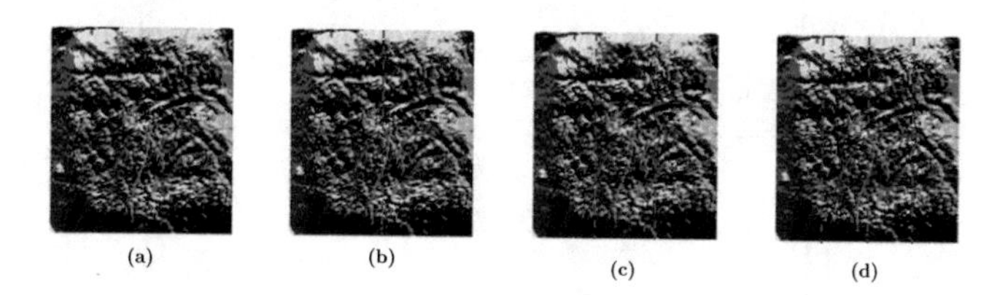

Fig. 2. Different grids that the characterization region can be divided into: (a) 1×1 (1 cell), (b) 2×2 (4 cells), (c) 3×3 (9 cells), (d) 4×4 (16 cells).

In Fig. 4, we illustrate the impact of CSLBP descriptor normalization on F-measure mean values for age estimation and gender classification. We note a significant improvement of 5.94% for age estimation and a slight increase of 0.22% for gender classification.

4.4 Performances Evaluation

In order to choose the optimal K value for KNN classifier, we test different values of K. The optimal K is selected based on cross-validation. In Table 1, we illustrate the mean values of F-measure and G-mean for age estimation and gender classification using KNN at different K values. For age estimation, KNN with K = 1 gives the best performance values: a 94.2% for F-measure and a

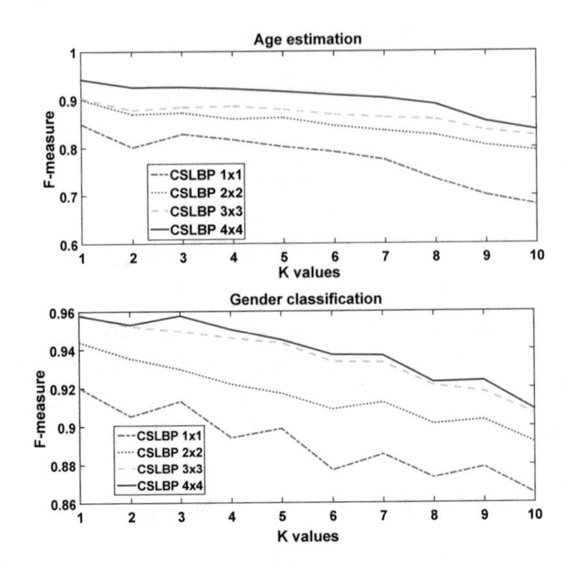

Fig. 3. F-measure mean values of age estimation and gender identification for different cell size of the CSLBP descriptor.

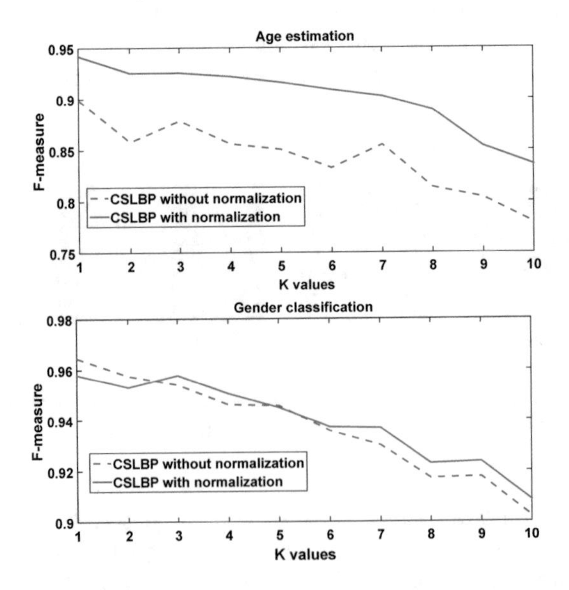

Fig. 4. Impact of CSLBP descriptor normalization on F-measure mean values for age estimation and gender classification.

94.4% for G-mean. When K = 10, performances decrease considerably: a 83.6% for F-measure and a 82.4% for G-mean. For gender classification, best results are achieved when K = 1: a 95.8% for F-measure and a 95.9% for G-mean. Moreover, performance metrics decrease with K = 10 yielding a rate of: a 90.8%

for F-measure and a 90.1% for G-mean. Thus, we find that by increasing the value of K, the performance metrics decrease. Moreover, the CSLBP descriptor is more discriminating for age estimation and gender classification respectively than the LBP descriptor.

Table 1. Age estimation and gender classification performance evaluation for different K values.

K	Age estimation				Gender classification			
	F-measure		G-mean		F-measure		G-mean	
	LBP	CSLBP	LBP	CSLBP	LBP	CSLBP	LBP	CSLBP
1	87.5	**94.2**	90	**94.4**	94.6	**95.8**	94.6	**95.9**
2	85.1	**92.5**	88.3	**93**	93.7	**95.3**	92.6	**94.6**
3	86.6	**92.6**	89.4	**93.1**	94	**95.8**	93.9	**95.7**
4	84.9	**92.2**	86.3	**91.9**	93.2	**95**	92.4	**94.3**
5	83.1	**91.6**	85.7	**91.6**	93.1	**94.5**	93	**94.4**
6	79.6	**90.9**	82	**90.5**	92.3	**93.7**	91.4	**93**
7	80.2	**90.2**	81.6	**90**	91.7	**93.7**	91.5	**93.4**
8	77	**88.9**	77.4	**88.2**	91.4	**92.3**	90.5	**91.5**
9	77.7	**85.3**	78.2	**84.9**	90.5	**92.3**	90.3	**92.2**
10	74.8	**83.6**	74.6	**82.4**	89.5	**90.8**	88.7	**90.1**
Mean	81.6	**90.2**	83.4	**90**	92.4	**93.9**	91.9	**93.5**

Table 2. Performances evaluation of KNN and WKNN for age estimation and gender classification.

K	Age estimation				Gender classification			
	F-measure		G-mean		F-measure		G-mean	
	WKNN	KNN	WKNN	KNN	WKNN	KNN	WKNN	KNN
1	**94.2**	**94.2**	**94.4**	**94.4**	**95.8**	**95.8**	**95.9**	**95.9**
2	92.4	**92.5**	**94.2**	93	94.6	**95.3**	**95.4**	94.6
3	**92.6**	**92.6**	**94**	93.1	**95.8**	**95.8**	**95.7**	**95.7**
4	**92.6**	92.2	**94**	91.9	94.7	**95**	**95.2**	94.3
5	**92.1**	91.6	**93.6**	91.6	**94.5**	**94.5**	**94.4**	**94.4**
6	**92**	90.9	**93.2**	90.5	**93.9**	93.7	**94.4**	93
7	**90.8**	90.2	**92.8**	90	**93.7**	**93.7**	**93.4**	**93.4**
8	**91.1**	88.9	**92.3**	88.2	**92.6**	92.3	**93**	91.5
9	**89.1**	85.3	**92.1**	84.9	**92.3**	**92.3**	**92.7**	92.2
10	**89.2**	83.6	**91.3**	82.4	**91.6**	90.8	**91.9**	90.1
Mean	**91.6**	90.2	**93.2**	90	**94**	93.9	**94.2**	93.5

WKNN leads to better performances for age estimation and gender classification, as illustrated in Table 2. When K = 1, KNN and WKNN results in the same performances for both age estimation and gender classification. By increasing the K value, when K = 10, performance metrics decrease slightly with WKNN compared to KNN respectively for age estimation and gender classification: a 89.2% and 91.6% for F-measure, a 91.3% and 91.9% for G-mean. Therefore, WKNN contributes to making the system less sensitive to the choice of parameter K.

5 Conclusion

In this paper, we propose a palm vein age and gender estimation system. For sake of accuracy in texture characterization, we apply CSLBP in feature extraction. We investigate two classification methods: KNN and WKNN, for good decision making about the gender and the age of palm veins. We adopt as performance metrics: F-measure and G-mean. In order to validate our system, we consider VERA palm vein database. Experimental results prove that WKNN provides a significant improvement in performance metrics compared to KNN. WKNN makes age estimation and gender classification system less sensitive to the choice of parameter K. Our proposed system leads to 94.2% and 95.8% for F-measure, 94.4% and 95.9% for G-mean, respectively for age estimation and gender classification. In future works, we aim to make our age estimation and gender classification system marketable in a variety of uses, such as a biometric identification system, a mobile application, a multimedia search system, an automatic distributor, a display board, etc.

References

1. Jain, A., Ross, A.A., Nandakumar, K.: Introduction to Biometrics. Springer Science & Business Media (2011)
2. Sun, Y., Zhang, M., Sun, Z., Tan, T.: Demographic analysis from biometric data: achievements, challenges, and new frontiers. IEEE Trans. Pattern Anal. Mach. Intell. **40**(2), 332–351 (2018)
3. Wang, Y., Zheng, H.: A preliminary analysis of the aging dorsal hand vein images. In: International Conference on Intelligent Human-Machine Systems and Cybernetics, vol. 2, pp. 271–274. IEEE (2013)
4. Damak, W., Trabelsi, R.B., Masmoudi, A.D., Sellami, D., Nait-Ali, A.: Age and gender classification from finger vein patterns. In: International Conference on Intelligent Systems Design and Applications, pp. 811–820. Springer (2016)
5. Lu, Y., Xie, S.J., Yoon, S., Wang, Z., Park, D.S.: An available database for the research of finger vein recognition. In: International Congress on Image and Signal Processing, vol. 1, pp. 410–415. IEEE (2013)
6. Zheng, H.G.: Static and dynamic analysis of near infra-red dorsal hand vein images for biometric applications. Ph.D. dissertation, University of Central Lancashire (2017)
7. Wang, J., Wang, G., Pan, Z.: Gender attribute mining with hand-dorsa vein image based on unsupervised sparse feature learning. IEICE Trans. Inf. Syst. **101**(1), 257–260 (2018)

8. Damak, W., Trabelsi, R.B., Damak, M.A., Sellami, D.: Dynamic ROI extraction method for hand vein images. IET Comput. Vis. **12**(5), 586–595 (2018)
9. Trabelsi, R.B., Masmoudi, A.D., Masmoudi, D.S.: Hand vein recognition system with circular difference and statistical directional patterns based on an artificial neural network. Multimed. Tools Appl. **75**(2), 687–707 (2016)
10. Trabelsi, R.B., Masmoudi, A.D., Masmoudi, D.S.: A novel biometric system based hand vein recognition. J. Test. Eval. **42**(4), 809–818 (2013)
11. Trabelsi, R.B., Masmoudi, A.D., Masmoudi, D.S.: A new multimodal biometric system based on finger vein and hand vein recognition. Int. J. Eng.Technol. **4**, 3175–3183 (2013)
12. Masmoudi, A.D., Trabelsi, R.B., Krid, M., Masmoudi, D.S.: Implementation of a fingervein recognition system based on improved gaussian matched filter. J. MAGNT Res. Rep. **2**(4), 251–260 (2014)
13. Masmoudi, A.D., Boukhris, R.T., Sellami, D.S.: A novel finger vein recognition system based on monogenic local binary pattern features. Int. J. Eng. Technol. **5**(6), 4528–4535 (2014)
14. Masmoudi, A.D., Trabelsi, R.B., Masmoudi, D.S.: A new biometric human identification based on fusion fingerprints and fingerveins using monoLBP descriptor. World Acad. Sci. Eng. Technol. **78**, 1658 (2013)
15. Trabelsi, R.B., Kallel, I.K., Masmoudi, D.S.: Person identification based on a new multimodal biometric system. Trans. Syst. Signals Devices **7**(3), 273–289 (2012)
16. Heikkilä, M., Pietikäinen, M., Schmid, C.: Description of interest regions with center-symmetric local binary patterns. In: Computer Vision, Graphics and Image Processing, pp. 58–69. Springer (2006)
17. Tan, S.: Neighbor-weighted k-nearest neighbor for unbalanced text corpus. Expert Syst. Appl. **28**(4), 667–671 (2005)
18. Tome, P., Vanoni, M., Marcel, S.: On the vulnerability of finger vein recognition to spoofing. In: International Conference of the Biometrics Special Interest Group, September 2014. http://publications.idiap.ch/index.php/publications/show/2910

Neuro-Evolutionary Feature Selection to Detect Android Malware

Silvia González[1], Álvaro Herrero[2(✉)], Javier Sedano[1], and Emilio Corchado[3]

[1] Instituto Tecnológico de Castilla y León, C/López Bravo 70, Pol. Ind. Villalonquejar, 09001 Burgos, Spain
{silvia.gonzalez, javier.sedano}@itcl.es

[2] Grupo de Inteligencia Computacional Aplicada (GICAP), Departamento de Ingeniería Civil, Escuela Politécnica Superior, Universidad de Burgos, Av. Cantabria s/n, 09006 Burgos, Spain
ahcosio@ubu.es

[3] Department of Computer Science and Automation, University of Salamanca, Plaza de la Merced, s/n, 37008 Salamanca, Spain
escorchado@usal.es

Abstract. Although great effort has been devoted to successfully detect Android malware, it still is a problem to be addressed. Its complexity increases due to the high number of features that can be obtained from Android apps in order to improve detection. Present paper proposes wrapper feature selection by applying a genetic algorithm and a Multilayer Perceptron. In order to validate this proposal, feature selection is performed on the well-known Drebin dataset on Apache Spark. Interesting results on the most informative features for the detection of existing Android malware have been obtained.

Keywords: Feature selection · Genetic algorithm · Multilayer Perceptron · Android malware

1 Introduction

Since the late 1990s, an increasing number of smartphones are sold every year and it is expected that the number of users pass the 2.7 billion mark by 2019 [1]. Among all of them, Google's Android still is the most widely-used one [1] and consequently, the number of Android users has permanently increased.

To fight against such a problem, it is required to understand the malware and its nature, given that this nature is constantly evolving as it happens with most software. Otherwise, it will not be possible to practically develop an effective solution [2]. One of the main problems is the high amount of features that can be extracted from Android apps in order to be considered for the detection of malware. This problem is addressed in present paper by analyzing a massive dataset on Apache Spark [3] for Feature Selection (FS).

FS methods are normally used to reduce the number of features considered in a classification task by removing irrelevant or noisy features [4, 5]. Filter methods

© Springer Nature Switzerland AG 2020
F. Martínez Álvarez et al. (Eds.): CISIS 2019/ICEUTE 2019, AISC 951, pp. 124–131, 2020.
https://doi.org/10.1007/978-3-030-20005-3_13

perform feature selection independently from the learning algorithm while wrapper models embed classifiers in the search model [6, 7].

In present research, wrapper FS is applied; as a result, different sets of features are generated by a genetic algorithm and the fitness score of each individual is calculated by training a Multi-Layer Perceptron (MLP) on this reduced set of features. The evolutionary search process is guided by crossover and mutation operators specific to the binary encoding and a fitness function that evaluates the quality of the encoded feature subset. To validate the proposed FS method, a real-life benchmark dataset [8, 9] has been analyzed in present research.

There are many advantages of feature selection for malware detection. Although many soft computing techniques have been previously applied to cybersecurity [10–12], little effort has been devoted until now to apply these methods of machine learning to deal with malware features [13]. In [14] just information gain is used to rank the 32 static and dynamic features from a self-generated malware dataset containing 14,794 instances, comprising 30 legitimate apps. Samples of malware come from five different families (GoldDream, PJApps, DroidKungFu2, Snake and Angry Birds Rio Unlocker). To rank the features, four machine learning classifiers (Naïve Bayes, RandomForest, Logistic Regression, and Support Vector Machine) were applied. The top 10 selected features were (in decreasing order of importance): Native_size, Native_shared, Other_shared, Vmpeak, Vmlib, Dalvik_RSS, Rxbytes, VmData, Send_SMS, and CPU_Usage. In a different work [15], 88 dynamic features from 43 apps were collected and then analysed to discriminate between games and tools. The underlying idea of this study was that distinguishing between games and tools would provide a positive indication about the ability of detection algorithms to learn and model the behavior applications and potentially detect malware. To do so, feature selection was applied to identify the 10, 20 and 50 best features, according to Information Gain, Chi Square, and Fisher Score. A similar analysis [16] by same authors proposed a selection from 22,000 static features about 2,285 apps to distinguish between games and tools apps once again. The following classifiers were applied: Decision Tree, Naïve Bayes, Bayesian Networks, PART, Boosted Bayesian Networks, Boosted Decision Tree, Random Forest, and Voting Feature Intervals. The obtained results shown that the combination of Boosted Bayesian Networks and the top 800 features selected using Information Gain yield an accuracy level of 0.918 with a False Positive Rate of 0.172.

More recently, static analysis of Android malware families was already proposed in [17], trying to identify the malware family of malicious apps based on their payload. On the other hand, filter FS has also been applied to Android malware [18], trying to characterize malware families.

The rest of this paper is organized as follows: the proposed evolutionary feature selection algorithm is described in Sect. 2 while the experiments for the Drebin dataset are presented in Sect. 3. Complementary, the obtained results are discussed in Subsect. 3.2 and the conclusions of the study are presented in Sect. 4.

2 Feature Selection

There is a big amount of features in the analysed dataset (see Sect. 3), that are to be selected in order to speed up detection of malware.

As defined by [19], FS involves a learning algorithm that approaches the problem of selecting a subset of features upon which it can focus its attention. The remaining features from the original dataset are not considered important and consequently ignored. There is also an induction algorithm that, as in general terms for supervised learning, tries to minimize classification error, being trained on different subsets of features taken from the original data. According to [20], two different levels of relevance (weak and strong) can be defined regarding each one of the features in the original dataset. One of such features is considered as strongly relevant if its removal causes a deterioration in the performance of the induction algorithm. In present paper, this is the underlying idea to identify key features that leads to a successful classification of Android malware.

There are mainly three different types of FS methods, namely: Wrapper, Filter and Embedded. In the former, as opposed to the other two, it is the learning algorithm that "wraps" the induction algorithm and may be equated with a "black box" and is run on different subsets of features taken from the original data. Wrapper FS [20] has been selected in present study.

As the induction algorithm under the Wrapper FS perspective, MLP are evaluated. To generate the different subsets of features that are provided to these two classifiers, standard Genetic Algorithms (GAs) [21] have been applied as preliminary results had suggested that they are a powerful mean of reducing the time for finding near-optimal subsets of features from large datasets [22]. These are computational methods based on natural selection and natural genetics, used for searching solutions to a given problem. More precisely, a GA can be seen as an heuristic-based method for global optimization.

In order to optimize solutions to a given problem, these are codified as binary strings. In present case, for feature selection, a bit is assigned to each one of the features in the original dataset; 1 means that the features is included in the given subset and 0 means that it is not included. As a result, vectors of length n (being n the number of features in the original dataset) are constructed as solutions to the feature-subset selection problem.

In GA, there is a fitness function that measures the "quality" of the generated. Its design is part of the modelling process of the whole optimization approach [23]. In present paper, as per feature selection, the fitness function for was defined as the highest negative error rate (lowest classification error) obtained by applying the above-mentioned classifiers to the generated subset of features, when trying to classify the testing data to forecast the internationalization decision. As usual, selection, mutation and crossover operators are applied, according to certain parameters (different values have been tested as described in Subsect. 3.2). All in all, the applied GA is defined as follows (Table 1):

Table 1. Pseudocode of the FS GA applied in present paper.

input: a feature selection problem	
1	set the generation counter $g = 0$
2	**for** *i := 1* to population size do
3	create a random combination of feature subset (solution)
4	**end for**
5	**while** the number of generations is not reached **do**
6	generate child solutions by applying the crossover operator (with a certain probability)
7	generate child solutions by applying the selection and crossover operator (with a certain probability)
8	compute fitness values by training and testing MLP on each individual (subset of features)
9	apply the mutation operator
10	selection of best child solutions (tournament)
11	replace the worst member of the population by the child solutions
12	g = g + 1
13	**end while**
output: the best subset of features for the given problem	

As previously stated, a widely-applied neural classifier (MLP) is considered in present research to implement the fitness function of the GA.

The MLP is an Artificial Neural Network consisting of several layers of nodes. There are weights associated to the connected nodes and output signals are generated by calculating the activation to the sum of the inputs. Its architecture consists of an input layer that pass the input vector to the other layers of the network. The terms "input vectors" and "output vectors" refer to the inputs and outputs of the MLP and are represented as single vectors [24]. Additionally, a MLP has one or more hidden layers, together with the output layer. MLPs are fully connected; that is, every node is connected to each one of the nodes in the previous and next layer.

During training, the update of weights is performed according to the learning rule. In present paper, the Broyden–Fletcher–Goldfarb–Shanno implementation [25] of this algorithm has been applied, that is an approximation to Newton's method.

3 Experimental Study

As previously explained, several approaches for features selection have been applied to perform a FS on Android malware. The analysed dataset is described in Subsect. 3.1 and the obtained results are introduced and described in Subsect. 3.2.

3.1 Drebin Dataset

The Drebin dataset [8, 9] is a collection of Android apps gathered from the Android official market (Google Play) and from some other un-official sources (alternative markets, websites, forums…) between 2010 and 2012. The gathered apps were analysed through the VirusTotal [26] service, being declared as malicious when more than one of the applied scanners identified the app as an anomalous one. As a result, the dataset contains 123,441 benign applications and 5,554 malicious applications (128,995 in total), being one of the largest publicly-available datasets containing legitimate and malicious Android apps.

Data were extracted from the manifest and the disassembled dex code of the apps, obtained by a linear sweep over the application's content [9]. Every sample in the dataset is associated to an analysed app and the values of the sample represent the given values of app characteristics, such as permissions, intents and API calls.

The following feature sets were extracted from the manifest file of every app [9]:

- Hardware components: contains information about the hardware components requested by the app.
- Requested permissions: contains information about the permission system, the main security mechanism of Android. Permissions declared by the app, and hence requested before installation, are taking into account.
- App components: contains information about the different types of components in the app, each defining different interfaces to the system.
- Filtered intents: contains information about intents (passive data structures exchanged as asynchronous messages for inter-process and intra-process communication).

Additionally, some other feature sets were extracted from the dex information extracted from the apk files of the apps [9]:

- Restricted API calls: contains information about the calls defined in the app to those APIs defined as critical. Although that information must be declared in the manifest file, exactly for being malware, some APIs may be accessed without declaring that in the manifest file (root exploits) and hence the information is double checked with the API calls from the dex code.
- Used permissions: contains information about the permissions that must be granted for the calls identified in previous feature subset. It is once again a way of double-checking the manifest file; the permissions in this case.
- Suspicious API calls: contains information about calls defined in the app to those APIs identified by the authors of the dataset as potentially dangerous. It includes calls for accessing sensitive data, communicating over the network, sending and receiving SMS messages, execution of external commands, and obfuscation.

- Network addresses: contains information about IP addresses, hostnames and URLs found in the dex code.

The previously defined features sets resulted in an initial set of 545,333 features for the analysed apps. Each one of these features takes binary values: 0 if the app does not contain such feature and 1 otherwise. To aggregate this information at a family level, feature data were summarized for each family, taking binary values as well: 1 if any app from the family does contain such feature and 0 otherwise.

3.2 Results

The previously described GA-based FS algorithm (see Sect. 2) has been applied to the Drebin dataset (see Subsect. 3.1). The obtained results can be found in Table 2, for the 5 experiments that have been carried out. For each one of them, different features have been selected by the FS algorithm and are compiled in the following tables as "Selected Features".

Table 2. Classification results for the different experiments on the Drebin dataset.

Exp.	Selected features	Accuracy	AUC	AUPR
1	1 from each subset	0.957902912621	0.492086667896	0.173056239954
2	4 from activity	0.957281553398	0.496649013461	0.039576263307
3	4 from url	0.956815533981	**0.516449492901**	0.242002312756
4	4 from service_receiver	**0.958213592233**	0.513517093859	**0.431448169338**
5	4 from call::Cipher	0.957281553398	0.474310455467	0.040550887106

Classification results are presented according to the following three metrics: Accuracy, Area under the Curve (AUC), and Area under Precision-Recall (AUPR). Accuracy has been calculated by means of a binary classifier, whereas AUC and AUPR have been calculated by means of a multiclass classifier.

From the results in Table 2 it can be said that features from the service_receiver category obtained the best value for both Accuracy and AUPR metrics, whereas the best AUC one was obtained by features from the url category. It is worth noting that features from call::Cipher category obtained the worst values for all the analyzed metrics.

For a fine-grained analysis of the different features, the selected ones in each one of the experiments are presented in Table 3.

It can be seen in Table 3 that the features selected when considering the 4 subsets are also selected as the most relevant ones when considering features from each one of them individually.

Table 3. Selected features on each one of the different experiments on Drebin dataset.

Exp.	Selected features
1	activity::.MainActivity call::Cipher(AES) service_receiver::.BatteryService url::http://techerrata.com/shinzul/hdmid/hdmid-
2	Activity::.MainActivity activity::.USArmyA activity::jp.hiro.android.Sewing22.USArmyU activity::.activity.MapViewActivity
3	url::http://techerrata.com/shinzul/hdmid/hdmid- url::http://10.0.0.172:80 url::http://82.165.128.81/services/uploads/ url::market
4	service_receiver::.BatteryService service_receiver::com.zanalytics.sms.SmsReceiverService service_receiver::.CalendarService service_receiver::com.sonyericsson.extras.liveview.plugins.PluginReceiver
5	call::Cipher(AES) call::Cipher(Blowfish) call::Cipher(password) call::Cipher(PBEWithSHA256And256BitAES-CBC-BC)

4 Conclusions and Future Work

By applying the proposed FS algorithms, the detection of Android malware is sped up. The main reason is that, key features to identify malware apps are identified, and hence least important ones could be discarded and not included in the dataset to be analyzed. Additionally, experimental results show that good classification results are obtained when reducing the amount of features to be analyzed.

Future work will focus on proposing further adaptations of feature selection algorithms to improve present results.

References

1. Global smartphone sales to end users from 1st quarter 2009. https://www.statista.com/statistics/266219/global-smartphone-sales-since-1st-quarter-2009-by-operating-system/
2. Yajin, Z., Xuxian, J.: Dissecting android malware: characterization and evolution. In: 2012 IEEE Symposium on Security and Privacy, pp. 95–109 (2012)
3. Apache Spark. https://spark.apache.org/
4. Guyon, I., Elisseeff, A.: An introduction to variable and feature selection. J. Mach. Learn. Res. **3**, 1157–1182 (2003)
5. Larrañaga, P., Calvo, B., Santana, R., Bielza, C., Galdiano, J., Inza, I., Lozano, J.A., Armañanzas, R., Santafé, G., Pérez, A.: Machine learning in bioinformatics. Brief. Bioinform. **7**(1), 86–112 (2006)

6. Ding, C., Peng, H.: Minimum redundancy feature selection from microarray gene expression data. J. Bioinform. Comput. Biol. **3**(02), 185–205 (2005)
7. Liu, H., Liu, L., Zhang, H.: Ensemble gene selection by grouping for microarray data classification. J. Biomed. Inform. **43**(1), 81–87 (2010)
8. Spreitzenbarth, M., Echtler, F., Schreck, T., Freling, F.C., Hoffmann, J.: Mobile-sandbox: having a deeper look into android applications. In: 28th International ACM Symposium on Applied Computing (SAC) (2013)
9. Arp, D., Spreitzenbarth, M., Hubner, M., Gascon, H., Rieck, K.: DREBIN: effective and explainable detection of android malware in your pocket. In: 21st Annual Network and Distributed System Security Symposium (2014)
10. Sánchez, R., Herrero, Á., Corchado, E.: Visualization and clustering for SNMP intrusion detection. Cybern. Syst. Int. J. **44**(6–7), 505–532 (2013)
11. Pinzón, C., Herrero, Á., De Paz, J.F., Corchado, E., Bajo, J.: CBRid4SQL: a CBR intrusion detector for SQL injection attacks, pp. 510–519. Springer, Heidelberg (2010)
12. Corchado, E., Herrero, Á., Baruque, B., Sáiz, J.M.: Intrusion detection system based on a cooperative topology preserving method. In: Ribeiro, B., Albrecht, R.F., Dobnikar, A., Pearson, D.W., Steele, N.C. (eds.) International Conference on Adaptive and Natural Computing Algorithms (ICANNGA 2005), pp. 454–457. Springer, Vienna (2005)
13. Feizollah, A., Anuar, N.B., Salleh, R., Wahab, A.W.A.: A review on feature selection in mobile malware detection. Digit. Investig. **13**, 22–37 (2015)
14. Hyo-Sik, H., Mi-Jung, C.: Analysis of android malware detection performance using machine learning classifiers. In: 2013 International Conference on ICT Convergence, pp. 490–495 (2013)
15. Shabtai, A., Elovici, Y.: Applying behavioral detection on android-based devices. In: Cai, Y., Magedanz, T., Li, M., Xia, J., Giannelli, C. (eds.) Mobile Wireless Middleware, Operating Systems, and Applications: Third International Conference, Mobilware 2010, Chicago, IL, USA, 30 June–2 July 2010, Revised Selected Papers, pp. 235–249. Springer, Heidelberg (2010)
16. Shabtai, A., Fledel, Y., Elovici, Y.: Automated static code analysis for classifying android applications using machine learning. In: 2010 International Conference on Computational Intelligence and Security, pp. 329–333 (2010)
17. Battista, P., Mercaldo, F., Nardone, V., Santone, A., Visaggio, C.: Identification of android malware families with model checking. In: 2nd International Conference on Information Systems Security and Privacy (2016)
18. Sedano, J., González, S., Chira, C., Herrero, A., Corchado, E., Villar, J.R.: Key features for the characterization of android malware families. Logic J. IGPL **25**(1), 54–66 (2017)
19. John, G.H., Kohavi, R., Pfleger, K.: Irrelevant features and the subset selection problem. In: 11th International Conference on Machine Learning, pp. 121–129. Morgan Kauffman, San Francisco (1994)
20. Kohavi, R., John, G.H.: Wrappers for feature subset selection. Artif. Intell. **97**(1–2), 273–324 (1997)
21. Goldberg, D.E.: Genetic Algorithms in Search, Optimization, and Machine Learning. Addison-Wesley, Reading (1989)
22. Siedlecki, W., Sklansky, J.: A note on genetic algorithms for large-scale feature selection. Pattern Recogn. Lett. **10**(5), 335–347 (1989)
23. Kramer, O.: Genetic Algorithm Essentials. Springer, Cham (2017)
24. Pal, S.K., Mitra, S.: Multilayer perceptron, fuzzy sets, and classification (2011)
25. Broyden, C.G., Dennis Jr., J.E., Moré, J.J.: On the local and superlinear convergence of Quasi-Newton methods. IMA J. Appl. Math. **12**(3), 223–245 (1973)
26. Virus Total. https://www.virustotal.com

Improving Blockchain Security Validation and Transaction Processing Through Heterogeneous Computing

Ciprian Pungila(✉) and Viorel Negru(✉)

Faculty of Mathematics and Informatics, Computer Science Department, West University of Timisoara, V. Parvan 4, Timisoara, Romania
{cpungila,vnegru}@info.uvt.ro

Abstract. We are presenting a new architectural model for enhancing blockchain processing of transactions and security validation through heterogeneous computing, by focusing and improving the consensus validation process, specifically the Merkle tree implementation and validation, both in terms of processing and storage efficiency. By using a heterogeneous architecture, coupled with an efficient storage model for instant transfers of data between memory and video memory, the validation process efficiency can be improved and offloaded from the CPU, with a direct result in increased transactional speeds and enhanced security.

Keywords: Blockchain · Consensus validation · High performance computing · Efficient storage · Merkle trees · GPU processing

1 Introduction

Blockchains have been widely spread lately with the surge of interest in the crypto-world, with Bitcoin [1] and Ethereum [2] taking the primary focus spots when it comes to interest from investors, crypto-enthusiasts and developers. With more than 2,000 coins in existence nowadays [3], and a market cap of more than 120 billion USD at the time of this writing for all of them, blockchains will have to keep focusing on improving security and transactional speeds, to a point where the technology becomes secure and gains more trust from the general public as well.

Blockchains however have practical purpose outside the crypto world as well, with multiple solutions emerging lately in the world of fintech [4], health systems [5], manufacturing [6], road safety [7], disaster prevention [8] and many more. Two main categories of blockchains exist: public blockchains (e.g. Bitcoin, Ethereum, etc.), where the principle of consensus allows validation of transactions by individual peers in a network, and private blockchains (e.g. Hyperledger Fabric [9]), where data is closed to the general public, but needs to be kept under strict control to avoid potential security breaches or vulnerabilities.

© Springer Nature Switzerland AG 2020
F. Martínez Álvarez et al. (Eds.): CISIS 2019/ICEUTE 2019, AISC 951, pp. 132–140, 2020.
https://doi.org/10.1007/978-3-030-20005-3_14

Keeping in mind the above, we are proposing an innovative architecture for improving the efficiency of the consensus validation process used in public blockchains like Bitcoin and others where a proof-of-work (PoW) system is based on hash validation, generally through a custom implementation of Merkle trees [10], and apply it to heterogeneous systems to discuss potential benefits. We discuss related work in Sect. 2 of the paper, describe our approach, with data structures, algorithm and methodology in Sect. 3, while Sect. 4 provides the necessary experimental results achieved during our testing.

2 Related Work

Public blockchains which have known increased public awareness lately, such as Bitcoin and Ethereum, are a constant target of attackers due to the high financial potential that they pose. Due to various flaws in the design of these blockchains, significants amounts of money are lost regularly, but as technology and security evolve, so do the blockchains that adopt them.

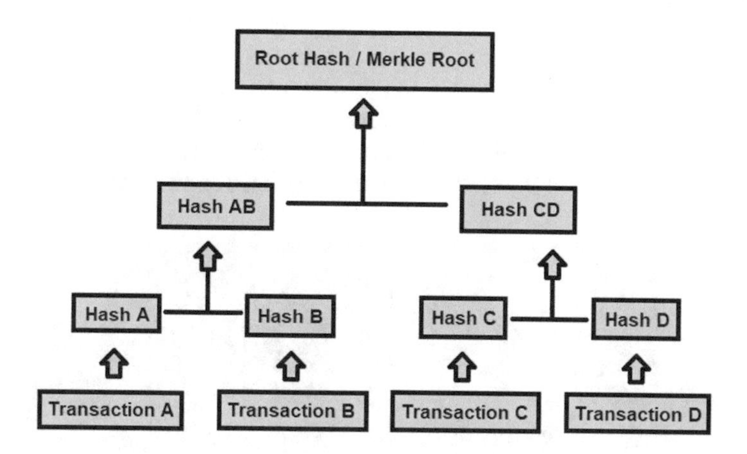

Fig. 1. A Merkle tree's typical layout.

In Proof-of-Work (PoW) systems such as Bitcoin and Ethereum, to name the most popular ones (but also applicable to thousands of other coins in existence nowadays), the principle of consensus validation is fundamental to the way blockchains function. In most such approaches, Merkle trees are being used as structures for efficient, secure validation of digital data in large amounts of information, as an essential mechanism to verify the consistency and authenticity of that content.

A Merkle tree is a structure that performs a security check of all the transactions included in a block of the blockchain, enabling user to verify whether or not a transaction is a part of a block. In Fig. 1, transactions A and B are hashed through a SHA-256 mechanism into a corresponding Merkle node, producing Hash A and Hash B, and then again both of these are hashed consecutively into a parent node, Hash AB. Following on the same logic for transactions C and D, and hashing in the final step nodes Hash AB with Hash CD, we obtain the Merkle root of the tree, with the unique property that should any of the underlying hashes (or transactions) change, at any depth, the root also changes adequately. This helps prevent malicious nodes in a blockchain inject fake or invalid transactions into the blockchain, and to a more marginal degree it helps prevent double-spend and denial-of-service attacks.

2.1 The CUDA Framework

As nVIDIA has introduced CUDA [11] back in 2007, the entire set of modern computing has changed. With graphics cards now capable of being put to use for other purposes except 3D rendering and gaming, a significant amount of research fields have emerged that benefit from the massively computational capabilities of GPGPU devices, including fluid dynamics, astronomical event modeling and numerous other data-parallel implementations of known or new algorithms. A warp in CUDA is a group of 32 threads, which is the minimum size of the data processed in SIMD fashion by a CUDA multiprocessor. The CUDA architecture works with blocks that can contain 64 to 512 threads. Blocks are organized into grids. Parallel portions of an application are executed on the device (GPU) as kernels, one kernel at a time, while many threads execute each kernel, with this aspect being an essential part of the architecture required to achieve high throughputs therefore. Figure 2 shows the CUDA architecture used in the implementation.

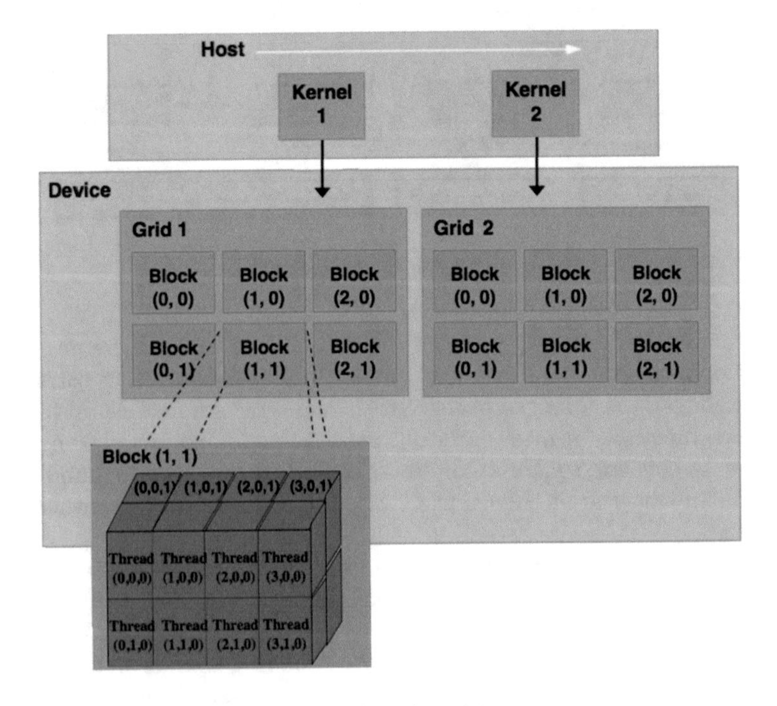

Fig. 2. The CUDA architecture.

Once the data is transferred in video memory, the CUDA parallel processing begins in a data-parallel manner. Most of the approaches split the input data into equally-sized chunks, which are being fed to the various threads being scheduled by the GPU kernel.

2.2 Tree Challenges in GPGPU Implementations

Because of the nature of trees in general, implementations in modern CPUs can be both fast and easy to achieve. In heterogeneous architectures however, there are several constraints related to the amount of video memory being available for use by the GPU, as well as the data to be processed requiring physical presence in video memory instead of RAM.

When working with dynamic trees in RAM memory, due to compiler optimizations and architecture limitations (or performance constraints), every allocation of new nodes will not follow a linear pattern of pointers in memory. We call this phenomenon "sparse pointers", since data structures are not perfectly aligned in memory, leading to gaps appearing between consecutive node allocations. While this may not be a problem usually, when dealing with large trees (with hundreds of thousands or millions of nodes), this can cause significant memory overhead (see Fig. 3a for an example of how sparse pointers are allocated; the ideal scenario, depicted in Fig. 3b, shows how memory should be aligned for maximum efficiency storage-wise). Eventually, such gaps can even double [12,13] the memory required to store such large trees.

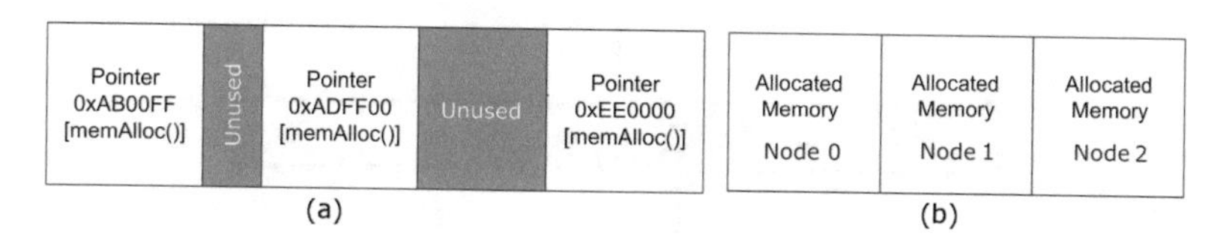

Fig. 3. (a) Sparse pointers in GPU memory for classic tree implementations; (b) The ideal layout of GPU memory for tree representation.

A significant amount of time however may be spent by modern computational systems validating transactions in a public blockchain that uses Merkle trees for example, which is why it would be desireable to attempt and offload this burden as much as possible. As a general rule of thumb, the less time spent on validating the blockchain's integrity, the more time we can spend on accepting more transactions, which could result in a higher number of transactions per second (Tx/s) that a blockchain could support.

3 Implementation

To implement our own model, we had to redesign the storage model used in Merkle trees, and optimize it for heterogeneous architectures. We have performed a serialization of the classic Merkle tree, by using a stack of nodes that have been linked together without any sparse pointer problems, by using a bitmapped representation.

To ensure a compact representation, we organize the compact Merkle tree in a form so that all children of any given node, are stored consecutively (without any interposing nodes between them) in the list of nodes. In addition to this, we use 2 bits in the bitmap for the left and right subtrees, an offset to indicate the position in the serialized tree where the first child node is stored, and the SHA-256 hash stored for this particular node, for validation purposes. The resulting node size is just 36 bytes, and the total maximum of supporting nodes is 2^{30} Fig. 4.

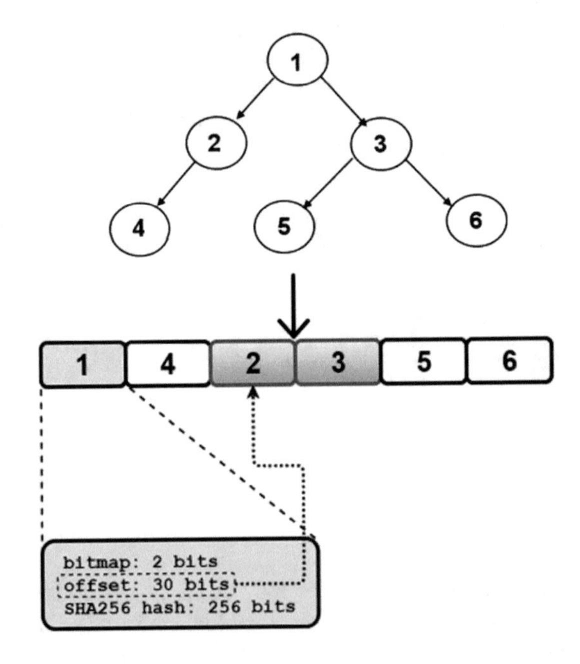

Fig. 4. Data structures used for individual nodes in our compact Merkle tree.

The compacting model for our Merkle tree, inspired by our earlier work in [12], is presented below, with "pos" starting at 0 and the "top" at 1:

```
void compactMerkleTree(merkleNode, pos)
{
    compactNode[pos].offset = top
    compactNode[pos].data = merkleNode.data
```

```
        old = top
        if (merkleNode.hasChild(left))
            setBit(compactNode[pos].bitmap, 0);
        if (merkleNode.hasChild(right))
            setBit(compactNode[pos].bitmap, 1);
        top += popCount(compactNode[pos].bitmap);
        if (merkleNode.hasChild(left))
            compactMerkleTree(merkleNode.child(left), old);
        if (merkleNode.hasChild(right))
            compactMerkleTree(merkleNode.child(right), old+1);
}
```

The algorithm proposed takes as input a classic Merkle tree and transforms it into its serialized, compact representation, which can be safely transferred between RAM and V-RAM in one single burst, without having to do individual transfers (e.g. one node at a time). To compare with, when performing individual node transfers for the same Merkle tree that had 1 million nodes, we required a total of 133 min to complete the transfer of the entire tree into V-RAM.

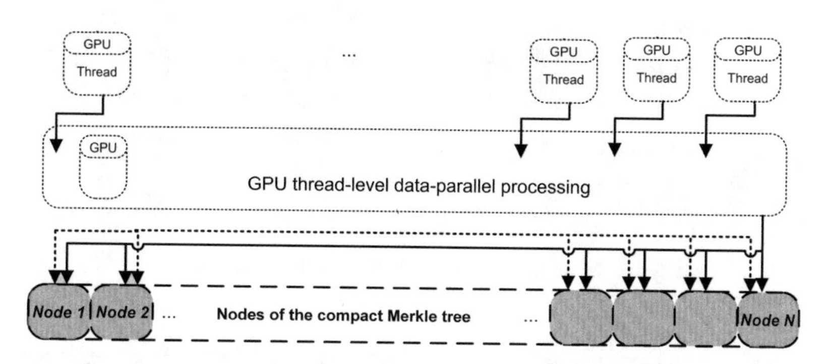

Fig. 5. A data parallel approach to Merkle tree processing using SIMD-driven, CUDA-based threads for GPGPU processing.

To allow validation of the tree in the GPU, as part of our heterogeneous processing architecture and model, we create and assign a single GPU thread to each specific node of the Merkle tree, for validation purposes, in a data parallel manner. Because of the way the compact tree is constructed, data parallel processing works very well with the CUDA capabilities of nVIDIA cards. If validation fails for any thread, the entire process stops of course. Figure 5 shows the input data being fragmented (dotted lines), with every chunk being sent to specific threads in order to achieve parallel processing of the input data.

4 Experimental Results

We have performed our testing using an i7 6700HQ CPU, backed up by 64 GB of RAM, and for heterogeneous processing we have used an nVIDIA Quadro

M1000M graphics card with 2 GB of DDR5 V-RAM The i7 has a memory bandwidth of 34.1 GB/s (for a single core), while the Quadro has 80 GB/s, meaning that a maximum theoretical speed-up for the GPU implementation vs CPU should see at most 2.34× higher throughput for the GPU. We have performed the experiments in three functional scenarios.

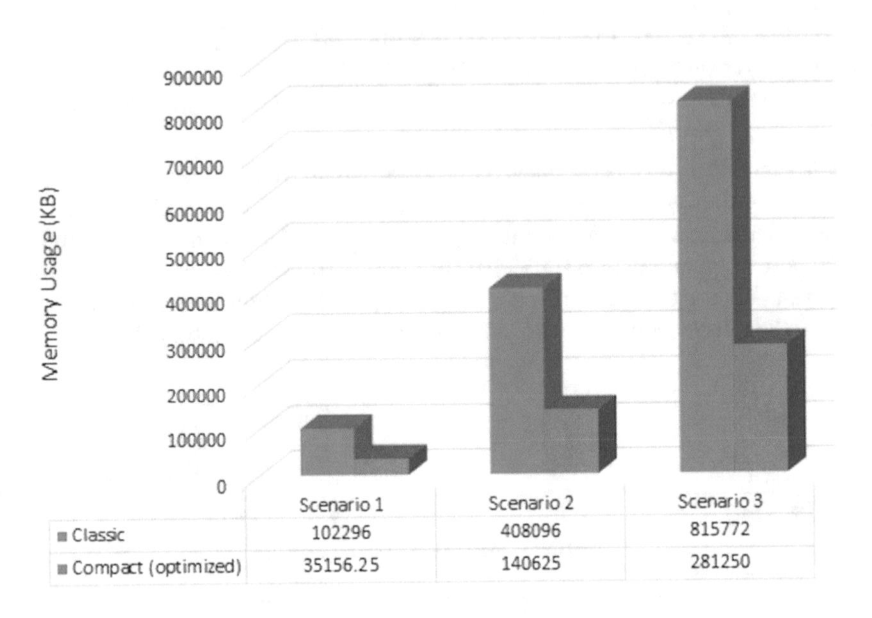

Fig. 6. Memory usage analysis of all test scenarios.

To perform the experiments, we have used real-world data from existing public blockchains, with real transaction information, and created the corresponding Merkle trees for three different scenarios, each with a purposely high number of transactions: we started with 1 million, followed by 4 million and, respectively, 8 million transactions. While they may not be achievable in practice at this very moment by most public blockchains (such as Bitcoin or Ethereum), there are numerous others that do claim a very high number of transactions - and we fully expect this side of the blockchain technology to mature even further in the years to come.

In the first stage, we have tested the single-CPU implementation, through a classic implementation of Merkle trees for all three scenarios outlined. We have focused on memory consumption and tree validation times, to emphasize the primary challenges that one would focus. The results are shown in Fig. 6.

In the second stage, we have tested the model obtained in the previous stage with a single-CPU implementation, to outline its capabilities and processing efficiency. We have observed that, first of all, the much lower memory footprint does not have a significant impact on the processing power of the newer, more compact storage model for our Merkle tree. The results are outlined in Fig. 7.

The memory footprint has been dramatically reduced in the optimized storage model, with the new approach producing a compression ratio of about 66% in all three scenarios tested by us. The processing time has not been dramatically affected by the new layout, in the first 2 scenarios we observed a small drop of performance of about 1.6% and 6% respectively, slightly compensated by a very subtle increase of performance of about 0.8% in the third scenario. Overall, the throughput on our test CPU topped at a little over 330,000 tx/s, a solid amount even for long-term ambitious blockchain development goals.

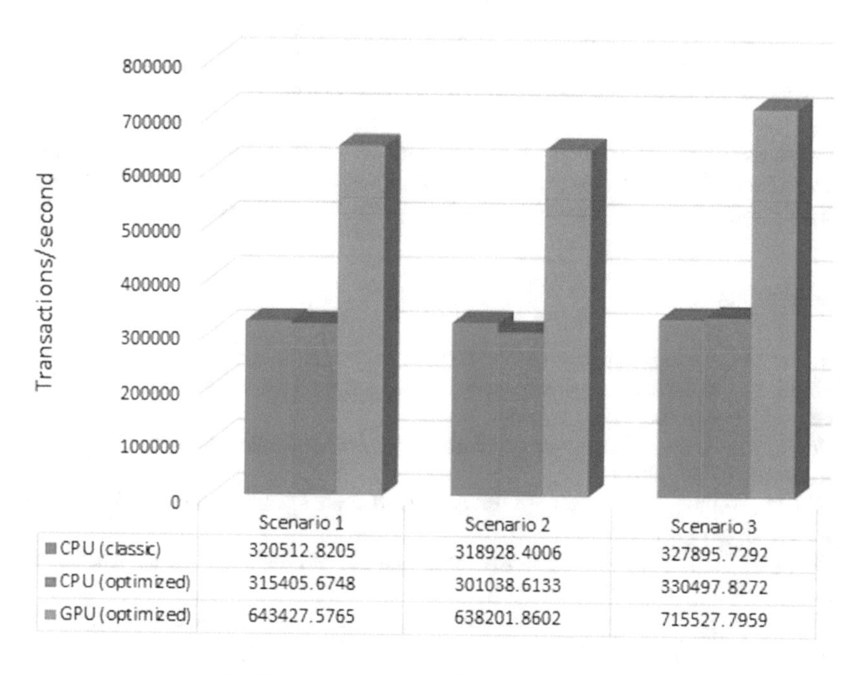

	Scenario 1	Scenario 2	Scenario 3
CPU (classic)	320512.8205	318928.4006	327895.7292
CPU (optimized)	315405.6748	301038.6133	330497.8272
GPU (optimized)	643427.5765	638201.8602	715527.7959

Fig. 7. Throughput analysis of our testing.

In the third and final stage, we have implemented our heterogeneous processing model in a CPU and GPU environment, with the consensus validation being performed by the GPU directly using the same memory footprint as tested in the previous stage. The results have shown that the proposed model is suitable for GPGPU processing, as data could be instantly transferred from RAM to V-RAM, and then processed through dedicated CUDA kernels, to ensure even higher throughputs: when offloading the computations to the GPU, we topped a little over 715,000 tx/s.

5 Conclusion

We have proposed a novel, efficient heterogeneous architecture for enhancing transactional speeds in blockchains based on proof-of-work consensus validation, specifically those using large Merkle trees, and have shown how we can significantly reduce the memory storage required for such trees, as well as improve the processing time overall, in both CPU (through increased cache locality) and heterogeneous (CPU and GPU) implementations. Our model has been enhanced to the point where it allows instant, full-duplex, full-bandwidth transfers (through the PCI Express architecture) of the entire Merkle tree required to be validated as part of the consensus process, with a significantly lower (up to 3 times lower) memory footprint than a classic implementation of such trees. In addition, the experiments performed with real-world data have shown that the model can be easily used in non-heterogeneous, single-CPU implementations, where it offers similar performance to classic implementations while retaining the same, much smaller memory footprint model for data representation.

Acknowledgment. This work was partially supported by InnoHPC - Interreg, Danube Transnational Programme grant. The views expressed in this paper do not necessarily reflect those of the corresponding projects consortium members.

References

1. Nakamoto, S.: Bitcoin whitepaper (2009). https://bitcoin.org/bitcoin.pdf
2. Wood, G.: Ethereum: a secure decentralized generalised transaction ledger (2015). https://gavwood.com/paper.pdf
3. CoinMarketCap: All Cryptocurrencies. https://coinmarketcap.com/all/views/all/. Accessed 15 Feb 2019
4. Zile, K., Renate, S.: Blockchain use cases and their feasibility. Appl. Comput. Syst. **23**(1), 12–20 (2018)
5. Holbl, M., et al.: A systematic review of the use of blockchain in healthcare. Symmetry **10**(10), 470 (2018)
6. Ko, T., et al.: Blockchain technology and manufacturing industry: real-time transparency and cost savings. Sustainability **10**, 4274 (2018)
7. PwC Report: Blockchain: The next innovation to make our cities smarter (2018). https://www.pwc.in/assets/pdfs/publications/2018/blockchain-the-next-innovation-to-make-our-cities-smarter.pdf
8. DH Network: Blockchain for the humanitarian sector: future opportunities (2016)
9. Androulaki, E., et al.: Hyperledger fabric: a distributed operating system for permissioned blockchains (2018). https://arxiv.org/pdf/1801.10228.pdf
10. Merkle, R.: Method of providing digital signatures, Patent, Patent US4309569A (1979). https://patents.google.com/patent/US4309569
11. NVIDIA: NVIDIA CUDA Compute Unified Device Architecture Programming Guide, version 4.1. http://developer.download.nvidia.com/compute/DevZone/docs/html/C/doc/CUDA_C_Programming_Guide.pdf
12. Pungila, C., Negru, V.: A highly-efficient memory-compression approach for GPU-accelerated virus signature matching. In: Information Security Conference (ISC) (2012)
13. Pungila, C., Negru, V.: Real-time polymorphic Aho-Corasick automata for heterogeneous malicious code detection. In: International Joint Conference SOCO17-CISIS17-ICEUTE17. AISC, vol. 239, pp. 439–448. Springer (2014)

Anomaly Detection on Patients Undergoing General Anesthesia

Esteban Jove[1,2(✉)], Jose M. Gonzalez-Cava[2], José-Luis Casteleiro-Roca[1], Héctor Quintián[1], Juan Albino Méndez-Pérez[2], and José Luis Calvo-Rolle[1]

[1] Department of Industrial Engineering, University of A Coruña, Avda. 19 de febrero s/n, 15405 Ferrol, A Coruña, Spain
esteban.jove@udc.es

[2] Department of Computer Science and System Engineering, Universidad de La Laguna, Avda. Astrof. Francisco Sánchez s/n, 38200 S/C de Tenerife, Spain
jamendez@ull.edu.es

Abstract. The importance of the infusion drug optimization in patients undergoing general anesthesia has led to the implementation of automatic control loops and models to predict the state of the patient. The appearance of any anomaly during the anesthetic process may lead, for instance, to incorrect drug administration. This could produce undesirable side effects that can affect the patient postoperative and also reduce the safety of the patient in the operating room. This study evaluates different one-class intelligent techniques to detect anomalies in patients undergoing general anesthesia. Due to the difficulty of obtaining data from anomaly situations, artificial outliers are generated to check the performance of each classifier. The final results give good performance in general terms.

Keywords: Anomaly detection · Outlier generation · Anesthesia

1 Introduction

During the last decades, one of the most important advances in the field of anesthesia has been related to the design of automatic control systems that decide the proper patient drug dose according to his/her needs [8,28].

There have been many researches focused on the anesthesia closed-loop control opposed to manual control [13]. The control process is a complex system that involves three main variables: muscular blockade, analgesia and hypnosis. In this sense, different automatic controllers [22,23,27] are proposed to optimize the right drug infusion rate depending on the needs of the patient. It is important to remark, that administering the correct dose of anesthetic drugs can avoid side effects during the postoperative process [5,6,17].

In this context, the appearance of any anomaly caused by multiple sources like: sensors misreadings [1], actuators malfunction, surgeon mistakes or patient

© Springer Nature Switzerland AG 2020
F. Martínez Álvarez et al. (Eds.): CISIS 2019/ICEUTE 2019, AISC 951, pp. 141–152, 2020.
https://doi.org/10.1007/978-3-030-20005-3_15

alterations during surgery, may probably lead to wrong control loop decisions. Then, if a proper drug delivery is sought, specially taking into consideration the safe-critical character of a surgery, an early detection of any kind of anomaly plays an important role [10,29,31].

From an analytical point of view, an anomaly is defined as a data pattern that has an unexpected behaviour in a specific application [7]. However, some important issues must be faced in an anomaly detection system, such as: the selection of a threshold between normal and anomaly data, the occurrence of noise in the data or the lack of data from anomaly operation [7,15].

Anomaly detection techniques are used on a wide variety of fields and applications, like intrusion detection in surveillance systems, fraud detection in bank accounts, fault detection in industrial process, and in medicine tasks [12,18,18,38].

This study proposes a method to detect anomalies in surgeries with real data from patients undergoing general anesthesia. Although the Bispectral Index (BIS, a variable that has influence on the hypnosis) is controlled directly with propofol, there are some studies that also uses remifentanil to reject small BIS deviations from the target [21]. This work is focused in the analysis of this correlations. Thus, three different variables will be analyzed during surgery: the Bispectral Index and the Electromyogram (EMG) signals, and the Remifentanil administered to the patient. Given the difficulty of obtaining data from faulty situations during a real surgery, the anomalies are artificially generated using the Boundary Value Method (BVM) [37].

Different one-class classifiers are implemented using Approximate Convex Hull (ACH) [3], Autoencoder [33] and Support Vector Machine (SVM) techniques [19]. Their performance is assessed, offering successful results.

This paper is structured as follows: after the introduction section, T the case of study is explained. Then, next section presents the model approach and the used techniques. Section 4 describes the experiments and results and finally, the conclusions and future works are shown in Sect. 5.

2 Case of Study

The main goal of an anesthetic process is to control the drug infusion according to the real needs of the patient. Two different drugs are administered in a Total Intravenous Anesthesia (TIVA): propofol for the hypnosis and remifentanil for the analgesia. Hence, it is necessary to have information about the current state of the patient. In this sense, two different signals are monitored during the surgery: the Bispectral Index (BIS), used to measure the hypnotic state of patient, and the Electromyogram (EMG) [30,34].

The anesthetic process is commonly divided in three phases: induction, maintenance and recovering. The induction part consists of administering the appropriate dose of drug to achieve adequate levels of analgesia, hypnosis and muscular relaxation to begin with the surgery [20]. This hypnotic state is reached by delivering a propofol (1%) intravenous bolus of 2 mg/Kg at the maximum pump rate. As a response to the propofol bolus dose, the patient enters the maintenance phase. In this phase the clinician should continuously infuse the proper propofol dose to control the BIS variable and the remifentanil dose to control the analgesia state. In this study, propofol was administered by using a closed-loop strategy and remifentanil was administered manually. Once the surgery finishes, the recovery phase starts and the drug infusion ceases.

The simple scheme representative of the case of study is shown in Fig. 1, where the inputs are the propofol and reminfentanil dose and the outputs are the BIS and EMG signals.

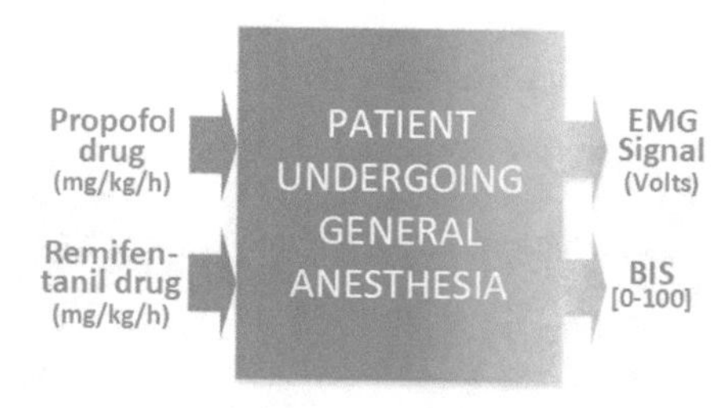

Fig. 1. Case of study. Input/Output representation

The scope of this study is the analysis of three of these variables: remifentanil, EMG and BIS. The study will be done for the maintenance phase where the hypothesis is that remifenanil variations affects the BIS signal. The aim is to detect anomalies in the evolution of these three variables.

2.1 Dataset Description

In the case under study, two Graseby 3500 intravenous pumps connected to a laptop via RS232 port are used for both remifentanil and propofosol infusion. Furthermore, the BIS and EMG signals are measured during the surgery with a 5 s sample rate.

This work is focused on detecting anomalies during surgery taking into account a real dataset obtained from 15 patients. For each patient, the values of EMG and BIS signals, and propofol and remifentanil rates are measured.

The initial dataset is pre-processed to detect null measurements, and replace them using linear interpolation.

With the dataset from the induction and maintenance phases, after the above mentioned pre-processing (19.941 samples), the classifier has been implemented.

3 Applied Techniques to Validate the Proposed Model

3.1 Model Approach

This work deals the assessment of three one-class classification techniques to detect anomalies of the anesthetic process in patients undergoing surgery. These techniques are briefly explained in Subsect. 3.2. To check the performance of the different one-class classifiers, the outliers are artificially generated according to the method shown in Subsect. 3.3.

The process followed to evaluate each one-class technique is shown in Fig. 2. First, the dataset from patients #2 to #15 is used to train the Classifier 1. This classifier is tested using data from patient #1, where 10% of the data is randomly selected to generate anomalies, and the 90% left remains unaltered. This test process is repeated 10 times, modifying for each iteration the data to be converted on anomaly. Once the results of Classifier 1 are calculated, the same process is conducted for each patient data. Consequently, fifteen different one-class classifiers are trained with fourteen patients and tested with one. Then, the average results and standard deviations of this classifiers are evaluated.

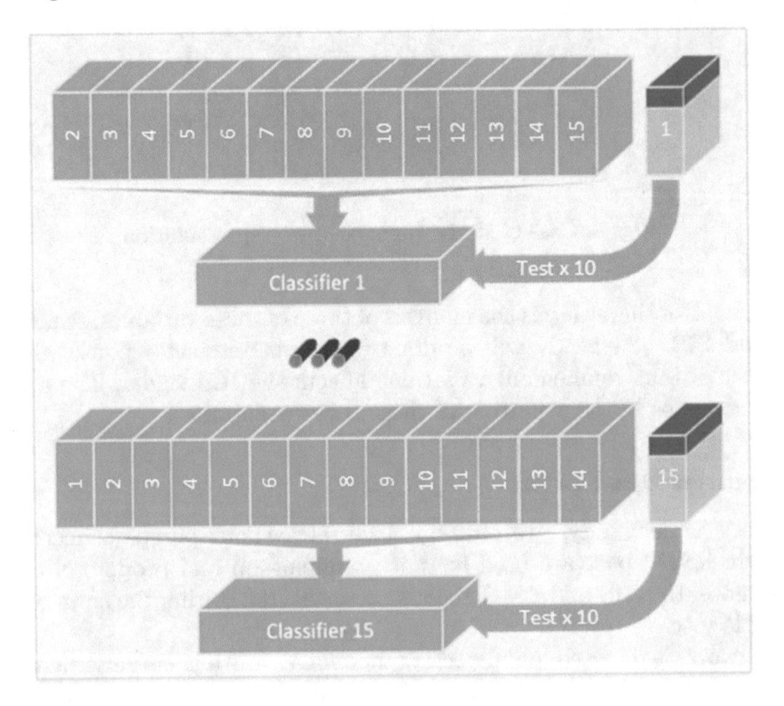

Fig. 2. Classifier train and test process

3.2 One-Class Techniques

Approximate Convex Hull. The Approximate Convex Hull (ACH) is a one-class classification technique, whose good performance has been proved in [4, 11].

The main idea of this technique is to estimate the boundaries of a dataset $A \in \mathbb{R}^n$ as its convex hull. Taking into consideration that, the convex limits of a given dataset with M samples and t variables, has a computational cost of $O(M^{(t/2)+1})$ [4], the convex hull can be approximated using n random projections on $2D$ planes, and identifying the convex limits on each projection. After obtaining the convex hull approximation using the n projections of the training data, if a new test sample arrives, it is considered as an anomaly, when it is out of the convex hull of any projection.

Furthermore, it is possible to define a parameter λ, that reduces or expands the convex limits from each centroid. If λ is set to a value higher than 1, there is an expansion in the convex limits. However, values lower than 1 leads to narrower boundaries.

The appearance of an anomaly point in a $\mathbb{R}^3$ space is shown in Fig. 3.

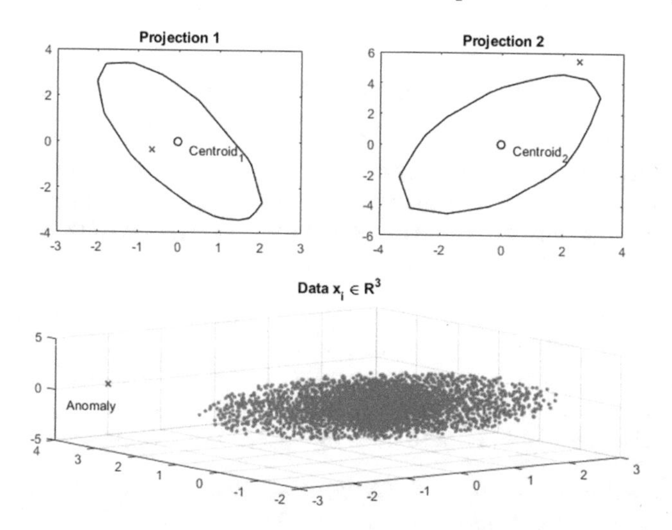

Fig. 3. Anomaly point in $\mathbb{R}^3$

Support Vector Machine. The Support Vector Machine (SVM) can be used for regression and classification tasks [9,16,19]. In case of one class classification, this technique maps the initial dataset into a high dimensional space using a kernel function. Then, in this high dimensional space, a hyper-plane that maximises the distance between the origin and the data is implemented [32].

After the training process, the criteria to identify if a test data must be considered as an anomaly, is based on the distance of that point to the hyper-plane. If the distance is positive, the data is considered as an anomaly, and it is inside the target class otherwise.

Artificial Neural Networks Autoencoder. The use of the Autoencoder configuration using Artificial Neural Networks (ANN) for one-class offer successful results in a wide variety of applications [14,33,36].

The Multilayer Perceptron (MLP) is the most used ANN for this task [39]. Its structure has an input layer, an output layer and a hidden layer. Each layer is composed by connected neurons with weighted links between layers and non-linear activation functions.

The aim of Autoencoder technique is to reconstruct the input patterns in the output using an intermediate nonlinear dimensional reduction in the hidden layer. Thus, the number n of input neurons and output neurons are the same as the variables of the dataset, and the hidden layer must have at least $(n-1)$ neurons.

The intermediate nonlinear reduction must eliminate the data, that is not consisted with the dataset. Then, anomaly data should lead to high reconstruction error, which is the difference between the input and the output estimated by the MLP.

3.3 Artificial Outlier Generation. The Boundary Value Method

In this work, the Boundary Value Method (BVM) [37] is applied to emulate anomalies during the anesthetic process. From an initial dataset of S samples and n attributes, this technique deals the anomaly generation shifting the samples out of the training dataset boundary. A detailed description of the steps followed to generate an artificial outlier from $a \in \mathbb{R}^n$ is described next:

1. Find in the initial dataset the minimum and maximum value of each attribute and save these values in two vectors $V_{min}, V_{max} \in \mathbb{R}^n$.
2. Select randomly two dimensions p and q of the n-dimensional dataset.
3. Replace the value of $a(p)$ randomly by $V_{max}(p)$ or $V_{min}(p)$.
4. Replace the value of $a(q)$ randomly by $V_{max}(q)$ or $V_{min}(q)$.

A simplified example of an anomaly conversion from a target class point is shown in Fig. 4. A point inside the target class (green point) in $\mathbb{R}^3$ is converted into an anomaly. The transformation Tx replaces the x coordinate by the maximum registered (yellow dot), and the transformation Ty changes the y value by its minimum (red dot). Then, it can be noticed that the data generated is out of the initial set.

4 Experiments and Results

This paper evaluates the performance of three one-class techniques on the task of detecting anomalies in the anesthetic process during a surgery, using data from fifteen patients undergoing general anesthesia. Given the difficulty of obtaining data from abnormal situations in this kind of processes, it is necessary to generate outliers artificially.

Fifteen different classifiers are implemented using fourteen patients for train and one for test. A random 10% of the test data is converted to outlier and this process is repeated ten times. The performance of each classifier is evaluated using the Area Under Curve (AUC) parameter, that relates false positives and true positives [2].

To generalize the performance of each one-class technique, the average value and standard deviation (SD) of the fifteen classifiers for each technique is presented in this work. Also, the computational cost in terms of training time is shown.

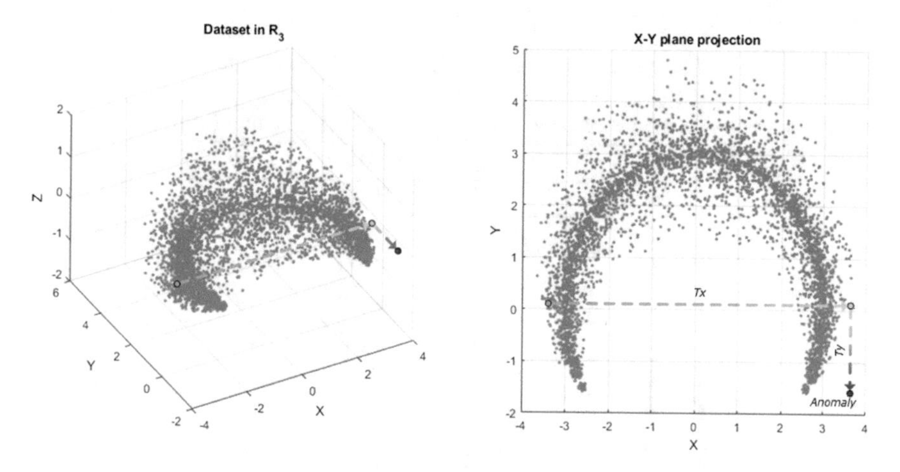

Fig. 4. Anomaly generation in $\mathbb{R}^3$

Two different experiments were performed: first, the four variables presented in Subsect. 2.1 are used as inputs. However, this configuration does not offer successful results and they are not included. The second experiment was made excluding the propofol signal. The next subsections show the results achieved for the second configuration.

4.1 Approximate Convex Hull Classifiers

Different classifiers where implemented using using 5, 10, 50, 100, 500 and 1000 $2D$ projections with a λ set as 0.9, 1 and 1.1. The AUC average value of fifteen classifiers, as well as the SD, and the average training time for each configuration can be seen in Table 1.

4.2 SVM Classifiers

To obtain the SVM one-class classifier the *fitcsvm* [25] function from Matlab is used. In this function, the outlier fraction of the training data is checked from 0 to 5 and the kernel function is set as Gaussian. The criteria to detect an anomaly is based on the distance of the point to the decision plane. This value is calculated using the *predict* [26] function. The results are shown in Table 2.

4.3 Artificial Neural Network Autoencoder Classifier

The Autoencoder classifiers were implemented using the *trainAutoencoder* Matlab function [24]. In this case, the number of neurons in the hidden layer varies from 1 to 2, as it must have a value lower than the number inputs.

The criteria to detect an anomaly is based on the reconstruction error. If this value is higher than the one obtained with 99% of the training set, it is considered as an anomaly. The obtained results are presented in Table 3.

Table 1. Average results with ACH classifiers

λ	Number of proj.	AUC (%)	STD (%)	Time (min)
0,9	5	83,22	10,03	0,05
	10	83,57	10,24	0,46
	50	85,78	10,25	0,50
	100	85,39	11,29	0,42
	500	87,59	10,76	0,38
	1000	88,60	8,83	0,38
1	5	67,90	9,43	0,49
	10	68,69	10,11	0,44
	50	71,38	9,89	0,35
	100	72,24	10,13	0,47
	500	73,93	9,73	0,41
	1000	74,56	9,85	0,47
1,1	5	51,53	3,93	0,49
	10	51,72	4,12	0,36
	50	56,59	6,48	0,37
	100	59,78	7,29	0,45
	500	65,23	7,19	0,49
	1000	66,28	7,85	0,48

Table 2. Average results with SVM classifiers

Outlier frac. (%)	AUC (%)	STD (%)	Time (min)
0	66,90	3,69	0,41
1	88,77	5,55	0,42
2	90,69	6,04	0,43
3	91,46	5,21	0,43
4	92,50	6,04	0,41
5	92,19	5,44	0,42

Table 3. Average results with Autoencoder classifiers

Number of neurons	AUC (%)	STD (%)	Time (min)
1	86,62	4,32	17,04
2	87,60	5,15	18,49

5 Conclusions and Future Works

This study is focused on detecting anomalies using different intelligent techniques validated with artificially generated outliers. The detection assessed in three variables: EMG, BIS and Remifentanil dose. From the clinical point of view the results obtained are important as they reflects that in the maintenance phase there are a correlation between the remifentanil dose and the BIS signal. The results shown the SVM as the best technique in terms of AUC, offering a 92,50%, with a significantly low SD between patients.

In this case, given the fact that, the classifier training can be done between surgeries, the computational cost is not a critical feature. Furthermore, the testing of a new data is far lower than 5 s for the three proposed techniques.

The achieved model can be used as valuable method to detect anomalies in the anesthetic system during surgeries. This can be used to avoid wrong drug infusion rates and adapt this values to the real needs of the patient and hence, decrease the possible side effects of patients undergoing general anesthesia.

In future works, the incorporation of more signals involved in the anesthetic process, like the Analgesia Nociception Index (ANI), could be taken into consideration. Also, the correlation between variables can be assessed using dimensional reduction [35]. To improve the performance of the anomaly detection, the possibility of achieving different classifiers depending on the patients nature (gender, weight,...) could be considered.

Acknowledgments. This research is partially supported through the "Fundación Canaria de Investigación Sanitaria" (FUNCANIS) [ref: PIFUN23/18].

Jose M. Gonzalez-Cava's research was supported by the Spanish Ministry of Education, Culture and Sport (www.mecd.gob.es), under the "Formación de Profesorado" grant FPU15/03347.

References

1. Baruque, B., Porras, S., Jove, E., Calvo-Rolle, J.L.: Geothermal heat exchanger energy prediction based on time series and monitoring sensors optimization. Energy **171**, 49–60 (2019)
2. Bradley, A.P.: The use of the area under the ROC curve in the evaluation of machine learning algorithms. Pattern Recognit. **30**(7), 1145–1159 (1997)
3. Casale, P., Pujol, O., Radeva, P.: Approximate convex hulls family for one-class classification. In: Sansone, C., Kittler, J., Roli, F. (eds.) Multiple Classifier Systems, pp. 106–115. Springer, Heidelberg (2011)
4. Casale, P., Pujol, O., Radeva, P.: Approximate convex hulls family for one-class classification. In: International Workshop on Multiple Classifier Systems, pp. 106–115. Springer (2011)
5. Casteleiro-Roca, J.L., Jove, E., Gonzalez-Cava, J.M., Pérez, J.A.M., Calvo-Rolle, J.L., Alvarez, F.B.: Hybrid model for the ANI index prediction using remifentanil drug and EMG signal. Neural Comput. Appl., 1–10 (2018)

6. Casteleiro-Roca, J.L., Pérez, J.A.M., Piñón-Pazos, A.J., Calvo-Rolle, J.L., Corchado, E.: Modeling the electromyogram (EMG) of patients undergoing anesthesia during surgery. In: 10th International Conference on Soft Computing Models in Industrial and Environmental Applications, pp. 273–283. Springer (2015)
7. Chandola, V., Banerjee, A., Kumar, V.: Anomaly detection: a survey. ACM Comput. Surv. (CSUR) **41**(3), 15 (2009)
8. Chang, J.J., Syafiie, S., Kamil, R., Lim, T.A.: Automation of anaesthesia: a review on multivariable control. J. Clin. Monit. Comput. **29**(2), 231–239 (2015)
9. Chen, Y., Zhou, X.S., Huang, T.S.: One-class SVM for learning in image retrieval. In: 2001 International Conference on Image Processing, Proceedings, vol. 1, pp. 34–37. IEEE (2001)
10. Chiang, L.H., Russell, E.L., Braatz, R.D.: Fault Detection and Diagnosis in Industrial Systems. Springer, London (2000)
11. Fernández-Francos, D., Fontenla-Romero, O., Alonso-Betanzos, A.: One-class convex hull-based algorithm for classification in distributed environments. IEEE Trans. Syst. Man Cybern. Syst. **99**, 1–11 (2018)
12. González, G., Angelo, C.D., Forchetti, D., Aligia, D.: Diagnóstico de fallas en el convertidor del rotor en generadores de inducción con rotor bobinado. Revista Iberoamericana de Automática e Informática industrial **15**(3), 297–308 (2018). https://polipapers.upv.es/index.php/RIAI/article/view/9042
13. Gonzalez-Cava, J.M., Reboso, J.A., Casteleiro-Roca, J.L., Calvo-Rolle, J.L., Méndez Pérez, J.A.: A novel fuzzy algorithm to introduce new variables in the drug supply decision-making process in medicine. Complexity **2018**, 15 (2018)
14. Goodfellow, I., Bengio, Y., Courville, A., Bengio, Y.: Deep Learning, vol. 1. MIT Press, Cambridge (2016)
15. Jove, E., Antonio Lopez-Vazquez, J., Isabel Fernandez-Ibanez, M., Casteleiro-Roca, J.L., Luis Calvo-Rolle, J.: Hybrid intelligent system to predict the individual academic performance of engineering students. Int. J. Eng. Educ. **34**(3), 895–904 (2018)
16. Jove, E., Gonzalez-Cava, J.M., Casteleiro-Roca, J.L., Méndez-Pérez, J.A., Antonio Reboso-Morales, J., Javier Pérez-Castelo, F., Javier de Cos Juez, F., Luis Calvo-Rolle, J.: Modelling the hypnotic patient response in general anaesthesia using intelligent models. Logic J. IGPL **27**, 189–201 (2018)
17. Jove, E., Gonzalez-Cava, J.M., Casteleiro-Roca, J.L., Pérez, J.A.M., Calvo-Rolle, J.L., de Cos Juez, F.J.: An intelligent model to predict ani in patients undergoing general anesthesia. In: Pérez García, H., Alfonso-Cendón, J., Sánchez González, L., Quintián, H., Corchado, E. (eds.) International Joint Conference SOCO'17-CISIS'17-ICEUTE'17 León, Proceeding, Spain, 6–8 September 2017, pp. 492–501. Springer, Cham (2018)
18. Moreno-Fernandez-de Leceta, A., Lopez-Guede, J.M., Ezquerro Insagurbe, L., Ruiz de Arbulo, N., Graña, M.: A novel methodology for clinical semantic annotations assessment. Logic J. IGPL **26**(6), 569–580 (2018). https://doi.org/10.1093/jigpal/jzy021
19. Li, K.L., Huang, H.K., Tian, S.F., Xu, W.: Improving one-class SVM for anomaly detection. In: 2003 International Conference on Machine Learning and Cybernetics, vol. 5, pp. 3077–3081. IEEE (2003)
20. Litvan, H., Jensen, E.W., Galan, J., Lund, J., Rodriguez, B.E., Henneberg, S.W., Caminal, P., Villar Landeira, J.M.: Comparison of conventional averaged and rapid averaged, autoregressive-based extracted auditory evoked potentials for monitoring the hypnotic level during propofol induction. J. Am. Soc. Anesthesiologists **97**(2), 351–358 (2002)

21. Liu, N., Chazot, T., Hamada, S., Landais, A., Boichut, N., Dussaussoy, C., Trillat, B., Beydon, L., Samain, E., Sessler, D.I., Fischler, M.: Closed-loop coadministration of propofol and remifentanil guided by bispectral index: a randomized multicenter study. Anesthesia Analgesia **112**(3), 546–557 (2011). www.refworks.com
22. Marrero, A., Méndez, J.A., Reboso, J.A., Martín, I., Calvo, J.L.: Adaptive fuzzy modeling of the hypnotic process in anesthesia. J. Clin. Monit. Comput. **31**(2), 319–330 (2017). https://www.scopus.com/inward/record.uri?eid=2-s2.0-84963700634&doi=10.1007%2Fs10877-016-9868-y&partnerID=40&md5=9d8d7b817499d3f41dacae54665a6af3
23. Marrero, A., Méndez, J.A., Reboso, J.A., Martín, I., Calvo, J.A.L.: Adaptive fuzzy modeling of the hypnotic process in anesthesia. J. Clin. Monit. Comput. **31**, 319–330 (2016)
24. MathWorks: Autoencoder. https://es.mathworks.com/help/deeplearning/ref/trainautoencoder.html. Accessed 29 Jan 2019
25. MathWorks: fitcsvm. https://es.mathworks.com/help/stats/fitcsvm.html. Accessed 29 Jan 2019
26. MathWorks: predict. https://es.mathworks.com/help/stats/classreg.learning.classif.compactclassificationsvm.predict.html. Accessed 29 Jan 2019
27. Mendez, J.A., Marrero, A., Reboso, J.A., Leon, A.: Adaptive fuzzy predictive controller for anesthesia delivery. Control Eng. Pract. **46**, 1–9 (2016)
28. Mendez, J.A., Leon, A., Marrero, A., Gonzalez-Cava, J.M., Reboso, J.A., Estevez, J.I., Gomez-Gonzalez, J.F.: Improving the anesthetic process by a fuzzy rule based medical decision system. Artif. Intell. Med. **84**, 159–170 (2018)
29. Miljković, D.: Fault detection methods: a literature survey. In: 2011 Proceedings of the 34th International Convention, MIPRO, pp. 750–755. IEEE (2011)
30. Pérez, J.A.M., Torres, S., Reboso, J.A., Reboso, H.: Estrategias de control en la práctica de anestesia. Revista Iberoamericana de Automática e Informática Industrial RIAI **8**(3), 241–249 (2011)
31. de la Portilla, M.P., Piñeiro, A.L., Sánchez, J.A.S., Herrera, R.M.: Modelado dinámico y control de un dispositivo sumergido provisto de actuadores hidrostáticos. Revista Iberoamericana de Automática e Informática industrial **15**(1), 12–23 (2017). https://polipapers.upv.es/index.php/RIAI/article/view/8824
32. Rebentrost, P., Mohseni, M., Lloyd, S.: Quantum support vector machine for big data classification. Phys. Rev. Lett. **113**, 130503 (2014). https://link.aps.org/doi/10.1103/PhysRevLett.113.130503
33. Sakurada, M., Yairi, T.: Anomaly detection using autoencoders with nonlinear dimensionality reduction. In: Proceedings of the MLSDA 2014 2nd Workshop on Machine Learning for Sensory Data Analysis, p. 4. ACM (2014)
34. Sánchez, S.S., Vivas, A.M., Obregón, J.S., Ortega, M.R., Jambrina, C.C., Marco, I.L.T., Jorge, E.C.: Monitorización de la sedación profunda. el monitor BIS. Enfermería Intensiva **20**(4), 159–166 (2009)
35. Segovia, F., Górriz, J.M., Ramírez, J., Martinez-Murcia, F.J., García-Pérez, M.: Using deep neural networks along with dimensionality reduction techniques to assist the diagnosis of neurodegenerative disorders. Logic J. IGPL **26**(6), 618–628 (2018). http://dx.doi.org/10.1093/jigpal/jzy026
36. Vincent, P., Larochelle, H., Lajoie, I., Bengio, Y., Manzagol, P.A.: Stacked denoising autoencoders: learning useful representations in a deep network with a local denoising criterion. J. Mach. Learn. Res. **11**, 3371–3408 (2010)

37. Wang, C.K., Ting, Y., Liu, Y.H., Hariyanto, G.: A novel approach to generate artificial outliers for support vector data description. In: IEEE International Symposium on Industrial Electronics, ISIE 2009, pp. 2202–2207. IEEE (2009)
38. Wojciechowski, S.: A comparison of classification strategies in rule-based classifiers. Logic J. IGPL **26**(1), 29–46 (2018). http://dx.doi.org/10.1093/jigpal/jzx053
39. Zeng, Z., Wang, J.: Advances in Neural Network Research and Applications, 1st edn. Springer, Heidelberg (2010)

Special Session: From the Least to the Least: Cryptographic and Data Analytics Solutions to Fulfill Least Minimum Privilege and Endorse Least Minimum Effort in Information Systems

An Internet Voting Proposal Towards Improving Usability and Coercion Resistance

Iñigo Querejeta-Azurmendi[1,2,3](✉), Luis Hernández Encinas[3], David Arroyo Guardeño[3], and Jorge L. Hernández-Ardieta[2]

[1] Minsait, Indra Sistemas S.A., Alcobendas, Spain
iquerejeta@minsait.com

[2] Universidad Carlos III de Madrid, Getafe, Spain
jlhernan@inf.uc3m.es

[3] Instituto de Tecnologías Físicas y de la Información, Consejo Superior de Investigaciones Científicas, Madrid, Spain
{luis,david.arroyo}@iec.csic.es

Abstract. This paper proposes a coercion-resistant internet voting protocol using a re-voting approach. It is not assumed for voters to own cryptographic keys prior to the election and the voting experience remains simple by only requiring voters to keep their authentication credentials. Furthermore, we reduce complexity in the filtering stage by leveraging the so-called *Millionaires Protocol*.

Keywords: Internet voting · Coercion-resistance · User privacy · Zero-knowledge proofs · Homomorphic encryption · Usability

1 Introduction

In on-site elections, voter's privacy and ballot integrity are protected by guaranteeing physical isolation when casting the ballot, and through a set of audit procedures by multiple and independent trustees. Internet voting (i-voting) eludes the call for physical presence of the agents involved in the voting process, which eases both the voting experience and the tallying process. Certainly, i-voting enables voters' ubiquity and the use of any personal device to cast a vote, and its digital nature makes it possible to conduct the whole election process without human intervention. Nonetheless, i-voting comes with several risks and pitfalls: the voter may cast the vote from anywhere, resulting in high possibility of coercion or vote-selling; the vote cast could go through an uncontrolled channel; remote authentication increases the chances of impersonation, and moreover, the voter casts a vote from her device, making it difficult to do a massive-scale, effective security assessment [14].

An electronic voting protocol must satisfy four main properties [4]: *universal verifiability* which allows an external party to verify that all steps of the election

© Springer Nature Switzerland AG 2020
F. Martínez Álvarez et al. (Eds.): CISIS 2019/ICEUTE 2019, AISC 951, pp. 155–164, 2020.
https://doi.org/10.1007/978-3-030-20005-3_16

went as expected, *ballot secrecy* which guarantees that a voters election remains secret, *eligibility verifiability* which allows verification that all counted votes were cast by an eligible voter, and, finally, *coercion resistance* which gives the voter the possibility of escaping coercion without the coercer noticing. Current literature presents several solutions offering coercion-resistance; however these are not usable in real-scale elections, either by the complication towards the voter or by the complexity of the solution. Proposed solutions in the non-hybrid i-voting setting are one of two types: (i) the voter ends up the registration phase owning many means of authentication (such as credentials and/or passwords) where only one is valid to cast a vote, enabling to cheat the coercer; (ii) the voter is capable to vote many times and the last vote counts, hence overwriting the coerced vote with the desired one.

One of the principal premises of our protocol is that it is not based in the requirement of the ownership of public/private key pairs by the users in order to foster the usability of a secure i-voting scheme. Requiring voters to transport, remember or store several credentials complicates voter's experience. For these reasons we have decided that the re-voting setting was preferable for our scheme.

The remainder of this work is organized as follows. Related work is presented in Sect. 1.1 followed by a short description of our contribution. Section 2 introduces the parties and the notation used in our protocol followed by a description of the latter in Sect. 3. Finally, we conclude our work and present possible lines of improvement in Sect. 4.

1.1 Related Work

Re-voting has been proposed in several constructions in current literature [16]. The main challenge that one finds when allowing multiple voting for coercion resistance is filtering, which consists in accepting only one vote per voter (normally the last one). It is a challenge because, in order to mitigate coercion attacks, voting must be *deniable*, meaning that the adversary must not be able to determine whether a specific voter re-cast her ballot, or more generally, not be able to know which of the voters re-voted. Concurrently, universal verification and proof that the process happened as expected must be provided.

The JCJ protocol [15] allows multiple voting, but the filtering phase is not deniable. One can determine whether her own vote has been filtered. Moreover, the filtering has a complexity of $\mathcal{O}(N^2)$, where N is the number of votes. The latter was most recently improved by Araújo et al. [2], still not satisfy deniability. The hybrid protocol presented by Spycher et al. [21] allows the voter to cast a vote in an electoral school, and therefore overwrite any previously cast votes. However, this results in the requirement of presence accessibility of the voter. Another used method is to do the filtering as a black box protocol. Trust is given to a server which will filter all votes and publish the result [11]. Such a method avoids any tracking or knowledge of which votes have been filtered, but verifiability is completely lost.

To the best of our knowledge, current filtering schemes that offer a deniable voting scheme which, at the same time have public verifiability, are the ones proposed by Achenbach *et al.* [1] and Locher *et al.* [18]. In a protocol where a Public Bulletin Board (PBB) is used in order to allow verifiability, the filtering process must be done after the mixing, otherwise, a voter (and thus the coercer) would know whether her vote was filtered or not. However, after the mixing, it is no longer possible to know which is the order of the votes, and therefore, before inserting the votes in the mix-net there must be some reference of their order. This solution faces this by performing Encrypted Plaintext Equivalence Tests (EPET) on the credentials for all votes against all lately cast votes before mixing. This consists in performing a PET (a cryptographic tool to compare two plaintexts without decrypting any of them) whose output is encrypted, and if any of the comparisons among the credentials is equivalent, then the output of the EPET hides a random number (alternatively, the encrypted number will be a one). Votes are then included in a mix-net. The filtering phase happens after mixing by decrypting the EPET, and only votes which output a 1 will be kept. These solutions are very interesting as they achieve the deniability property with no trust in any of the servers. However, to offer this there is a high increase in the complexity as the EPET have to be performed for each pair of votes, resulting in a complexity of $\mathcal{O}(N^2)$ prior to the mixing, and in $\mathcal{O}(N)$ decryptions and zero-knowledge proofs of correct decryption during the filtering.

These filtering schemes are implemented by comparing the voting credentials, requiring voters to maintain voting credentials throughout the whole election.

1.2 Our Contribution

In this paper, we present an i-voting scheme to achieve a reasonable balance between security and usability. We slightly increase the trust assumptions on some entities in the different stages of the voting process. On this basis, we show that it is not necessary to sacrifice usability to implement means for granting coercion-resistance, election verifiability, and ballot secrecy. As a matter of fact, our protocol provides a user-friendly interface for voters and, simultaneously, reduces the computational cost of the tallying phase with respect to existing solutions [15,17,18]. Our construction generates anonymous digital certificates each time a user casts a vote so that she does not need to remember, transport or possess any kind of cryptographic keys and/or secret apart from her authentication credentials. We mitigate coercion by using the solution of re-voting. Our construction achieves the deniability property by using concepts of the 'Millionaires Protocol' [6].

2 Parties and Building Blocks

The parties involved in the voting protocol are the following: (i) the electorate, given by the set $V = \{V_1, V_2, \dots, V_n\}$ of n actual voters among n_e potential voters; (ii) the t trustees $T = \{T_1, T_2, \dots, T_t\}$, which hold a share of the master voting key. In order to decrypt a ciphertext encrypted with this key, at least $k \leq t$ trustees must cooperate; (iii) the Certificate Authority, CA; (iv) the Voting

Server, VS, which validates and posts the votes; (v) the Tallying Server, TS, which performs the filtering and counting of votes and (vi) the Public Bulletin Board, PBB, which acts as an append-only public ledger where votes, proofs (to prove re-encryption, shuffle and decryption), and all intermediate steps until the tally are posted. A specification of a possible construction is explained in [13].

The building blocks for the voting protocol are the following.

Elliptic Curve Cryptography (ECC): Let $\mathcal{E}$ denote an elliptic curve defined over a finite field $\mathbb{F}_p$, being p a prime number, and G a generator of $\mathcal{E}$, with prime order, q. Without loss of generality, we assume that $p = q$. The key generation algorithm, $\texttt{KeyGen}(\mathcal{E}, G, q)$ outputs an asymmetric public-private key pair (pk, sk) such that $sk \in_R \mathbb{Z}_q^*$ and $pk = sk \cdot G$. Encryption function $\texttt{Enc}(pk, M)$ takes as input a public key pk and an elliptic curve point $M \in \mathcal{E}$ and returns a ciphertext $\gamma = (\gamma_1, \gamma_2) = (r \cdot G, M + r \cdot pk)$ for $r \in_R \mathbb{Z}_q^*$. The decryption function $\texttt{Dec}(sk, \gamma)$ takes as input a private key sk and a ciphertext γ, and returns the elliptic curve point $M = \gamma - sk \cdot \gamma_1 = M + r \cdot pk - sk \cdot r \cdot G$. Finally, the signature function $\texttt{Sign}(sk, M)$ takes as input a private key sk and a message $M \in \mathcal{E}$, and outputs a signature on M, following [19]. For the sake of simplicity, when using $\texttt{Sign}(sk, M)$, we refer both to the message and the signature of the latter.

The cryptosystem is probabilistic and additively homomorphic, i.e., given two ciphertexts $\texttt{Enc}(pk, M_1)$ and $\texttt{Enc}(pk, M_2)$, then

$$\begin{aligned}\texttt{Enc}(pk, M_1) \oplus \texttt{Enc}(pk, M_2) &= (r_1 G, M_1 + r_1 pk) + (r_2 G; M_2 + r_2 pk)\\ &= ((r_1 + r_2)G, M_1 + M_2 + (r_1 + r_2)pk) = \texttt{Enc}(pk, M_1 \oplus M_2),\end{aligned}$$

where $\oplus$ is the operation defined in ciphertext and plaintext spaces.

Threshold Cryptography: Due to the homomorphic nature of the cryptosystem we use, the voting private key is generated in a distributed manner by T, and thus decryption also demands the collaboration of all trustees. In other words, we impede any single trustee to access any private key [10].

Non-Interactive Zero-Knowledge Proofs (NIZKP): We will use the Fiat-Shamir heuristic [9] (using a hash function modeled as the random oracle) in order to make Zero-Knowledge Proofs non-interactive. Different constructions of NIZKP, to prove correct ballot generation [8], correct decryption of the final result and correct re-encryption of a ciphertext [20] are used.

Anonymous Credentials: Our scheme uses anonymous credentials during the registration phase and vote cast. The only requirement of these credentials is that they certify certain attributes which are used to group voters by electoral colleges and filter votes cast by the same voter [3,5,7].

Verifiable Re-Encryption Shuffles: In order to offer coercion resistance in a private way, we use verifiable shuffles [12]. These allow an entity to produce a shuffle of a list of homomorphic ciphertexts, in such a way that it is infeasible for a computationally bounded adversary to relate ciphertexts in the input list with ciphertexts in the output. Moreover, this is done in a verifiable manner.

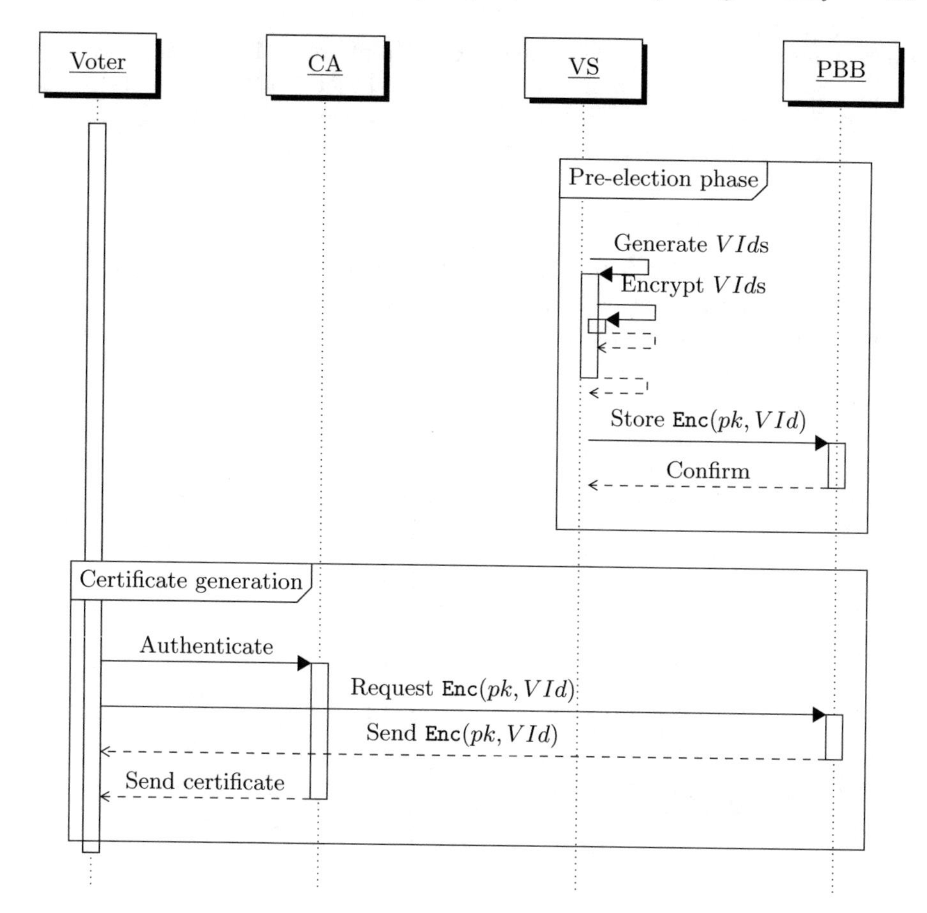

Fig. 1. Pre-election and certificate generation

3 Description of Our Protocol

This section presents an overview of the protocol using Figs. 1 and 2 followed with a more exhaustive explanation of the three parts of our protocol: *pre-election, election* and *tally* phases. Note that the *election* phase is divided into the *certification, voting,* and *verification phase.*

3.1 Pre-election Phase

This phase begins by generating, with $\texttt{KeyGen}(\mathcal{E}, G, q)$, the different keys of the servers. The keys of CA, VS, and TS are generated by each authority. The master voting keypair, (π, σ), is distributively generated amongst the trustees. Each trustee T_h, $1 \leq h \leq t$, ends up with the pair (π, σ_h), being σ_h the share of T_h of key σ. Such protocols allow a set of trustees to compute a public key pair. This public key is directly computed from the different shares of the private key. Note that the 'full' private key is never computed [10]. Then, VS, with its keypair, (pk_{VS}, sk_{VS}), takes as input n_e and randomly generates different voting identifiers, $VId_i \in \mathcal{E}$, $1 \leq i \leq n_e$. It then commits to these values by encrypting all of the VId_is, $VC_i = \texttt{Enc}(pk_{VS}, VId_i)$. These values are published in the PBB together with the signature of VS for VC_i, $\texttt{Sign}(sk_{VS}, VC_i)$.

3.2 Election Phase

This phase comprises all steps that are taken while the election process is open. Note that a voter needs to follow a certification phase for each vote cast, which allows to avoid coercion without a high increase in computational complexity whilst simplifying the task for voters to cast votes multiple times even from different devices.

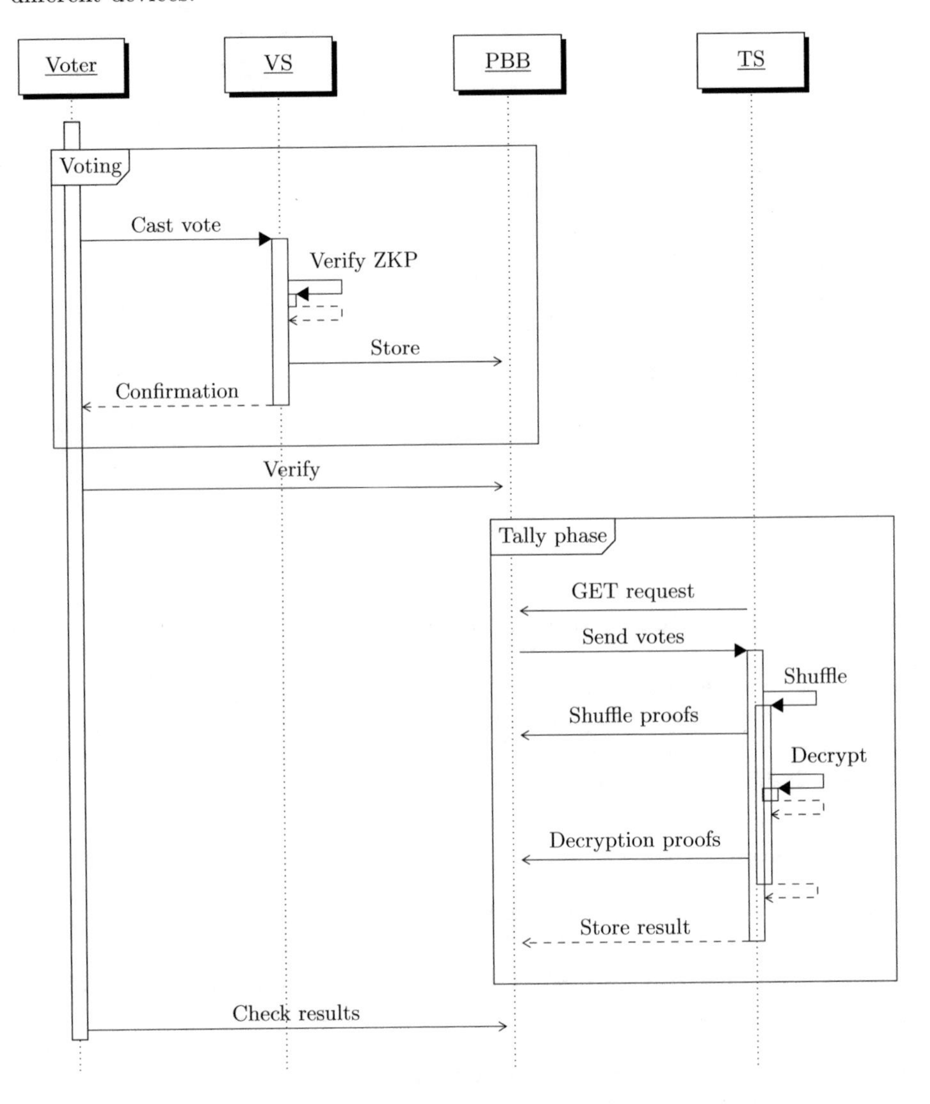

Fig. 2. Vote and tallying phases

Certification: The voter authenticates to the CA, which responds with a one-time use anonymous certificate with a re-encryption, VC'_i, of the respective VC_i. With the certificate, CV_i, CA includes a NIZKP of correct re-encryption of VC_i.

Voting Phase: The cast of the vote involves V_i in a 'regular vote-cast protocol'. The voter selects a candidate, encrypts her choice, v_i, with the master voting public key, π, signs it using the voter private key sk_{CV_i}, and creates a NIZKP, θ_i, proving that the vote has a correct form, together with VC'_i. She then sends it to VS. The latter verifies correctness and publishes this information in the PBB together with VC'_i: $(CV_i, VC'_i, \mathtt{Sign}(sk_{CV_i}, \mathtt{Enc}(\pi, v_i)), \theta_i)$. The elements in the PBB will have assigned a counter, which orders the entries by the time of reception (a counter between 1 and the total number of votes would suffice).

Verification Phase: The voter is now able to check that the vote has indeed been recorded as cast using a search engine in the PBB with a reference of the certificate, which is unique per vote cast. Any third party is able to check that all votes recorded in the PBB come from a certified voter, the VC'_i is related to the certificate, and the votes have a correct format. Note that voters may repeat the protocol until this step as many times as they wish from different devices.

3.3 Tallying Phase

At this point, the election process is closed and all votes from the voters have been stored. There is possibly more than one vote from some voters; hence, before proceeding to the tallying, it is necessary to do the filtering. If the filtering is made in an unprovable manner, high levels of trust is put in the VS.

On the other hand, if all the counters (or timestamps, any reference of the time when the vote was cast) together with each one of the votes are published, coercion could easily happen, as this information could be individually identified. We therefore make use of a proof determining whether $a > b$, with a, b being the counters of the objects in the PBB, without giving any other information of a, b.

Public Filtering: To do the filtering in a secret and provable way, we use the following result which we derive from a work presented in [6]. For an affine point R in the elliptic curve $\mathcal{E}$ and $C_1, C_2 \in \mathbb{Z}^+$ whose expressions in bits are denoted by $C_h^{(j)}$, $1 \leq j \leq k$, $1 \leq h \leq 2$, let

$$A_j = R + \gamma_2^{(1,j)} - \gamma_2^{(2,j)} + \sum_{d=j+1}^{k} 2^d \left(\gamma_2^{(1,d)} - \gamma_2^{(2,d)}\right),$$

$$B_j = \gamma_1^{(1,j)} - \gamma_1^{(2,j)} + \sum_{d=j+1}^{k} 2^d \left(\gamma_1^{(1,d)} - \gamma_1^{(2,d)}\right),$$

$$\gamma^{(i,j)} = (\gamma_1^{(i,j)}, \gamma_2^{(i,j)}) = \mathtt{Enc}(pk, (2 - C_i^{(j)}) \cdot R).$$

Then, C_1 is greater than C_2 if and only if there exists the infinity point of the elliptic curve $v_j = \infty \in \mathcal{E}$ with $v_j = A_j - dB_j$, for some $j \in [1, \lfloor \log_2(\max(C_1, C_2)) \rfloor]$.

This result allows us to go through the filtering without revealing any information to a coercer and without having to trust any entity. The role of the TS is merely for completeness purposes, as it has to add a proof of correctness. We now describe the steps taken in the filtering phase. Once the election is closed, the information displayed in the PBB is:

$$(C_i, CV_i, VC'_i, \mathtt{Sign}(sk_{CV}, \mathtt{Enc}(\pi, (v_i)), \theta_i),$$

where C_i is simply an increasing counter in the PBB. TS begins by removing redundant information for this step. That is, the only information of interest for the filtering are the counters, VC'_i, and $\mathtt{Enc}(\pi, v_i)$ (without the signature). TS, by using its public key, generates another table and publishes it in the PBB:

$$\left(\mathtt{Enc}(pk, C_i^{(j)})_{j=1}^k, \mathtt{Enc}(\pi, v_i), VC'_i\right).$$

The encryption of the bits of the counters is done with randomness zero, to make it easily verifiable. Next step consists in producing a verifiable re-encryption shuffle, where the resulting list of ciphertexts, $\left(\mathtt{Enc}'(pk_{TS}, C_i^{(j)})_{j=1}^k, \mathtt{Enc}'(\pi, v_i), VC''_i\right)$, is published in the PBB. After the mixing, TS requests VS to decrypt the VC''_i including a proof of correct decryption, Θ_i^1, resulting in:

$$\left(\mathtt{Enc}'(pk_{TS}, C_i^{(j)})_{j=1}^k, \mathtt{Enc}'(\pi, v_i), VId_i, \Theta_i^1\right).$$

Now it is possible to proceed to the filtering phase. Note that the TS has access to the counters (it can decrypt each one of them) and therefore, proceeds to do the filtering by comparing VId's. It decrypts each counter (not publicly) and publishes only the last vote cast by each VId_i. It now needs to prove that the process was done in an honest manner for which it adds, for each VId_i, a NIZKP, Θ_l^2, that there exists v_l, as it was mentioned above, that decrypts to ∞ for each of the filtered VId_j with $VId_i = VId_j$ for some $l \in \{1, \ldots, k\}$. Let S denote the subset of the filtered votes and S_i denote the subset of votes having the same VId as VId_i. The TS, after filtering, publishes:

$$\left(\mathtt{Enc}'(pk_{TS}, C_i^{(j)})_{i=1}^k, \mathtt{Enc}'(\pi, v_j), VId_j, \left(\Theta_j^2\right)_{s\in S_j}\right)_{j\in S}.$$

Any third party can compute all values v_l for each pair of counters, as there is the encryption of each of the bits published in the PBB. Together with the proof of correct decryption to ∞, it is universally verifiable that the filtering phase has been done in an honest way, and only the votes with the higher counter (and therefore, the last vote cast by a certain voter) are accepted and proceed to the tallying phase. Furthermore, this result is reached without the need of doing PETs between each pair of votes, as presented in [1] or [18], simplifying the computational cost of the filtering phase asymptotically, as our construction, in comparison to the $\mathcal{O}(N^2)$ complexity of these protocols, uses a filtering that has complexity $\mathcal{O}(n \cdot n_v)$, where n_v is the maximum number of votes cast by a single voter. Moreover, our scheme only requires the interaction of two entities and a single proof of correct decryption for each pair of votes sharing the same VId, while the proper execution of PETs is done in a distributed manner.

Tallying: Finally, the tallying phase happens in a straight forward way for schemes using threshold homomorphic encryption. TS calculates the full encrypted result by adding all ciphertexts accepted after the filtering

$$\mathtt{Enc}(\pi, v_{\mathtt{Final}}) = \sum_{i=1}^{n} \mathtt{Enc}(\pi, v_i) = \mathtt{Enc}\left(\pi, \sum_{i=1}^{n} v_i\right).$$

Then, the result goes through the group of trustees T holding the different shares of the private key, ending with the decrypted result. With this, a proof of correct decryption is added. Any auditor or third party can calculate the sum of all the ciphertexts and verify that the result is a proper decryption of such a product.

4 Conclusion and Future Work

We have presented a protocol which is coercion-resistant, with low requirements for deployment, and with a friendly voting experience. The complexity is also improved in comparison to existing schemes, and, while the assumptions are not optimal in comparison to some other electronic voting schemes (trust is needed in the CA and VS to offer coercion resistance), these allow an organization to employ remote voting elections in a secure, deployable and verifiable manner without the requirement of voters having cryptographic keys. Our current lines of investigation are directed towards a formal proof of the security requirements of our model together with an implementation of the protocol to test performance evaluations of the later.

Acknowledgements. This research has been partially supported by Ministerio de Economía, Industria y Competitividad (MINECO), Agencia Estatal de Investigación (AEI), and Fondo Europeo de Desarrollo Regional (FEDER, UE) under project COP-CIS, reference TIN2017-84844-C2-1-R, and by Comunidad de Madrid (Spain) under project reference P2018/TCS-4566-CM (CYNAMON), also co-funded by European Union FEDER funds.

References

1. Achenbach, D., Kempka, C., Löwe, B., Müller-Quade, J.: Improved coercion-resistant electronic elections through deniable re-voting. USENIX J. Election Technol. Syst. (JETS) **3**(2), 26–45 (2015)
2. Araújo, R., Barki, A., Brunet, S., Traoré, J.: Remote electronic voting can be efficient, verifiable and coercion-resistant. In: Financial Cryptography and Data Security - 2016 International Workshops, BITCOIN, VOTING, and WAHC. Lecture Notes in Computer Science, vol. 9604, pp. 224–232. Springer (2016)
3. Arroyo, D., Diaz, J., Rodriguez, F.B.: Non-conventional digital signatures and their implementations—a review. In: International Joint Conference, pp. 425–435. Springer (2015)

4. Benaloh, J.: Simple verifiable elections. In: Proceedings of the USENIX/Accurate Electronic Voting Technology Workshop 2006 on Electronic Voting Technology Workshop, EVT 2006. USENIX Association, Berkeley, CA, USA (2006)
5. Brands, S.A.: Rethinking Public Key Infrastructures and Digital Certificates: Building in Privacy. MIT Press, Cambridge (2000)
6. Brandt, F.: Efficient cryptographic protocol design based on distributed ElGamal encryption. In: Information Security and Cryptology-ICISC 2005: 8th International Conference, pp. 32–47. Springer (2006)
7. Camenisch, J., Lysyanskaya, A.: A signature scheme with efficient protocols, pp. 268–289. Springer, Heidelberg (2003)
8. Cramer, R., Damgård, I., Schoenmakers, B.: Proofs of partial knowledge and simplified design of witness hiding protocols. In: Advances in Cryptology-CRYPTO 1994, LNCS, pp. 174–187. Springer, Heidelberg (1994)
9. Fiat, A., Shamir, A.: How to prove yourself: practical solutions to identification and signature problems. In: Advances in Cryptology-CRYPTO 1986, Proceedings, Lecture Notes in Computer Science, vol. 263, pp. 186–194, Santa Barbara, California, USA. Springer (1986)
10. Gennaro, R., Jarecki, S., Krawczyk, H., Rabin, T.: Secure distributed key generation for discrete-log based cryptosystems. J. Cryptol. **20**(1), 51–83 (2007)
11. Gjøsteen, K.: Analysis of an internet voting protocol. Technical report, IACR Cryptology ePrint Archive (2010) https://eprint.iacr.org/2010/380.pdf
12. Groth, J.: A verifiable secret shuffle of homomorphic encryptions. J. Cryptol. **23**(4), 546–579 (2010). https://doi.org/10.1007/s00145-010-9067-9
13. Heather, J., Lundin, D.: The append-only web bulletin board. In: Formal Aspects in Security and Trust, pp. 242–256. Springer, Heidelberg (2009)
14. Hernandez-Ardieta, J.L., Gonzalez-Tablas, A.I., De Fuentes, J.M., Ramos, B.: A taxonomy and survey of attacks on digital signatures. Comput. Secur. **34**, 67–112 (2013)
15. Juels, A., Catalano, D., Jakobsson, M.: Coercion-resistant electronic elections. In: Proceedings of the 2005 ACM Workshop on Privacy in the Electronic Society, WPES 2005, pp. 61–70. ACM, New York, NY, USA (2005)
16. Kulyk, O., Volkamer, M.: Efficiency comparison of various approaches in e-voting protocols. In: Proceedings of International Conference on Financial Cryptography and Data Security, pp. 209–223. Springer (2016). https://link.springer.com/chapter/10.1007/978-3-662-53357-4_14
17. Locher, P., Haenni, R.: Verifiable internet elections with everlasting privacy and minimal trust. In: E-Voting and Identity, pp. 74–91. Springer International Publishing (2015)
18. Locher, P., Haenni, R., Koenig, R.E.: Coercion-resistant internet voting with everlasting privacy. In: Financial Cryptography and Data Security, pp. 161–175. Springer, Heidelberg (2016)
19. Pornin, T.: Deterministic Usage of the Digital Signature Algorithm (DSA) and Elliptic Curve Digital Signature Algorithm (ECDSA). RFC 6979, August 2013. https://rfc-editor.org/rfc/rfc6979.txt
20. Schoenmakers, B., Veeningen, M.: Universally verifiable multiparty computation from threshold homomorphic cryptosystems. In: Malkin, T., Kolesnikov, V., Lewko, A.B., Polychronakis, M. (eds.) Applied Cryptography and Network Security, pp. 3–22. Springer International Publishing (2015)
21. Spycher, O., Haenni, R., Dubuis, E.: Coercion-resistant hybrid voting systems. In: Krimmer, R., Grimm, R. (eds.) 4th International Workshop on Electronic Voting, pp. 269–282 (2010)

Linearization of Cryptographic Sequences

Sara D. Cardell[1(✉)] and Amparo Fúster-Sabater[2]

[1] Instituto de Matemática, Estatística e Computação Científica, University of Campinas (UNICAMP), Campinas, Brazil
sdcardell@ime.unicamp.br
[2] Instituto de Tecnologías Físicas y de la Infomación, Consejo Superior de Investigaciones Científicas, Madrid, Spain
amparo@iec.csic.es

Abstract. The generalized self-shrinking generator (or generalized generator) produces binary sequences (generalized sequences) with good cryptographic properties. On the other hand, the binomial sequences can be obtained considering infinite successions of binomial coefficients modulo 2. It is possible to see that the generalized sequences can be computed as a finite binary sum of binomial sequences. Besides, the cryptographic parameters of the generalized sequences can be studied in terms of the binomial sequences.

Keywords: Binary sequences · Binomial coefficients · Generalized sequences

1 Introduction

The Internet of Things (IoT) is one of the buzzwords in computer science and information technology. In the near future, IoT will be used more and more to connect devices of many different types. Some of them use powerful processors and can be expected to perform the same cryptographic algorithms as those of standard PCs. However, many other devices use extremely low power micro-controllers that can hardly devote a small fraction of their computing power to security. It is inside this kind of applications where stream ciphers, the simplest and fastest among all the encryption procedures, play a leading part.

The main concern in stream cipher design is to generate from a short and truly random key a long and pseudorandom sequence called *keystream* sequence. In emission, the sender performs the bitwise XOR (exclusive-OR) operation among the bits of the plaintext and the keystream sequence. The result is the ciphertext that is sent to the receiver. In reception, the receiver generates the same keystream sequence, performs the bitwise XOR operation between ciphertext and keystream sequence and recovers the original plaintext.

Most keystream generators are based on maximal-length Linear Feedback Shift Registers (LFSRs) [9]. Inside the class of LFSR-based cryptographic generators, the irregularly decimated generators are some of the most popular [8,13,14].

© Springer Nature Switzerland AG 2020
F. Martínez Álvarez et al. (Eds.): CISIS 2019/ICEUTE 2019, AISC 951, pp. 165–174, 2020.
https://doi.org/10.1007/978-3-030-20005-3_17

As examples of this family, we can enumerate: (a) the *shrinking generator* [5] that includes two LFSRs, (b) the *self-shrinking generator* [12] involving only one LFSR and (c) the most representative element of this family, the *generalized self-shrinking generator* or family of generators [10], that includes the self-shrinking generator as one of its members. Irregularly decimated generators produce sequences that exhibit good cryptographic properties: long periods, excellent run distribution and self-correlation, balancedness [6], simplicity of implementation, etc. The underlying idea of this type of generators is the irregular decimation of a PN-sequence according to the bits of another. The decimation result is a sequence that will be used as keystream sequence in the encryption/decryption procedure. This work focuses on the generalized self-shrinking generators and their output sequences the so-called *generalized self-shrunken sequences* (GSS sequences) or simply generalized sequences.

On the other hand, the binomial sequences are a family of binary sequences whose terms are binomial numbers reduced modulo 2. More precisely, the binomial sequences correspond to the diagonals of the Sierpinski's triangle modulo 2. In this way, the binomial sequences exhibit many attractive properties that can be very useful in the analysis and generation of keystream sequences. In this work, it is shown the close relationship between generalized sequences and the class of binomial sequences, in the sense that these binomial sequences can be used to analyse the cryptographic properties of those generalized sequences.

The work is organized as follows. In Sect. 2, we introduce the concept of binomial sequence and binomial representation of sequences. Next in Sect. 3, we remind the concept of generalized self-shrinking generator. The representation and computation of generalized sequences in terms of binomial sequences is the subject of Sect. 4. Finally, conclusions in Sect. 5 end the paper.

2 Binomial Sequences

The binomial coefficient $\binom{n}{i}$ is the coefficient of the power x^i in the polynomial expansion of $(1+x)^n$. For every positive integer n, it is a well-known fact that $\binom{n}{0} = 1$ and $\binom{n}{i} = 0$ for $i > n$. Moreover, it is worth noticing that if we arrange these binomial coefficients into rows for successive values of $n = 0, 1, 2, \ldots$, then the generated structure is the Pascal's triangle (see Fig. 1a). On the other hand, if we color the odd numbers of the Pascal's triangle and shade the other ones, we can find the Sierpinski's triangle (see Fig. 1b).

The binomial coefficients reduced modulo 2 allow us to introduce the concept of binomial sequence.

Definition 1. *Given a fixed integer $k \geq 0$, the sequence $\{b_n^k\}_{n\geq 0}$ given by:*

$$b_n^k = \begin{cases} 0 & \text{if } n < k \\ \binom{n}{k} \bmod 2 & \text{if } n \geq k \end{cases}$$

is known as the ***binary k-th binomial sequence****.*

Table 1. Binomial coefficients, binomial sequences, periods and complexities

Coefficient	Sequence	Period	Linear complexity
$\binom{n}{0}$	1 1 1 1 1 1 1 1 1 1 1 1 1 1 1 1 ...	$T_0 = 1$	$LC_0 = 1$
$\binom{n}{1}$	0 1 0 1 0 1 0 1 0 1 0 1 0 1 0 1 ...	$T_1 = 2$	$LC_1 = 2$
$\binom{n}{2}$	0 0 1 1 0 0 1 1 0 0 1 1 0 0 1 1 ...	$T_2 = 4$	$LC_2 = 3$
$\binom{n}{3}$	0 0 0 1 0 0 0 1 0 0 0 1 0 0 0 1 ...	$T_3 = 4$	$LC_3 = 4$
$\binom{n}{4}$	0 0 0 0 1 1 1 1 0 0 0 0 1 1 1 1 ...	$T_4 = 8$	$LC_4 = 5$
$\binom{n}{5}$	0 0 0 0 0 1 0 1 0 0 0 0 0 1 0 1 ...	$T_5 = 8$	$LC_5 = 6$
$\binom{n}{6}$	0 0 0 0 0 0 1 1 0 0 0 0 0 0 1 1 ...	$T_6 = 8$	$LC_6 = 7$
$\binom{n}{7}$	0 0 0 0 0 0 0 1 0 0 0 0 0 0 0 1 ...	$T_7 = 8$	$LC_7 = 8$

Table 1 shows the first bits of the binomial sequences and the values of their periods and linear complexities, denoted by T_k and LC_k, respectively, for the first binomial coefficients $\binom{n}{k}$, $k = 0, 1, \ldots, 7$.

Recall that the successive binomial sequences correspond to the successive diagonals of the Sierpinski's triangle (see Fig. 1b) reduced modulo 2.

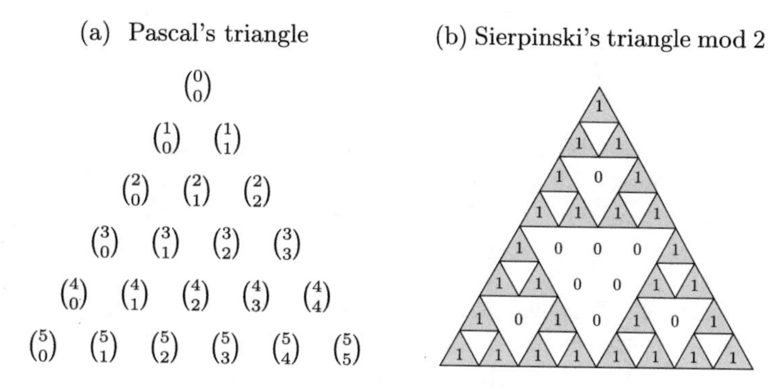

Fig. 1. Pascal's triangle and Sierpinski's triangle

Next, the relation between binomial sequences and every binary sequence with period a power of 2 appears in the following result.

Theorem 1. *Let $\{a_n\}$ be a binary sequence with period 2^L, with L a positive integer. Then, $\{a_n\}$ can be written as a linear combination of binomial sequences.*

Proof. Since the period of $\{a_n\}$ is a power of 2, then the next difference equation holds:

$$(E^{2^L} + 1)a_n = (E + 1)^{2^L} a_n = 0,$$

where E is the shifting operator that acts on the terms of a sequence $\{a_n\}$, that is: $E^k a_n = a_{n+k}$ for all integer $k \geq 0$.

Algorithm 1. Constructing the binomial representation of a given sequence

Input:
$\boldsymbol{s}$: Sequence of period 2^L

```
01: r = {0 0 0 0 ... };
02: n = length(v);
03: for j = 0 to n − 1 do
04:    if s_j ≠ r_j then
05:       r = r + (n choose j);
06:    endif
07: endfor
```

Output:
$\boldsymbol{r}$: Binomial representation of the intercepted bits

The characteristic polynomial of the previous equation is $(x+1)^m$ with $m = 2^L$, that is $x = 1$ is the unique root of the polynomial $(x+1)$ but with multiplicity m. Therefore, the binary solutions of this equation are given [8] by the expression:

$$a_n = c_0\binom{n}{0} + c_1\binom{n}{1} + \cdots + c_{T-1}\binom{n}{T-1} \quad \text{for } n \geq 0, \tag{1}$$

where the coefficients $c_i \in \mathbb{F}_2$ and $\binom{n}{i}$ are binomial coefficients reduced modulo 2. When n takes successive values $n = 0, 1, 2, \ldots$, then each binomial coefficient modulo 2 defines a different binomial sequence. Thus, the sequence $\{a_n\}$ is just the bit-wise XOR of such binomial sequences weighted by binary coefficients c_i. □

Different choices of c_i will produce different sequences $\{a_n\}$ with distinct characteristics and properties, but all of them with period 2^l, $0 \leq l \leq L$.

According to Theorem 1, a binary sequence of period power of 2 is the bit-wise XOR of binomial sequences. Therefore, we introduce the following definition.

Definition 2. *The set of binomial sequences necessary to obtain a binary sequence of period power of* 2 *is called the* ***binomial representation*** *of such a sequence.*

Given a sequence $\boldsymbol{s}$ of period 2^L, Algorithm 1 introduces a method to recover the binomial representation of such a sequence. In round j, the algorithm compares $\boldsymbol{s}_j$ with the corresponding bit in the sequence represented by $\sum_{i=0}^{j}\binom{n}{i}$. If they match, then $\binom{n}{j}$ is part of the binomial representation. Otherwise, the term $\binom{n}{j}$ is discarded and the algorithm continues. This method is based on the fact that the first j bits of the sequence represented by $\binom{n}{j}$ are 0s.

Let us introduce now an illustrative example.

Example 1. Consider the sequence $\boldsymbol{s} = \{1\ 1\ 0\ 1\ \ldots\}$ of period 4. The first two bits ($\boldsymbol{s}_0 = 1$ and $\boldsymbol{s}_1 = 1$) match with the first two bits of the sequence $\binom{n}{0} = \{\mathbf{1}\ \mathbf{1}\ 1\ 1\ \ldots\}$. This means that one of the binomial representations of the sequence starts with $\sum_{i=0}^{1} c_i \binom{n}{i} = \binom{n}{0}$ $(c_0 = 1, c_1 = 0)$.

The bit $\boldsymbol{s}_2 = 0$ matches with the corresponding bit of the sequence $\binom{n}{0} + \binom{n}{2}$:

$$\begin{array}{l} \binom{n}{0} : 1\,1\,1\,1\,1\,1\,1\,1\ldots \\ \binom{n}{2} : 0\,0\,1\,1\,0\,0\,1\,1\ldots \\ \hline \quad\;\; 1\,1\,\mathbf{0} \end{array}$$

Then, the binomial representation we are considering starts with $\sum_{i=0}^{2} c_i \binom{n}{i} = \binom{n}{0} + \binom{n}{2}$ $(c_0 = 1, c_1 = 0, c_2 = 1)$.

Finally, the bits $\boldsymbol{s}_3 = 1$ match with the corresponding bits of the sequence $\binom{n}{0} + \binom{n}{2} + \binom{n}{3}$:

$$\begin{array}{l} \binom{n}{0} : 1\,1\,1\,1\,1\,1\,1\,1\ldots \\ \binom{n}{2} : 0\,0\,1\,1\,0\,0\,1\,1\ldots \\ \binom{n}{3} : 0\,0\,0\,1\,0\,0\,0\,1\ldots \\ \hline \quad\;\; 1\,1\,0\,\mathbf{1}\,\mathbf{1}\,\mathbf{1}\,\mathbf{0}\,\mathbf{1} \end{array}$$

Therefore, the binomial representation of our sequence is $\binom{n}{0} + \binom{n}{2} + \binom{n}{3}$. ■

Several cryptographic generators produce sequences with periods that are powers of 2 [4,10–12]. Next, we study on of the most representative element in the class of irregularly decimated generators, the generalized self-shrinking generator [10].

3 The Generalized Self-shrinking Generator

The generalized self-shrinking generator is a particularization of the shrinking generator [3,5] that includes the sequences generated by: the self-shrinking generator [1,12], the modified self-shrinking generator [2,11] and the t-modified self-shrinking generator [4]. The family of generalized sequences is described as follows:

Definition 3. *Let $\{a_n\}$ $(n = 0, 1, 2, \ldots)$ be a PN-sequence generated by a maximal-length LFSR with an L-degree characteristic polynomial. Let t be an integer and $\{v_n\}$ $(n = 0, 1, 2, \ldots)$ be an t-position left shifted version of $\{a_n\}$ with $(t = 0, 1, 2, \ldots, 2^L - 2)$. The decimation rule is very simple:*

1. *If $a_n = 1$, then v_n is output.*
2. *If $a_n = 0$, then v_n is discarded and there is no output bit.*

Thus, for each t an output sequence $\{s_0\, s_1\, s_2 \ldots\}$ denoted by $\{S(t)_n\}$ $(n \geq 0)$ is generated. Such a sequence is called the generalized self-shrunken sequence (GSS-sequence) (or simply generalized sequence) associated with the shift t.

Table 2. GSS-sequences for $p(x) = 1 + x + x^4$

p	α^p	$\{v_n\}$ sequences	GSS-sequence
0	1	111100010011010	11111111
1	α	111000100110101	11100100
2	α^2	110001001101011	11000011
3	α^3	100010011010111	10001101
4	$1+\alpha$	000100110101111	00011011
5	$\alpha+\alpha^2$	001001101011110	00100111
6	$\alpha^2+\alpha^3$	010011010111100	01001110
7	$1+\alpha+\alpha^3$	100110101111000	10010110
8	$1+\alpha^2$	001101011110001	00111100
9	$\alpha+\alpha^3$	011010111100010	01101001
10	$1+\alpha+\alpha^2$	110101111000100	11011000
11	$\alpha+\alpha^2+\alpha^3$	101011110001001	10101010
12	$1+\alpha+\alpha^2+\alpha^3$	010111100010011	01010101
13	$1+\alpha^2+\alpha^3$	101111000100110	10110001
14	$1+\alpha^3$	011110001001101	01110010
		111100010011010	

Recall that $\{a_n\}$ remains fixed while $\{v_n\}$ is the sliding sequence or left-shifted version of $\{a_n\}$. When t ranges in the interval $t \in [0, 1, 2, \ldots, 2^L - 2]$, then the family of $2^L - 1$ generalized sequences is obtained. For each possible sequence $\{v_n\}$ and after the application of the decimation rule, a new generalized sequence is generated. The GSS-sequence family includes the $2^L - 1$ generalized self-shrunken sequences plus the identically null sequence. Some important facts extracted from [7,10] are enumerated:

- This family always includes [7] the sequence $\{111111\ldots\}$ for $t = 0$ and the sequences $\{101010\ldots\}$ and $\{010101\ldots\}$ for $t = m, m+1$, respectively, where m is an integer corresponding to the power $\alpha^m \in \mathbb{F}_{2^L}$ satisfying $\alpha^{m+1} = \alpha^m + 1$.
- All the sequences in this family are balanced except for sequences $\{0000\ldots\}$ and $\{1111\ldots\}$, [10, Theorem 1].
- By construction, the family of generalized self-shrinking sequences consists of 2^L sequences of 2^{L-1} bits each of them [10, Section I]. Consequently, the period of each one of these sequences is a factor of 2^{L-1}.
- The family of generalized self-shrinking sequences has structure of Abelian group whose group operation is the bit-wise addition mod 2, the neutral element is the sequence $\{0000\ldots\}$ and the inverse element of each sequence is the own sequence, [10, Theorem 2].
- The self-shrinking sequence is a member of the GSS-sequence family [10, Section I] with shift $t = 2^{L-1}$.

Example 2. For an LFSR with characteristic polynomial $p(x) = 1 + x + x^4$ and initial state $\{1\ 1\ 1\ 1\}$, we get the generalized sequences (GSS-sequences) shown in Table 2. The bits in bold in the different sequences $\{v_n\}$ are the digits of the corresponding GSS-sequences associated to their corresponding p. The PN-sequence $\{a_n\}$ with period $T = 2^4 - 1$ is written at the bottom of the table. Note that the sequences corresponding to $t = 2, 8$ are the same but starting at different terms. The same holds for the sequences corresponding to $t = 1, 5, 6, 14$ and $t = 3, 4, 10, 13$. In brief, inside the family of generalized sequences there are shifted versions of the same sequence. ■

Since the generalized sequences have periods that are powers of 2, then Theorem 1 can be directly applied to the GSS-sequence family. In this way, every generalized sequence can be written as a linear combination of binomial sequences as shown in the Eq. (1).

4 Binomial Representation of Generalized Sequences

In this section, we analyze the binomial representation of the different sequences in the class of GSS-sequences. In order to do that, we study the additive subgroups of $(\mathbb{F}_{2^L}, +)$, which will help us to divide the set of generalized sequences associated to a given polynomial into separated groups.

Let $\mathbb{F}_{2^L}$ be an additive group, then we can construct the group isomorphism:

$$\begin{aligned} \Phi : (\mathbb{F}_{2^L}, +) &\longrightarrow (\mathcal{S}, +) \\ \alpha^t &\longrightarrow \{S(t)\}, \end{aligned} \tag{2}$$

which defines a relationship between α^t (α being a root of $p(x)$, the LFSR characteristic polynomial) and the generalized sequence $\{S(t)\}$, see Table 2 for the isomorphism (2) corresponding to the polynomial $p(x) = 1 + x + x^4$.

Now, we are going to write the most simple sequences of the generalized sequence family in terms of their binomial representations. In fact, we have seen in Sect. 3 that there exists an integer $m \in \{0, 1, \ldots, 2^{L-2}\}$ such that $\alpha^{m+1} = 1 + \alpha^m$. The generalized sequences corresponding to α^m and α^{m+1} are $\{1\ 0\ 1\ 0\ 1\ 0\ 1\ 0 \ldots\}$ and $\{0\ 1\ 0\ 1\ 0\ 1\ 0\ 1 \ldots\}$ and their binomial representations are $\binom{n}{0} + \binom{n}{1}$ and $\binom{n}{1}$, respectively. In the same way, for α^0 its corresponding generalized sequence is $\{1\ 1\ 1\ 1\ 1\ 1\ 1\ 1 \ldots\}$ with binomial representation $\binom{n}{0}$. Coming back to Example 2, we have now that $m = 11$ and consequently

$$\Phi(\alpha^{11}) = \{S(11)\} = \{1\ 0\ 1\ 0\ 1\ 0\ 1\ 0 \ldots\} = \binom{n}{0} + \binom{n}{1}$$

$$\Phi(\alpha^{12}) = \{S(12)\} = \{0\ 1\ 0\ 1\ 0\ 1\ 0\ 1 \ldots\} = \binom{n}{1}$$

$$\Phi(\alpha^{0}) = \{S(0)\} = \{1\ 1\ 1\ 1\ 1\ 1\ 1\ 1 \ldots\} = \binom{n}{0}$$

These three basic binomial representations allow one to determine the binomial representation for the rest of generalized sequences.

Table 3. Binomial representation of the GSS-sequences $\{S(p)\}$ of Example 2

p	$\{S(p)\}$	Binomial representation
0	1 1 1 1 1 1 1 1	$\binom{n}{0}$
1	1 1 1 0 0 1 0 0	$\binom{n}{0}+\binom{n}{3}+\binom{n}{4}+\binom{n}{5}$
2	1 1 0 0 0 0 1 1	$\binom{n}{0}+\binom{n}{2}+\binom{n}{4}$
3	1 0 0 0 1 1 0 1	$\binom{n}{0}+\binom{n}{1}+\binom{n}{2}+\binom{n}{3}+\binom{n}{5}$
4	0 0 0 1 1 0 1 1	$\binom{n}{3}+\binom{n}{4}+\binom{n}{5}$
5	0 0 1 0 0 1 1 1	$\binom{n}{2}+\binom{n}{3}+\binom{n}{5}$
6	0 1 0 0 1 1 1 0	$\binom{n}{1}+\binom{n}{3}+\binom{n}{4}+\binom{n}{5}$
7	1 0 0 1 0 1 1 0	$\binom{n}{0}+\binom{n}{1}+\binom{n}{2}+\binom{n}{4}$
8	0 0 1 1 1 1 0 0	$\binom{n}{2}+\binom{n}{4}$
9	0 1 1 0 1 0 0 1	$\binom{n}{1}+\binom{n}{2}+\binom{n}{4}$
10	1 1 0 1 1 0 0 0	$\binom{n}{0}+\binom{n}{2}+\binom{n}{3}+\binom{n}{5}$
11	1 0 1 0 1 0 1 0	$\binom{n}{0}+\binom{n}{1}$
12	0 1 0 1 0 1 0 1	$\binom{n}{1}$
13	1 0 1 1 0 0 0 1	$\binom{n}{0}+\binom{n}{1}+\binom{n}{3}+\binom{n}{4}+\binom{n}{5}$
14	0 1 1 1 0 0 1 0	$\binom{n}{1}+\binom{n}{2}+\binom{n}{3}+\binom{n}{5}$

Next, we continue considering the additive subgroup of $\mathbb{F}_{2^L}$ of order 4 given by:

$$\mathcal{S}_4 = \{0, 1, \alpha^m, \alpha^{m+1}\}.$$

If we denote the zero sequence by $\mathbf{0}$, the corresponding sequences through the isomorphism defined in Eq. (2) are:

$$\Phi(\mathcal{S}_4) = \left\{\mathbf{0}, \binom{n}{0}, \binom{n}{1}, \binom{n}{0} + \binom{n}{1}\right\}$$

Consider now every subgroup of order 8 that contains $\mathcal{S}_4$. They are of the form

$$\{0, 1, \alpha^m, \alpha^{m+1}, \alpha^{m_1}, \alpha^{m_2}, \alpha^{m_3}, \alpha^{m_4}\},$$

such that,

$$\begin{aligned} \alpha^{m_1} + \alpha^{m_2} &= 1, \\ \alpha^{m_1} + \alpha^{m_3} &= \alpha^m, \\ \alpha^{m_1} + \alpha^{m_4} &= \alpha^{m+1}. \end{aligned}$$

Clearly, the binomial representation of the corresponding sequences will have a common part $\Lambda = \sum_{i=2}^{2^{L-1}-(L-2)} a_i\binom{n}{i}$, with $a_i \in \mathbb{F}_2$ and they will have the form:

$$\left\{\mathbf{0}, \binom{n}{0}, \binom{n}{1}, \binom{n}{0} + \binom{n}{1}, \Lambda, \Lambda + \binom{n}{0}, \Lambda + \binom{n}{1}, \Lambda + \binom{n}{0} + \binom{n}{1}\right\}$$

Example 3. Consider again the Example 2. In Table 3, we can see the binomial representation of each GSS-sequence produced by $p(x) = 1+x+x^4$. For instance, consider the four sequences in bold:

$$\left\{\binom{n}{5}+\binom{n}{4}+\binom{n}{3},\binom{n}{5}+\binom{n}{4}+\binom{n}{3}+\binom{n}{0},\binom{n}{5}+\binom{n}{4}+\binom{n}{3}+\binom{n}{1},\right.$$
$$\left.\binom{n}{5}+\binom{n}{4}+\binom{n}{3}+\binom{n}{1}+\binom{n}{0}\right\}$$

Their common part is $\Lambda = \binom{n}{5}+\binom{n}{4}+\binom{n}{3}$ and they correspond to α, α^4, α^6 and α^{13}. In this case, $m = 11$ and the corresponding additive subgroup is given by: $\{0, 1, \alpha^{11}, \alpha^{12}, \alpha, \alpha^4, \alpha^6, \alpha^{13}\}$. ■

As a consequence of what we have seen before, we can divide the set of GSS-sequences into sets of four elements each of them sharing the same common part. Therefore, given a primitive polynomial of degree L, we can generate 2^L GSS-sequences that can be divided into 2^{L-2} groups (including the trivial group $\{\mathbf{0}, \binom{n}{0}, \binom{n}{1}, \binom{n}{0}+\binom{n}{1}\}$).

In addition, it is possible to determine the linear complexity and period of the different generalized sequences in terms of their binomial representations.

The linear complexity of a generalized sequence is $LC = k+1$, where k is the greatest integer of the sequence $\binom{n}{k}$ in the binomial representation of the generalized sequence.

The period of a generalized sequence is the period of the binomial sequence $\binom{n}{k}$ where k is the greatest integer in the binomial representation of the generalized sequence. In Example 2, for sequences in bold we have linear complexity $LC = 6$ and period $T = 8$.

For the class of generalized sequences associated to a characteristic polynomial $p(x)$, it can be checked that there are $L-2$ different linear complexities LC_i among the generalized sequences of the same family (apart from the trivial ones $0, 1, 2$) satisfying:

$$LC_1 > LC_2 > \cdots > LC_{L-3} > LC_{L-2},$$

with $LC_i > 2^{L-2}$ [10]. Indeed, there are $2^{L-(i+2)}$ groups (containing four sequences each of them) with linear complexity LC_i. Let us consider the primitive polynomial $p(x) = 1+x+x^6$ of degree 6. Such a polynomial generates 64 GSS-sequences divided into 16 groups of 4 sequences each group: 8 groups with $LC_1 = 28$, 4 groups with $LC_2 = 27$, 2 groups with $LC_3 = 26$, 1 group with $LC_4 = 25$ and the group $\{\mathbf{0}, \binom{n}{0}, \binom{n}{1}, \binom{n}{0}+\binom{n}{1}\}$.

5 Conclusions

The binomial sequences are binary sequences able to construct all the binary sequences whose periods are powers of 2; therefore, they are able to construct the generalized sequences. Recall that the generalized sequences are binary sequences that exhibit good cryptographic properties such as long periods and large linear complexity. The cryptographic parameters of these sequences can be determined via the binomial sequences. At the same time, a family of generalized sequences can be analyzed and partitioned in terms of its binomial representation. Moreover, the binomial sequences allow one to generate the whole generalized family from a minimum number of its elements.

Acknowledgements. Research partially supported by Ministerio de Economía, Industria y Competitividad, Agencia Estatal de Investigación, and Fondo Europeo de Desarrollo Regional (FEDER, UE) under project COPCIS (TIN2017-84844-C2-1-R) and by Comunidad de Madrid (Spain) under project CYNAMON (P2018/TCS-4566), also co-funded by European Union FEDER funds. The first author was supported by CAPES (Brazil).

References

1. Cardell, S.D., Fúster-Sabater, A.: Linear models for the self-shrinking generator based on CA. J. Cell. Automata **11**(2–3), 195–211 (2016)
2. Cardell, S.D., Fúster-Sabater, A.: Recovering the MSS-sequence via CA. Procedia Comput. Sci. **80**, 599–606 (2016)
3. Cardell, S.D., Fúster-Sabater, A.: Modelling the shrinking generator in terms of linear CA. Adv. Math. Commun. **10**(4), 797–809 (2016)
4. Cardell, S.D., Fúster-Sabater, A.: The t-Modified self-shrinking generator. In: Shi, Y., et al. (eds.) ICCS 2018. Lecture Notes in Computer Science, vol. 10860, pp. 653–663. Springer, Cham (2018)
5. Coppersmith, D., Krawczyk, H., Mansour, Y.: The shrinking generator. In: Proceedings of CRYPTO 1993, Lecture Notes in Computer Science, vol. 773, pp. 22–39. Springer (1994)
6. Fúster-Sabater, A., García-Mochales, P.: A simple computational model for acceptance/rejection of binary sequence generators. Appl. Math. Model. **31**(8), 1548–1558 (2007)
7. Fúster-Sabater, A., Caballero-Gil, P.: Chaotic modelling of the generalized self-shrinking generator. Appl. Soft Comput. **11**(2), 1876–1880 (2011)
8. Fúster-Sabater, A.: Generation of cryptographic sequences by means of difference equations. Appl. Math. Inf. Sci. **8**(2), 1–10 (2014)
9. Golomb, S.W.: Shift Register-Sequences. Aegean Park Press, Laguna Hill (1982)
10. Hu, Y., Xiao, G.: Generalized self-shrinking generator. IEEE Trans. Inf. Theor. **50**(4), 714–719 (2004)
11. Kanso, A.: Modified self-shrinking generator. Comput. Electr. Eng. **36**(1), 993–1001 (2010)
12. Meier, W., Staffelbach, O.: The self-shrinking generator. In: Cachin, C., Camenisch, J. (eds.) Advances in Cryptology – EUROCRYPT 1994, Lecture Notes in Computer Science, vol. 950, pp. 205–214. Springer (1994)
13. Menezes, A.J., et al.: Handbook of Applied Cryptography. CRC Press, New York (1997)
14. Paar, C., Pelzl, J.: Understanding Cryptography. Springer, Berlin (2010)

On-the-Fly Testing an Implementation of Arrow Lightweight PRNG Using a LabVIEW Framework

Alfonso Blanco Blanco, Amalia Beatriz Orúe López(✉), Agustin Martín Muñoz, Victor Gayoso Martínez, Luis Hernández Encinas, Oscar Martínez-Graullera, and Fausto Montoya Vitini

Institute of Physical and Information Technologies (ITEFI), Spanish National Research Council (CSIC), C/ Serrano 144, Madrid, Spain
{alfonso,amalia.orue,agustin,victor.gayoso,luis,fausto}@iec.csic.es
oscar.martinez.graullera@csic.es

Abstract. This work proposes a LabVIEW framework suitable for simulating and on-the-fly testing a hardware implementation of the Arrow lightweight pseudorandom generator. Its aim is twofold. The first objective is to provide a framework to simulate the pseudorandom generator behavior in a personal computer, allowing to modify dynamically the configuration parameters of the generator. Moreover, to visualize the randomness of the output sequences useful techniques like the chaos game and return maps are used. The second objective is to generate an architecture implementing the Arrow algorithm which can be downloaded into a real Complex Programmable Logic Device or a Field-Programmable Gate Array. Plots are shown which demonstrate the usefulness of the proposed framework.

Keywords: Lightweight random number generation · Software tools · Hardware simulation · Internet of Things

1 Introduction

It is widely agreed that Internet of Things (IoT) has become an important technology which promotes the creation of systems by interconnecting ubiquitous and highly heterogeneous networked objects –things– such as sensors, actuators, smartphones, etc. It has applications in many fields, from industrial developments to users' daily life (smart home objects). An interesting example of technological development linked to the evolution of IoT is e-health, including remote patient monitoring through body sensors networks which collect data of different relevance as heart activity data or the body temperature [1,2].

The pervasive use of IoT in such sensitive fields requires IoT to address new security challenges [3,4]. Many of the devices typically employed in IoT have limited resources in terms of memory and computational capabilities and, thus, common cryptographic standards are not suitable [5, Sect. 4.1]. Indeed, the

© Springer Nature Switzerland AG 2020
F. Martínez Álvarez et al. (Eds.): CISIS 2019/ICEUTE 2019, AISC 951, pp. 175–184, 2020.
https://doi.org/10.1007/978-3-030-20005-3_18

ISO/29192 standards aim to provide lightweight cryptography for constrained devices, including block and stream ciphers, hash functions, random number generators, and asymmetric mechanisms, for securing IoT [6,7].

The European Union Agency for Network and Information Security (ENISA) recommends to use hardware that incorporates security features to strengthen the protection and integrity of devices in IoT environments, for example, specialized security chips/coprocessors that integrate security in the own processor, that provide, among other features, encryption and random number generators (RNGs) [8].

Many solutions have been proposed to generate on-board pseudo RNG (PRNG) to secure Radio Frequency Identification (RFID) and wireless sensor networks systems [6]. Among the wide portfolio of existing PRNG, Orúe *et al.* have proposed Arrow, a lightweight PRNG with a hardware complexity that fits IoT demands and that has demonstrated a good performance on Atmel 8-bit AVR and Intel Curie™ 32-bit microcontrollers [9].

This work proposes a framework suitable for simulating a hardware implementation of Arrow PRNG in a PC using LabVIEW. The LabVIEW simulation allows to generate an Arrow architecture to integrate it on a real Field-Programmable Gate Array (FPGA) or on a Complex Programmable Logic Device (CPLD), which are suitable for IoT. The simulation framework allows to adjust dynamically the parameters of the PRNG and on-the-fly testing the randomness of the output sequences by means of techniques like the chaos game or the return map.

The rest of the paper is organized as follows. In Sect. 2 related work about the implementation and testing of RNGs in hardware devices like FPGAs is reviewed. Sect. 3 briefly describes the characteristics of the generator under study. The proposed framework is explained in Sect. 4, highlighting its main characteristics and showing some figures which demonstrate its utility. Finally, conclusions are presented in Sect. 5.

2 Related Work

Different contributions have been published proposing the design of True RNG (TRNG) and PRNG, some of them implemented in hardware devices such as FPGAs (see [10] for the main characteristics of the PRNG and TRNG).

In order to evaluate their design quality and their implementation, it is important to use several statistical tests such as the SP 800-22Rev.1a [11] and Diehard statistical tests [12], which help us to certify the randomness of the sequences of the generator; however, note that there is no set of complete statistical tests that gives this work finished. For cryptographic applications, it is imperative to analyze the unpredictability of the generated sequence and a study of the state of the art of the attacks to this type of cryptographic primitives.

Santoro *et al.* pointed out in [13] that, usually, when an RNG is evaluated, designers put a huge bit stream into memory and then submit it to software tests. If the bit stream successfully passes a certain number of statistical tests, the RNG is said to be sufficiently random. These tests are specifically designed for deterministic generators (PRNGs). In order to evaluate a TRNG under the same conditions, the generators must be implemented in the same chip. Since TRNG analysis via statistical tests software is difficult and time-consuming, Santoro *et al.* proposed a methodology to estimate TRNG quality by using a hardware version of a proven and recognized suite of statistical tests.

Bhaskar and Gawande showed, in [14], an implementation of a PRNG based on Blum Blum Shub, XOR Shift, Fibonacci series, and Galois LFSR methods, where the design was specified in VHDL (Very High Speed Integrated Circuit Hardware Description Language) and was implemented on an Altera FPGA device. The authors demonstrated how the introduction of application specificity in the architecture could deliver huge performance in terms of area and speed, but they did not provide information about how they assessed the randomness of the obtained output sequences.

Random number generation of Linear Feedback Shift Registers (LFSR) based stream encryption algorithms and their hardware implementations were presented in [15]. In that instance, LFSR-based stream encryption algorithms were implemented on an Altera's FPGA development board using the VHDL language. The quality of the obtained random numbers was assessed by using the NIST statistical tests.

In [16], an efficient implementation of a PRNG based on Microcontroller Unit internal and external resources was proposed, so it could be used in cryptosystems with small size memory that do not need high levels of consumption or any additional hardware, which makes it suitable for IoT. The proposed approach was implemented on a new development kit made by Texas Instruments and was tested using the NIST statistical test suite.

In [17], a comprehensive survey on well-known PRNGs and TRNGs implemented on FPGAs including extensive technical details about their implementations and definitions is presented. The authors made a comparison between the different hardware implementations analyzed taking into account both the speed and the area occupied by them. They also analyzed the results obtained when applying standard test batteries such as the NIST and Diehard statistical tests for each of the cases studied.

It is important to point out that, unlike the works reviewed, our framework allows to quickly perform different statistical tests for different values of the parameters, thus guaranteeing the use of the appropriate ones before generating the code that will be embedded into the selected device.

3 A Description of the Generator Under Study: Arrow PRNG

The structure of Arrow [9], consists of two coupled Lagged Fibonacci Generators (LFG), which are mutually scrambled through the perturbation of the most and least significant bits of the lagged samples, before the sum of their outputs module m. The generator is defined by the following equations:

$$x_n = \left(x_{n-r_1} \oplus (y_{n-s_2} \ll d_1) + x_{n-s_1} \oplus (y_{n-r_2} \gg d_3)\right) \mod m,$$
$$y_n = \left(y_{n-r_2} \oplus (x_{n-s_1} \ll d_2) + y_{n-s_2} \oplus (x_{n-r_1} \gg d_4)\right) \mod m,$$

where d_1, d_2, d_3, d_4 are constants, $0 < d_i < N$; $\oplus$ is the bitwise exclusive-or; $\gg$ and $\ll$ are the right-shift and left-shift operators in the C/C++ language.

The output sequence of Arrow is $w_n = x_n \oplus y_n$, were w_n is the output sample of Arrow generator at the moment n and x_n, y_n are the output samples of the two generators, respectively. Arrow is suitable to secure the great majority of IoT applications, like those used in smart cards, RFID tags, and wireless sensor nodes. A scheme of Arrow is shown in the screenshot of the developed framework (see Fig. 1).

As explained in [9], the best results for Arrow PRNG were reached when the cross perturbation is achieved with only half the bit size of the samples, $d_i = N/2$, N being the word length. The sample x_{n-r_1} is right-shifted d_4 bits and then it is combined with y_{n-s_2} by means of a bitwise exclusive-or, while the sample x_{n-s_1} is left-shifted d_3 bits and then it is combined with y_{n-r_2}, also using a bitwise exclusive-or. The samples y_{n-r_2} and y_{n-s_2} are perturbed similarly.

The architecture of Arrow can be built with simple hardware or software and it is flexible enough to allow an assortment of implementations with different word lengths on different platforms. In [9], the Arrow algorithm was first implemented using C++, to verify its correct operation, and then it was implemented on two different platforms: an Arduino UNO –a low-power 8-bit microcontroller– and an Arduino 101, based on a 32-bit Intel Quark™ microcontroller.

4 LabVIEW Framework

The general objective of this work is to generate a flexible set of tools to analyze, check, optimize, and implement a PRNG algorithm on different hardware architectures. Then, it is possible to port it on specific hardware and to select the most suitable architecture, test the implementation, monitor and choose the parameters iteratively, verifying that the deployed architecture fulfills the requirements. Once the optimum configuration has been defined, the LabVIEW framework allows us to choose and generate an architecture which includes the correct parameters defined in the previous step and which is going to be used in a specific hardware context through the selected device programming environment.

Natively, LabVIEW is a dataflow-oriented language instead of an instruction-flow language and allows to perform multiple parallel tasks simultaneously. It is a natural approximation to a hardware/physical realization. The proposed framework also takes advantage of the state machines [18], queue and FIFO offered by LabVIEW.

With the LabVIEW platform, the programming is basically the same for any target. Thus, we have developed and then deployed our design using desktop/Windows-based PCs and an FPGA. First, the algorithm of Arrow generator was simulated in the LabVIEW-PC (running on PC's CPU). Then, a framework was designed with a set of utilities to evaluate the randomness of the output sequences of Arrow. Finally, we have implemented the algorithm in the Labview-FPGA (running on an FPGA). From the implementation in the FPGA, the quality of the sequences of the generator is also checked using the aforementioned framework.

As mentioned above, the objective of the framework is to simulate the behavior of Arrow PRNG, assessing in real time whether any bias can be observed in the output sequences, without needing to obtain a sequence long enough to use the common randomness statistical tests.

4.1 Real-Time Simulation

A front panel simulating the Arrow PRNG structure that allows to modify the configuration parameters interactively has been generated with LabVIEW (see Fig. 1). Additionally, the framework includes the graphs of the time series, the FFT and Walsh transforms, the chaos game, the return map, and the autocorrelation function, which allows a user to quickly –on the fly– evaluate the quality of the generator output sequence for the parameters shown in the configuration tables.

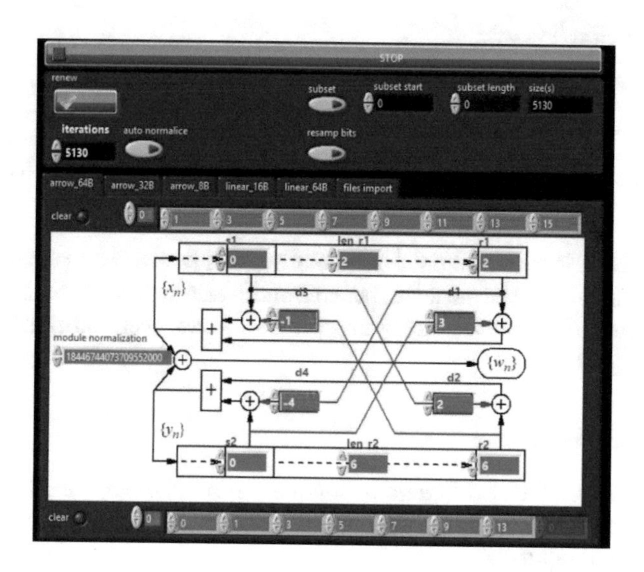

Fig. 1. Screenshot of the front panel of the PRNG Arrow simulation.

As it can be seen in Fig. 1, the simulated framework incorporates hardware implementations of Arrow for 64-, 32-, and 8-bit words, as well as implementations of 64- and 16-bit word linear generators, as shown in the different tabs. A sixth tab allows to import other implementations from a file. The different parameters of the generator can be selected in the configuration boxes, and the output sequence $\{w_n\}$ is obtained. The quality of the sequence is then assessed visually by means of the return map and the chaos game.

Note that this framework allows us to obtain the code that will be downloaded to a CPLD or an FPGA as an alternative to the algorithm which is being simulated in the PC. Figure 2(a) shows a LabVIEW block diagram code for Arrow simulation and Fig. 2(b) shows a LabVIEW block diagram code to implement Arrow on an FPGA. As it can be seen, the block diagram is quite similar to the one which is running on the PC (see Fig. 1).

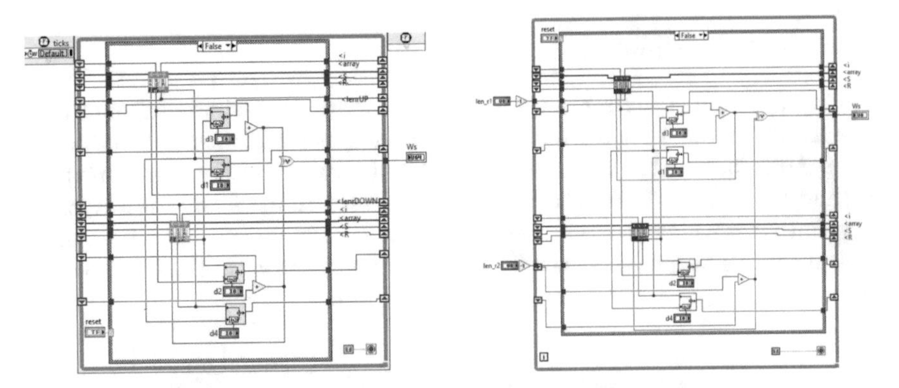

(a) Partial block diagram code to simulate Arrow PRNG.

(b) Partial block diagram code for FPGA implementation of Arrow.

Fig. 2. LabVIEW partial block diagram code for Arrow PRNG.

Observe that the aforementioned code fragments are only part of the complete code. Many relevant aspects like communication management, subsystems (FIFO devices), constraints, clock domain, etc., have been omitted here due to space limitations.

4.2 On-the-Fly Testing

The framework includes different techniques to assess the randomness in real time: time series representation, Walsh transform, histogram, autocorrelation, return map, and chaos game.

Below, the return map and the chaos game are explained, as methods for examining the randomness of a PRNG that are especially decisive for quickly discarding the configurations that lead to non-random sequences.

Return Map: The return map is a powerful framework for the cryptanalysis of chaotic systems. It was first used in [19] to break two schemes based on the Lorenz chaotic system.

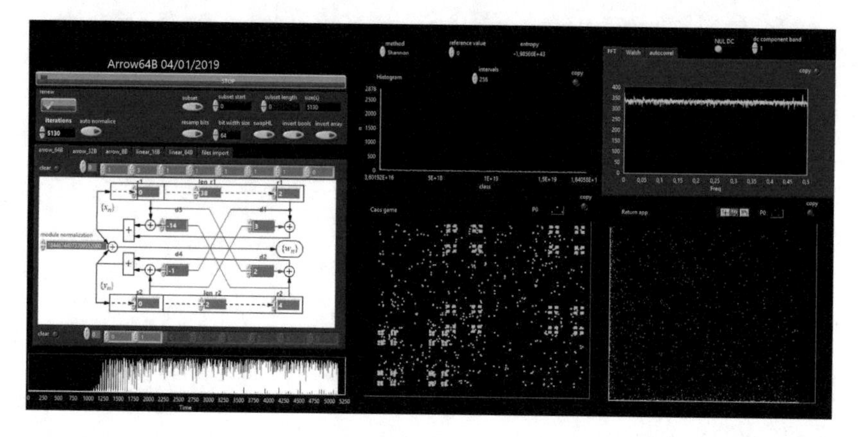

Fig. 3. Screenshot of the framework with a set of parameters which produce a non-random sequence.

The return map of a generator for a given set of parameters is a plot of a time series as a function of the current and the previous values.

Taking into account the time series $i(n)$ for $n = 1, 2, \ldots, N$, a m-dimensional return map is obtained by representing the vector $v(n) = [i(n), i(n-\tau), i(n-2\tau), \ldots, i(n-m\tau)]^T$, where T is an integer representing the delay.

In the analysis of chaotic systems, if $i(n)$ is the output of a dynamical system and the observation time N is large enough, the characteristics of the phase space can be derived from the return map. Similarly, if a pseudorandom sequence is analyzed, that plot will provide information about the possible internal structures of the sequence and, in certain cases, it could be possible to reconstruct the values of the parameters used in the generator to obtain the sequence; thus, the non-randomness of the PRNG becomes apparent. Indeed, a defined pattern in the map (or the fact that it is not completely filled), denotes a lack of entropy and a high level of determinism [20].

Chaos Game: The chaos game is mathematically described through an Iterated Functions System (IFS) [21], and provides a suitable framework to study the transition to chaos associated with fractals. Given an IFS $\{w_i\}_{i=1}^N$, a non-zero probability $p_i \in (0, 1)$ is associated with each application w_i such that $\sum_{i=1}^N p_i = 1$, according to the following algorithm:

Finally, the chaos game plot is built by drawing the points $x_0, x_1, \ldots, x_n$, thus obtaining an image of the set which is the attractor of the IFS.

When analyzing a PRNG, the chaos game representation is a visual representation of the structure of its output sequence. Non-uniformity of the distribution of subsequences produces non-uniformity in the chaos game representation [22].

As an example of the use of the chaos game in the presented simulation framework, Fig. 3 shows, among others, the plot of the chaos game obtained with a 64-bit word Arrow PRNG for 5130 iterations. As it can be observed, there is

Algorithm 1. Chaos game

1: Randomly select a point $x_0 \in X$
2: **for** $i = 1$ **to** N **do**
3: Select k: output of the PRNG(P).
4: Select w_i according to k
5: Compute $x_{n+1} = w_i(x_n)$
6: Return to step 4 and repeat with x_{n+1} replacing x_n
7: **end for**

a geometrical structure in the space, demonstrating the non-randomness of the output of the generator for the used parameters.

The front-end of the simulation framework includes all figures mentioned along the different sections, enabling a user to quickly –on the fly– assess the quality of the output for the parameters shown in the configuration boxes.

Figures 3 and 4 are examples of the effect of the right selection of the parameters. As it can be seen, parameter r_1 in Fig. 3 has a value $r_1 = 2$ while, in Fig. 4, it has a value $r_1 = 3$.

It can be observed that, for $r_1 = 2$, different patterns (geometrical structures) are generated in the chaos game. This implies a clear non-random behavior of the sequence. Similarly, the distribution of points in the return map shows the existence of a certain ordering –many points are concentrated on the left vertical and bottom horizontal axes– which can enable some kind of inverse engineering if a cryptanalyst knows the architecture of the generator.

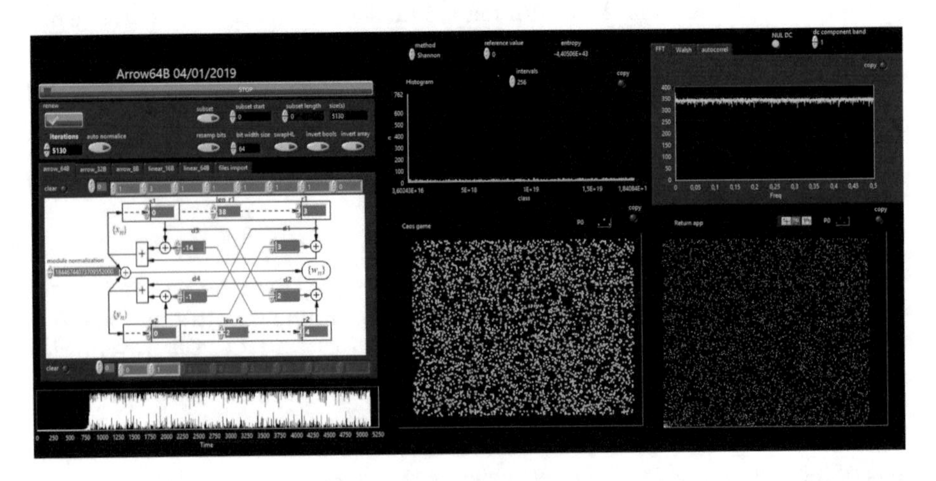

Fig. 4. Screenshot of the framework with a set of parameters which produce a random sequence.

Figure 4 shows that, for $r_1 = 3$, the points are uniformly spread in both the chaos game representation and the return map. In this case, the generator does not show any evident weakness or leakage.

Once the quality of the simulated generator has been checked, we can generate the structure that will be downloaded into a real FPGA.

5 Conclusion

In this work, a LabVIEW framework suitable for simulating and on-the-fly testing a hardware implementation of the Arrow PRNG has been presented. The framework simulates the PRNG behavior in a PC. It allows to modify the configuration parameters of the generator iteratively and to visualize, in real time, the randomness of the output sequences. This is done by means of the chaos game, the return map, the histogram, etc. This way, a first assessment of the quality of the generator can be made in order to select its optimum configuration parameters. Finally, the LabVIEW simulation allows to generate an Arrow architecture which can be downloaded into a real FPGA.

It is important to point out that we have carried out the simulations and verifications of the Arrow generator using Windows-based PCs and the verification of its implementation in an FPGA under a single and seamless integrated LabVIEW framework. Additionally, the framework lets us to add complementary instrumentation (for example, to measure FPGA emissions when generating the sequence). Note that this LabVIEW framework can be used for the analysis and implementation of other PRNGs.

Acknowledgments. This research has been partially supported by Ministerio de Economía, Industria y Competitividad (MINECO), Agencia Estatal de Investigación (AEI), and Fondo Europeo de Desarrollo Regional (FEDER, EU) under project COPCIS, reference TIN2017-84844-C2-1-R, and by the Comunidad de Madrid (Spain) under the project CYNAMON (P2018/TCS-4566), co-financed with FSE and FEDER EU funds.

References

1. Minoli, D., Sohraby, K., Occhiogrosso, B., Kouns, J.: Security considerations for IoT support of e-Health applications. In: Hassan, Q.F., ur Rehman Khan, A., Madani, S.A. (eds.) Internet of Things. Challenges, Advances, and Applications, pp. 321–346. Chapman and Hall/CRC (2018). Chap. 16
2. Sagahyroon, A., Aburukba, R., Aloul, F.: The Internet of Things and e-Health: remote patients monitoring. In: Hassan, Q.F., ur Rehman Khan, A., Madani, S.A. (eds.) Internet of Things. Challenges, Advances, and Applications, pp. 303–321. Chapman and Hall/CRC (2018). Chap. 15
3. Sfar, A.R., Natalizio, E., Challal, Y., Chtourou, Z.: A roadmap for security challenges in the Internet of Things. Digit. Commun. Netw. **4**, 118–137 (2018)
4. Grammatikis, P.I.R., Sarigiannidis, P.G., Moscholios, I.D.: Securing the Internet of Things: challenges, threats and solutions. Internet Things J. **5**, 41–70 (2019)
5. Misra, S., Maheswaran, M., Hashmi, S.: Security Challenges and Approaches in Internet of Things. Springer (2017)
6. Li, S., Xu, L.D.: Securing the Internet of Things, 1st edn. Syngress-Elsevier, Cambridge (2017)

7. Biryukov, A., Perrin, L.: State of the art in lightweight symmetric cryptography. IACR Cryptology ePrint Archive, p. 511 (2017)
8. ENISA: Good practices for security of internet of things in the context of smart manufacturing. European Union Agency for Network and Information Security. Technical report (2018). https://www.enisa.europa.eu/publications/good-practices-for-security-of-iot
9. Orúe López, A.B., Hernández Encinas, L., Martín Muñoz, A., Montoya Vitini, F.: A lightweight pseudorandom number generator for securing the Internet of Things. IEEE Access **5**, 27800–27806 (2017)
10. Schindler, W.: Random number generators for cryptographic applications. In: Koç Ç.K. (eds.) Cryptographic Engineering, pp. 2–23 (2009). Cap. 2
11. Bassham, L., Rukhin, A., Soto, J., Nechvatal, J., Smid, M., Barker, E., Leigh, S., Levenson, M., Vangel, M., Banks, D., Heckert, N., Dray, J.: NIST special publication 800–22 Rev 1a. A statistical test suite for random and PRNG for cryptographic applications. U.S., Department of Commerce/NIST (2010)
12. Marsaglia, G.: Diehard, a battery of tests for RNGs (2002). https://archive.is/IrySf
13. Santoro, R., Sentieys, O., Roy, S.: On-the-fly evaluation of FPGA-based true random number generator. In: Proceedings of the 2009 IEEE Computer Society Annual Symposium on VLSI, pp. 55–60 (2009)
14. Bhaskar, P., Gawande, P.D.: A survey on implementation of random number generator in FPGA. Int. J. Sci. Res. **4**(3), 1590–1592 (2015)
15. Tuncer, T., Avaroğlu, E.: Random number generation with LFSR based stream cipher algorithms. In: Proceedings of the 40th International Convention on Information and Communication Technology, Electronics and Microelectronics (MIPRO), pp. 171–175, May 2017
16. Souaki, G., Halim, K.: Random number generation based on MCU sources for IoT application. In: Proceedings of the 2017 International Conference on Advanced Technologies for Signal and Image Processing (ATSIP), pp. 1–6, May 2017
17. Bakiri, M., Guyeux, C., Couchot, J.-F., Oudjida, A.K.: Survey on hardware implementation of random number generators on FPGA: theory and experimental analyses. Comput. Sci. Rev. **27**, 135–153 (2018)
18. National Instruments: Tutorial: State machines (2018). http://www.ni.com/tutorial/7595/en/
19. Pérez, G., Cerdeira, H.A.: Extracting messages masked by chaos. Phys. Rev. Lett. **74**, 1970–1973 (1995)
20. Orúe, A.B., Montoya, F., Hernández Encinas, L.: Trifork, a new pseudorandom number generator based on lagged Fibonacci maps. J. Comput. Sci. Eng. **2**(2), 46–61 (2010)
21. Peitgen, H.-O., Jürgens, H., Saupe, D.: Chaos and Fractals. New Frontiers of Science. Springer, New York (2004)
22. Jeffrey, H.: Chaos game visualization of sequences. Comput. Graph. **16**(1), 25–33 (1992)

A Novel Construction for Streaming Networks with Delays

Joan-Josep Climent(✉), Diego Napp(✉), and Verónica Requena(✉)

Department of Mathematics, Universitat d'Alacant, Alacant, Spain
{jcliment,diego.napp,vrequena}@ua.es

Abstract. In this paper we study convolutional codes tailor made for fast decoding over burst erasure channels. This class of streaming codes are suitable for multimedia streaming applications where a stream of source packets must be transmitted in strict delay constraints. We show that in the case of dealing with burst erasure channels it is possible to come up with very simply constructions of encoders of convolutional codes that admit the fastest possible decoding delay to correct all bursts of a given length with a fixed rate. An explicit class of such encoders is presented. The last part of the paper is devoted to treat isolated errors. We propose the use of MDP convolutional codes to recover this kind of losses.

Keywords: Network coding · Convolutional codes · Burst erasure channel delay · Isolated errors

1 Introduction

When transmitting a stream of packets we can regard each packet as an element or sequence of elements from a large alphabet. In applications, in order to protect the packets, Cyclic Redundancy Check (CRC) codes are used so that the receiver can verify whether the packet has been received correctly or it is corrupted [11,18]. Moreover, due to the headers and number sequences describing their position within a given stream, the receiver knows where the erasures have happened. Packet sizes are upper bounded by $12,000$ bits - the maximum that the Ethernet protocol allows. In this way, the channels that use packets to transmit the information can be considered as erasure channels and each sent symbol is either correctly received or was corrupted or simply not received. Another important feature of this type of channels, such as the Internet, is that erasures tend to occur in bursts and are typically modeled by means of the Gilbert-Elliot channel model [8], which is a Markov model with two states: a good-state and a bad state (that represents a burst loss event).

In many important streaming applications the main problem observed by the receiver is the delay experienced in the transmission, for instance the Internet video, cloud computing, mobile gaming, etc. The use of CRC codes in this type of applications is not very efficient to reduced this delay, as if an erasure is

© Springer Nature Switzerland AG 2020
F. Martínez Álvarez et al. (Eds.): CISIS 2019/ICEUTE 2019, AISC 951, pp. 185–194, 2020.
https://doi.org/10.1007/978-3-030-20005-3_19

detected, one needs to ask for the missing packets and wait for the new packet to arrive. Another interesting possibility to address this problem that has attracted much of attention recently among several researchers is the use of convolutional error correcting codes [5–7, 19, 20]. In real-time applications source packets arrive sequentially at the destination, and the transmission must be reproduced (such as video streaming) using only the packets available up to that time within a designed maximum delay. Thus, these codes can be used to recover missing packets from the correctly received ones.

Streaming codes for such applications have not been properly studied and the large body of the existing literature in error correcting codes cannot be directly used to this context. Optimal codes of streaming applications need to be investigated. Recently, important advances towards the construction and implementation of these codes have been proposed in the literature. A class of codes that seems to have great potential in this context are convolutional codes [2–4, 7, 13, 14, 16]. Within this thread of research lies the work presented in [10], where a novel and very simple construction of convolutional codes tailor made to correct burst of erasures was presented. This construction had the advantage that the polynomial encoder representing the convolutional code could be defined with entries over the binary field. However, such construction did not performed optimally when many isolated erasures occur since that we need a long delay. In this paper we aim at complementing our previous work [10] and present codes to deal with isolated erasures only.

Hence, we continue the research in streaming codes and propose a convolutional encoders that perform efficiently when only isolated erasures occur, i.e., they admit the shortest possible decoding delay to correct isolated erasures. Here, isolated erasures means that the whole packet is lost and retain previous and later packets arrived correctly. We model each packet as an element in $\mathbb{F}^n$ where $\mathbb{F} = GF(q)$ be a finite field of size q.

2 Preliminaries

Let $\mathbb{F}[D]$ be the ring of polynomials with coefficients in $\mathbb{F}$.

Convolutional encoders take a stream of information bits and convert it into a stream of transmitted bits (by means of shift registers) and therefore they are very suitable for streaming applications. Due to their rich structure, convolutional codes have an interesting property called sliding window property that allows adaptation of the correction process to the distribution of the erasure pattern. If we introduce a variable D, called the *delay operator*, to indicate the time instant in which each information arrived or each codeword was transmitted, then we can represent the sequence message $(\boldsymbol{v}_0, \boldsymbol{v}_1, \cdots, \boldsymbol{v}_\mu)$ as a polynomial sequence $v(D) = \boldsymbol{v}_0 + \boldsymbol{v}_1 D + \cdots + \boldsymbol{v}_\mu D^\mu$.

A rate k/n **convolutional code** $\mathcal{C}$ [17, Definition 2.3] is an $\mathbb{F}[D]$-submodule of $\mathbb{F}[D]^n$ with rank k given by

$$\mathcal{C} = \mathrm{im}_{\mathbb{F}[D]} G(D) = \{\boldsymbol{v}(D) \in \mathbb{F}[D]^n \mid \boldsymbol{v}(D) = \boldsymbol{u}(D) G(D), \text{ with } u(D) \in \mathbb{F}[D]^k\}$$

where $G(D) \in \mathbb{F}[D]^{k\times n}$ is a right invertible matrix called *encoder matrix or generator matrix.*

Note that if $\boldsymbol{v}(D) = \boldsymbol{u}(D)G(D)$, with

$$\boldsymbol{u}(D) = \boldsymbol{u}_0 + \boldsymbol{u}_1 D + \boldsymbol{u}_2 D^2 + \cdots \quad \text{and} \quad G(D) = \sum_{j\geq 0} G_j D^j$$

then,

$$\boldsymbol{v}_0 + \boldsymbol{v}_1 D + \boldsymbol{v}_2 D^2 + \cdots = \boldsymbol{u}_0 G_0 + (\boldsymbol{u}_1 G_0 + \boldsymbol{u}_0 G_1)\, D + (\boldsymbol{u}_2 G_0 + \boldsymbol{u}_1 G_1 + \boldsymbol{u}_0 G_2)\, D^2 + \cdots$$

The maximum degree of all polynomials in the j-th column of $G(D)$ is denoted by δ_j. The degree δ of $\mathcal{C}$ is defined as the maximum degree of the full size minors of $G(D)$. We say that $\mathcal{C}$ is an (n, k, δ) convolutional code [17].

Now, we can define the **j-th column distance** [12] as

$$d_j^c(\mathcal{C}) = \min\left\{\mathrm{wt}\left(\boldsymbol{v}_{[0,j]}(D)\right) | \boldsymbol{v}(D) \in \mathcal{C} \text{ and } \boldsymbol{v}_0 \neq \boldsymbol{0}\right\},$$

where $\boldsymbol{v}_{[0,j]}(D) = \boldsymbol{v}_0 + \boldsymbol{v}_1 D + \cdots + \boldsymbol{v}_j D^j$ represents the j-th truncation of the codeword $\boldsymbol{v}(D) \in \mathcal{C}$ and

$$\mathrm{wt}(\boldsymbol{v}_{[0,j]}(D)) = \mathrm{wt}(\boldsymbol{v}_0) + \mathrm{wt}(\boldsymbol{v}_1) + \cdots + \mathrm{wt}(\boldsymbol{v}_j)$$

where $\mathrm{wt}(\boldsymbol{v}_i)$ is the Hamming weight of $\boldsymbol{v}_i$ which determines the number of nonzero components of $\boldsymbol{v}_i$, for $i = 1, \ldots, j$. For simplicity, we use d_j^c instead of $d_j^c(\mathcal{C})$.

The j-th column distance is upper bounded [9] by

$$d_j^c \leq (n-k)(j+1) + 1,$$

and the maximality of any of the column distances implies the maximality of all the previous ones, that is, if $d_j^c = (n-k)(j+1)+1$ for some j, then $d_i^c = (n-k)(i+1)+1$ for all $i \leq j$. For the value

$$M = \left\lfloor \frac{\delta}{k} \right\rfloor + \left\lfloor \frac{\delta}{n-k} \right\rfloor, \tag{1}$$

this bound is achieved and an (n, k, δ) convolutional code $\mathcal{C}$ with $d_M^c = (n-k)(M+1)+1$ is called a *maximum distance profile* (MDP) code [9].

Assume that $G(D) = \sum_{j=0}^{\ell} G_j D^j, G_j \in \mathbb{F}^{k\times n}, G_\ell \neq 0$, and consider the associated **sliding matrix**

$$G_j^c = \begin{pmatrix} G_0 & G_1 & \cdots & G_j \\ & G_0 & \cdots & G_{j-1} \\ & & \ddots & \vdots \\ & & & G_0 \end{pmatrix} \tag{2}$$

with $G_j = 0$ when $j > \ell$, for $j \in \mathbb{N}$.

If $G(D)$ is a basic generator matrix [17], then the code $\mathcal{C}$ can be equivalently described using a full rank polynomial *parity-check matrix* $H(D) \in \mathbb{F}[D]^{(n-k)\times n}$ defined by

$$\mathcal{C} = \ker_{\mathbb{F}[D]} H(D) = \left\{ v(D) \in \mathbb{F}[D]^n \mid H(D)v(D)^T = 0 \right\}.$$

The associated **sliding matrix** of $H(D) = \sum_{i=0}^{m} H_i D^i$ is

$$H_j^c = \begin{pmatrix} H_0 & & & \\ H_1 & H_0 & & \\ \vdots & \vdots & \ddots & \\ H_j & H_{j-1} & \cdots & H_0 \end{pmatrix} \tag{3}$$

with $H_j = 0$ when $j > m$, for $j \in \mathbb{N}$, and we called m to the memory of $H(D)$.

The MDP convolutional codes can be characterized from the associated sliding matrices of $G(D)$ and $H(D)$ as follows:

Theorem 1 (Theorem 2.4 in [9]). *Let G_j^c and H_j^c be the matrices defined in (2) and (3), respectively. Then the following statements are equivalent:*

1. $d_j^c = (n-k)(j+1)+1$*;*
2. *every $(j+1)k \times (j+1)k$ full size minor of G_j^c formed from the columns with the indices $1 \leq t_1 \leq \cdots \leq t_{(j+1)k}$, where $t_{ik+1} \leq in$, for $i = 1, 2, \ldots, j$ is nonzero;*
3. *every $(j+1)(n-k) \times (j+1)(n-k)$ full size minor of H_j^c formed from the columns with the indices $1 \leq r_1 \leq \cdots \leq r_{(j+1)(n-k)}$, where $r_{i(n-k)} \leq in$, for $i = 1, 2, \ldots, j$ is nonzero.*

In particular, when $j = M$, $\mathcal{C}$ is an MDP code.

2.1 Burst Erasure Channel

A **burst erasure channel** is a communication channel model where sequential symbols $\boldsymbol{v}_i$ of the codeword $\boldsymbol{v}(D)$ are either received correctly or completely lost. The receiver knows that a symbol has not been received and its location; moreover, losses can occur in bursts or as isolated losses. In this paper, we are primarily interested in building encoders that allow to recover these types of losses as soon as possible. We define different process for each kind of them. We follow previous approaches [3,19] and consider the symbols $\boldsymbol{v}_i$ as packets and consider that losses occur on a packet level.

Firstly, we assume that we have been able to correctly decode up to an instant i and a burst of length L is received at time instant i, i.e., one or more packets are lost from the sequence $(\boldsymbol{v}_i, \boldsymbol{v}_{i+1}, \ldots, \boldsymbol{v}_{i+L-1})$. To recover this burst of lost information, we need to receive information correctly after instant i. Then, we say that the **decoding delay** is T if the decoder can reconstruct each source packet with a delay of T source packets, i.e., we can recover $\boldsymbol{u}_{i+j}$ (for $j \in \{0, 1, \ldots, L-1\}$) once $\boldsymbol{v}_{i+L}, \boldsymbol{v}_{i+L+1}, \ldots, \boldsymbol{v}_{i+j+T}$ are received. To know if a burst erasure code is optimal, Martinian presented in [16] the following result which gives us a trade-off between delay and redundancy.

Theorem 2 (Theorem 1 in [16]). *If a rate R encoder enables correction of all erasure bursts of length L with decoding delay at most T, then,*

$$\frac{T}{L} \geq \max\left\{1, \frac{R}{1-R}\right\}. \tag{4}$$

Error-correcting convolutional codes are widely used in communications, owing to their efficient encoding and optimal yet simple sequential decoding algorithms. However, they require, in general, large finite fields and long delays [1]. Even though these type of codes have been proposed for applications over erasure channels, see [20], they do not generally achieve the best trade-off between delay, redundancy, field size and burst correction. The most common distances for convolutional codes, free distance and column distance, are not the best option in burst erasure channels. An indicator of the error-burst-correction capabilities of an encoder is the notion of column span introduced by Martinian in [16].

Definition 1. *The column span of G_j^c is defined as*

$$CS(j) = \min_{\substack{\boldsymbol{u} = (\boldsymbol{u}_0, \boldsymbol{u}_1, \ldots, \boldsymbol{u}_j), \\ \boldsymbol{u}_0 \neq 0}} \operatorname{span}(\boldsymbol{u} G_j^c)$$

where the span of a vector equals $j - i + 1$, with j the last index where the vector is nonzero and i the first such index.

If a burst of maximum length L occurs within a window of length $W + 1$, then, it can be corrected if and only if $CS(W) > L$, see also [7, Lemma 1.1] for a similar result. In this paper we are interested in convolutional codes with high erasure-burst-correcting capabilities but that also admit low latency decoding.

3 Convolutional Codes over the Erasure Channel with Low Delay

Convolutional codes have more advantages than block codes over a erasure channel, since that the received information can be grouped with flexibility, that is the blocks are not fixed as in the block code case. This property is due to the *sliding window* characteristic of convolutional codes. Moreover, the dependencies between the convolutional blocks allow us to recover the lost information sequentially.

3.1 Burst of Erasures

Consider a systematic encoder $G(D) = \begin{pmatrix} I_k & \widehat{G}(D) \end{pmatrix}$, $\widehat{G}(D) = \sum_{j \geq 0} \widehat{G}_j D^j$. If $\boldsymbol{v}(D) = \boldsymbol{u}(D)G(D)$, using constant matrices and the systematic form of $G(D)$, one can write

$$
(\boldsymbol{v}_0, \boldsymbol{v}_1, \boldsymbol{v}_2, \cdots) = (\boldsymbol{u}_0, \boldsymbol{u}_1, \boldsymbol{u}_2, \cdots) \begin{pmatrix} I_k & \widehat{G}_0 & O & \widehat{G}_1 & O & \widehat{G}_2 & \dots \\ & & I_k & \widehat{G}_0 & O & \widehat{G}_1 & \dots \\ & & & & I_k & \widehat{G}_0 & \dots \\ & & & & & & \ddots \end{pmatrix}, \tag{5}
$$

where we can express the vector $\boldsymbol{v}_i = (\boldsymbol{u}_i \ \widehat{\boldsymbol{v}}_i)$, with $\widehat{\boldsymbol{v}}_i = \sum_{j=0}^{i} \boldsymbol{u}_{i-j} \widehat{G}_j$ due to the systematic form of $G(D)$. The systematic form of the matrix $G(D)$ facilitate us, not only a direct decoding, but also a simple recovery of lost information as we show in the next example.

Due to this simple form of the system (5), we can reduce it to

$$
(\widehat{\boldsymbol{v}}_0, \widehat{\boldsymbol{v}}_1, \widehat{\boldsymbol{v}}_2, \cdots) = (\boldsymbol{u}_0, \boldsymbol{u}_1, \boldsymbol{u}_2, \cdots) \begin{pmatrix} \widehat{G}_0 & \widehat{G}_1 & \widehat{G}_2 & \dots \\ & \widehat{G}_0 & \widehat{G}_1 & \dots \\ & & \widehat{G}_0 & \dots \\ & & & \ddots \end{pmatrix}. \tag{6}
$$

In our recovering method, we will consider a submatrix of the sliding matrix $\widehat{G}_j^c$, with $j \geq L + T$, denoted by

$$
\widehat{G}_{L+T}^{trunc} = \begin{pmatrix} \widehat{G}_L & \widehat{G}_{L+1} & \cdots & \widehat{G}_T \\ \widehat{G}_{L-1} & \widehat{G}_L & \cdots & \widehat{G}_{T-1} \\ \vdots & \vdots & \cdots & \vdots \\ \widehat{G}_1 & \widehat{G}_2 & \cdots & \widehat{G}_{T-L+1} \end{pmatrix}, \tag{7}
$$

of size $Lk \times (T - L + 1)(n - k)$, extracted from the system given in (6).

This matrix will play an important role in the construction of good burst correction convolutional codes with low delay. This fact will be evident in the Example 1 in Subsect. 3.2 where we explain how to recover the lost packets in a burst of erasures. Firstly, we present a systematic encoder convolutional code for burst erasure correction with low delay.

3.2 Our Construction

Suppose that in the channel only bursts of erasures of length L occur. We first consider the case $k > n - k$, say, $(n - k)\lambda + \gamma = k$ for some integer λ and $\gamma < n - k$. Let $G(D) = (I_k \ \ \widehat{G}(D)) \in \mathbb{F}^{k \times n}$ be a systematic encoder, with $\widehat{G}(D) = \sum_{j \geq 0} \widehat{G}_j D^j$ given by

$$
\widehat{G}_{iL} = \begin{pmatrix} O_{(i-1)(n-k) \times (n-k)} \\ I_{n-k} \\ O_{k-i(n-k) \times (n-k)} \end{pmatrix}, \text{ for } i = 1, 2, \dots, \lambda.
$$

If $(n-k) \nmid k$, i.e., $\gamma \neq 0$, then, we also define

$$\widehat{G}_{(\lambda+1)L} = \begin{pmatrix} O_{(k-\gamma)\times\gamma} & O_{(k-\gamma)\times(n-k-\gamma)} \\ I_\gamma & O_{\gamma\times(n-k-\gamma)} \end{pmatrix}.$$

The remaining coefficients of $\widehat{G}(D)$ are null matrices.

Suppose that a burst of erasure of length L occurs at time j. Then, one can verify that at time instant $j+iL$, we recover $n-k$ coordinates of $\boldsymbol{u}_j$ for $i=1,\dots,\lambda$, and wait until time $j+(\lambda+1)L$ to retrieve the remaining part of $\boldsymbol{u}_j$, if required. Then, the delay to recover $\boldsymbol{u}_j$ is $T=\lceil\frac{k}{n-k}\rceil$. Furthermore, due to the block-Toeplitz structure of the sliding matrix it follows that T is also the delay for decoding all the remaining erasures of $\boldsymbol{v}_s$, $s=j+1,\dots,j+L-1$. Assume now, for simplicity, that $\gamma=0$ and we see that the bound in (4) is achieved with equality. First note that $\frac{R}{1-R}=\lambda$ for the selected parameters $(n-k)\lambda=k$ and $R=k/n$. Now, it is easy to verify that $T=\lambda L$ and therefore $\frac{T}{L}=\lambda=\frac{R}{1-R}$.

Thus, the proposed construction admits an optimal delay decoding when only bursts of erasures of length up to L occur. Note that this construction requires only binary entries, whereas previous contributions (see for instance [15,16]) require larger finite fields. As a consequence, the decoding of our construction is computationally more efficient. The case $k \leq n-k$ readily follows by considering

$$\widehat{G}_L = \begin{pmatrix} O \; I_k \end{pmatrix}$$

and the remaining coefficients of $G(D)$, G_j, $j \notin \{0, L\}$ null matrices.

Example 1. Let $n=9, k=5$ be the parameters of our convolutional encoder and consider $L=2$ the maximum length of the burst and $T=4$ the decoding delay satisfying inequation (4).

Suppose without lost of generality that we have received a burst of erasures in the first two packets of our message

$$\boldsymbol{v} = (\bigstar, \bigstar, \boldsymbol{v}_2, \boldsymbol{v}_3, \boldsymbol{v}_4, \boldsymbol{v}_5),$$

that is, this burst of erasures starts at instant $t=0$; since that the decoding delay is $T=4$, we need to receive correctly until the information packet $\boldsymbol{v}_5$. Hence, in order to recover the packets $\boldsymbol{u}_0$ and $\boldsymbol{u}_1$, we consider them as unknowns $\boldsymbol{x}_0$ and $\boldsymbol{x}_1$.

First, we will recover $\boldsymbol{u}_0$, and from the system (6) and the submatrix defined in (7), we only need to solve the following linear system of equations

$$(\boldsymbol{x}_0, \boldsymbol{x}_1)\widehat{G}_6^{trunc} = (\boldsymbol{x}_0, \boldsymbol{x}_1) \begin{pmatrix} \widehat{G}_2 & \widehat{G}_3 & \widehat{G}_4 \\ \widehat{G}_1 & \widehat{G}_2 & \widehat{G}_3 \end{pmatrix} = (\boldsymbol{b}_0^{(0)}, \boldsymbol{b}_1^{(0)}, \boldsymbol{b}_2^{(0)}), \tag{8}$$

where

$$\left(\boldsymbol{b}_0^{(0)}, \boldsymbol{b}_1^{(0)}, \boldsymbol{b}_2^{(0)}\right) = \left(\widehat{\boldsymbol{v}}_2 - \boldsymbol{u}_2\widehat{G}_0, \widehat{\boldsymbol{v}}_3 - \boldsymbol{u}_3\widehat{G}_0 - \boldsymbol{u}_2\widehat{G}_1, \widehat{\boldsymbol{v}}_4 - \boldsymbol{u}_4\widehat{G}_0 - \boldsymbol{u}_3\widehat{G}_1 - \boldsymbol{u}_2\widehat{G}_2,\right)$$

is a vector with information received correctly.

The construction described above yields the following matrices, from

$$\widehat{G}_1 = \widehat{G}_3 = O_{5\times 4}, \text{ and } \widehat{G}_2 = \begin{pmatrix} 1\,0\,0\,0 \\ 0\,1\,0\,0 \\ 0\,0\,1\,0 \\ 0\,0\,0\,1 \\ 0\,0\,0\,0 \end{pmatrix},$$

and, as we have that $\gamma \neq 0$, we compute

$$\widehat{G}_4 = \begin{pmatrix} 0\,0\,0\,0 \\ 0\,0\,0\,0 \\ 0\,0\,0\,0 \\ 0\,0\,0\,0 \\ 1\,0\,0\,0 \end{pmatrix}.$$

Now, we have

$$\widehat{G}_6^{trunc} = \left(\begin{array}{c|c|c} 1\,0\,0\,0 & 0\,0\,0\,0 & 0\,0\,0\,0 \\ 0\,1\,0\,0 & 0\,0\,0\,0 & 0\,0\,0\,0 \\ 0\,0\,1\,0 & 0\,0\,0\,0 & 0\,0\,0\,0 \\ 0\,0\,0\,1 & 0\,0\,0\,0 & 0\,0\,0\,0 \\ 0\,0\,0\,0 & 0\,0\,0\,0 & 1\,0\,0\,0 \\ \hline 0\,0\,0\,0 & 1\,0\,0\,0 & 0\,0\,0\,0 \\ 0\,0\,0\,0 & 0\,1\,0\,0 & 0\,0\,0\,0 \\ 0\,0\,0\,0 & 0\,0\,1\,0 & 0\,0\,0\,0 \\ 0\,0\,0\,0 & 0\,0\,0\,1 & 0\,0\,0\,0 \\ 0\,0\,0\,0 & 0\,0\,0\,0 & 0\,0\,0\,0 \end{array}\right).$$

We can check that in the first part of the system given in (8)

$$\begin{aligned} &(x_{00}, x_{01}, x_{02}, x_{03}, x_{04}, x_{10}, x_{11}, x_{12}, x_{13}, x_{14})\, \widehat{G}_6^{trunc} \\ &= (u_{00}\ u_{01}\ u_{02}\ u_{03}, u_{10}\ u_{11}\ u_{12}\ u_{13}, u_{04}\ 0\ 0\ 0)\,. \end{aligned}$$

We recover directly the vector $\boldsymbol{u}_0$ completely, at time instant 6, from the known information of $\widehat{\boldsymbol{v}}_2$ and $\widehat{\boldsymbol{v}}_4$. In order to recover the packet $\boldsymbol{u}_1$, we repeat the arguments.

We have to solve the system

$$(\boldsymbol{x}_1, \boldsymbol{v}_2)\widehat{G}_6^{trunc} = (\boldsymbol{b}_1^{(1)}, \boldsymbol{b}_2^{(1)}, \boldsymbol{b}_3^{(1)}),$$

where

$$\begin{aligned} \left(\boldsymbol{b}_1^{(1)}, \boldsymbol{b}_2^{(1)}, \boldsymbol{b}_3^{(1)}\right) = \Big(&\widehat{\boldsymbol{v}}_3 - \boldsymbol{u}_3\widehat{G}_0 - \boldsymbol{u}_2\widehat{G}_1 - \boldsymbol{u}_0\widehat{G}_3, \widehat{\boldsymbol{v}}_4 - \boldsymbol{u}_4\widehat{G}_0 - \boldsymbol{u}_3\widehat{G}_1 - \boldsymbol{u}_2\widehat{G}_2 - \boldsymbol{u}_0\widehat{G}_4, \\ &\widehat{\boldsymbol{v}}_5 - \boldsymbol{u}_5\widehat{G}_0 - \boldsymbol{u}_4\widehat{G}_1 - \boldsymbol{u}_3\widehat{G}_2 - \boldsymbol{u}_2\widehat{G}_3 - \boldsymbol{u}_0\widehat{G}_5\Big) \end{aligned}$$

is a vector of information received correctly.

4 Isolated Erasures

In order to recover isolated erasures, we could use the previous method for burst erasures with the value $L = 1$. The problem is that, in this case, the method is not optimal when values k and n are close, since that value of T increases fastly. Next, we follow an example.

Our aim is to recover the maximum number of erasures in a received sequence. With the method to recover bursts and isolated erasures presented in this paper, there are some unsolved cases. For instance, if we have two bursts of length L separated by a delay period T, and within of this delay period appears an isolated erasure, we would not be able to recover the first burst with our construction; or if the delay that we have until the next burst allows us to recover the isolated erasure with our method. However, if we could first recover the isolated error and then use our method to recover the bursts, we had solved this problem.

Our main idea to recover isolated erasures is the use of MDP convolutional codes, since that have the property their column distances are as large as possible [12]. These codes have optimal recovery rate [19] for windows of a certain length over an erasure channel. So, we will use these codes as tool to recover these type of erasures.

If we construct a concatenation of two convolutional codes, one outer code to recover the bursts of erasures (method presented in Subsect. 3.2) and a second inner code, a MDP code which recover the isolated erasures, we could recover many unsolved cases.

Our future work is focused on this line, determine a concatenation of codes to solve these difficult cases.

5 Conclusion

In this paper, we have present a systematic encoder for burst erasure correction with low delay, from convolutional codes. We have built an specific encoder in binary field. In addition, we have introduced a way to recover isolated erasures using MDP convolutional codes. In both cases, the computation of erasures require only lineal algebra.

Our future work consists on combining both methods presented to recover the maximum number of erasures. Other way is the use of reverse-MDP convolutional codes to recover packages for which do not work the previous methods. Moreover, we will study simulations of our models over erasure channels variating the rates of the convolutional codes.

Acknowledgment. This work was partially supported by Spanish grant AICO/2017/128 of the Generalitat Valenciana and the University of Alicante under the project VIGROB-287.

References

1. Almeida, P., Napp, D., Pinto, R.: Superregular matrices and applications to convolutional codes. Linear Algebra Appl. **499**, 1–25 (2016)
2. Arai, M., Yamaguci, A., Fukumpto, S., Iwasaki, K.: Method to recover lost internet packets using (n, k, m) convolutional code. Electron. Commun. Jpn. Part 3 **88**(7), 1–13 (2005)
3. Badr, A., Khisti, A., Tan, W.T., Apostolopoulos, J.: Robust streaming erasure codes based on deterministic channel approximations. In: 2013 IEEE International Symposium on Information Theory, pp. 1002–1006 (2013)
4. Badr, A., Khisti, A., Tan, W.-T., Apostolopoulos, J.: Layered constructions for low-delay streaming codes. IEEE Trans. Inf. Theory **63**(1), 111–141 (2017)
5. Climent, J.J., Napp, D., Pinto, R., Simões, R.: Decoding of 2D convolutional codes over the erasure channel. Adv. Math. Commun. **10**(1), 179–193 (2016)
6. Costello Jr., D.: A construction technique for random-error-correcting convolutional codes. IEEE Trans. Inf. Theory **15**(5), 631–636 (1969)
7. Deng, H., Kuijper, M., Evans, J.: Burst erasure correction capabilities of $(n, n-1)$ convolutional codes. In: 2009 IEEE International Conference on Communications, pp. 1–5 (2009)
8. Gilbert, E.N.: Capacity of a burst-noise channel. Bell Syst. Tech. J. **39**, 1253–1265 (1960)
9. Gluesing-Luerssen, H., Rosenthal, J., Smarandache, R.: Strongly MDS convolutional codes. IEEE Trans. Inf. Theory **52**(2), 584–598 (2006)
10. Napp, D., Climent, J.-J., Requena, V.: Block toeplitz matrices for burst-correcting convolutional codes. Rev. Real Acad. Cienc. Exactas Fsicas Nat. Ser. A Mat (submitted)
11. Jain, S., Chouhan, S.: Cyclic Redundancy Codes: Study and Implementation. Int. J. Emerg. Technol. Adv. Eng. **4**, 2250–2459 (2014)
12. Johannesson, R., Zigangirov, K.Sh.: Fundamentals of Convolutional Coding. IEEE Press, New York (2015)
13. Kuijper, M., Bossert, M.: On (partial) unit memory codes based on reed-solomon codes for streaming. In: 2016 IEEE International Symposium on Information Theory (ISIT), pp. 920–924 (2016)
14. Mahmood, R., Badr, A., Khisti, A.: Streaming-codes for multicast over burst erasure channels. IEEE Trans. Inf. Theory **61**(8), 4181–4208 (2015)
15. Mahmood, R., Badr, A., Khisti, A.: Convolutional codes with maximum column sum rank for network streaming. IEEE Trans. Inf. Theory **62**(6), 3039–3052 (2016)
16. Martinian, E., Sundberg, C.E.W.: Burst erasure correction codes with low decoding delay. IEEE Trans. Inf. Theory **50**(10), 2494–2502 (2004)
17. McEliece, R.J.: The algebraic theory of convolutional codes. In: Handbook of Coding Theory, vol. 1, pp. 1065–1138. Elsevier Science Publishers (1998)
18. Peterson, W.W., Brown, D.T.: Cyclic codes for error detection. Proc. IRE **49**(1), 228–235 (1961)
19. Tomás, V., Rosenthal, J., Smarandache, R.: Decoding of MDP convolutional codes over the erasure channel. In: Proceedings of the 2009 IEEE International Symposium on Information Theory (ISIT 2009), Seoul, Korea, pp. 556–560. IEEE, June 2009
20. Tomás, V., Rosenthal, J., Smarandache, R.: Decoding of convolutional codes over the erasure channel. IEEE Trans. Inf. Theory **58**(1), 90–108 (2012)

Hyot: Leveraging Hyperledger for Constructing an Event-Based Traceability System in IoT

Jesús Iglesias García[1(✉)], Jesus Diaz[2], and David Arroyo[3]

[1] Escuela Politécnica Superior, Universidad Autónoma de Madrid, Madrid, Spain
jesusgiglesias@gmail.com
[2] BBVA Next Technologies, Madrid, Spain
jesus.diaz.vico.next@bbva.com
[3] Instituto de Tecnologías Físicas y de la Información,
Consejo Superior de Investigaciones Científicas, Madrid, Spain
david.arroyo@csic.es

Abstract. In this work it is introduced Hyot, a blockchain solution for activity registration in the Internet of Things. Specifically, a permissioned blockchain is used to record anomalous occurrences associated with sensors located on a Raspberry Pi 3. Likewise, a web system is provided to consume the information collected in real time.

Keywords: Blockchain · Hyperledger · Internet of Things · Raspberry Pi · Security · Sensor · Smart Contracts · Privacy

1 Introduction

The Internet has changed everyday life and our daily activity depends more and more on the information obtained from the Internet. Therefore, we should develop mechanisms for data protection and information reliability. Traditionally, this issue has been resolved through some Trusted Third Party (TTP) [12, Ch. 13]. However, the participation of intermediaries also presents some disadvantages, such as the possible degradation of privacy if the TTP accesses, without the express consent, to sensitive and personal information [7].

One possible solution is the decentralization of information management. The blockchain (BC) technology is intended to protect information integrity without resorting to any type of TTP. The BC is a type of Decentralised Ledger of Transactions (DLT) where data recording demands the collaboration between the components of a *Peer-to-Peer* (P2P) network. As a result, any piece of data in the BC cannot be either written or deleted by a unique entity. Furthermore, any user with access to the BC can obtain an exact replica of the consolidated information, which paves the way for auditability and data transparency.

In its origin, the BC was devised to support Bitcoin [14], the first decentralised cryptocurrency which is not issued by a central bank. In the case of

© Springer Nature Switzerland AG 2020
F. Martínez Álvarez et al. (Eds.): CISIS 2019/ICEUTE 2019, AISC 951, pp. 195–204, 2020.
https://doi.org/10.1007/978-3-030-20005-3_20

the BC of Bitcoin, the goal was on validating financial transactions without the participation of any sort of TTP. However, the Bitcoin's data model enables the storing of arbitrary information and the creation of communications protocols [9, Ch. 12]. This being the case, there are plenty of initiatives harnessing the Bitcoin's BC to conduct business in areas as finance, logistics or healthcare [2].

The impact of the BC technology has been further propelled by the irruption of the so-called smart contracts [18], which enable the possibility of writing software programs onto the BC. This pieces of software are executed automatically, i.e., without human-intervention and only upon the fulfillment of a set of conditions and requirements. The popularity of smart contracts is clearly correlated with the deployment of the Ethereum platform, specially in the context of crowdfunding initiatives associated with Initial Coin Offerings (ICOs) [4].

In fact, Ethereum's smart contracts are powerful tools to construct Decentralised Autonomous Organisations [11]. They draw a scenario where transparency is endorsed by the public nature of the Ethereum's BC and automation is fuelled as a result of the automatic execution of the smart contracts that are stored in the BC. Therefore, smart contracts can be interpreted as key components in the design and implementation of governance structures for Cyber-Physical Systems (CPS). In more detail, the management and control of CPS is conducted through the information that is obtained from the outside physical world and codified according to specific sensors. Once this information has been properly treated and digitalised, a control law is executed to correspondingly exert an action. A critical concern in this loop is information and code integrity, since any data distortion could lead to undesirable outcomes from the control algorithm. Taking into account that BC guarantees data immutability and smart contracts foster automatic code execution, it represents a very attractive option as the base of new protocols for the control and management of CPS.

Among the different BC-based applications in CPS, in this work we are mainly concerned with those in the field of the Internet of Things (IoT). IoT is a specific implementation approach [13] of CPS, and it refers to the interconnection of numerous physical objects in the real world -from a TV, a vehicle to a refrigerator- which are provided with an Internet connection and interact by exchanging data. Each device works differently and has different levels of security implemented. As a consequence, there is an urge to standardise monitoring procedures and techniques. The BC has been exploited as bottom line of this effort to extract, model, and process information from the heterogeneous ecosystem derived from the variability of IoT in terms of information interfaces and communication protocols [5].

In this paper, the main goal is on the deployment of event gathering procedures to buttress digital audit and forensics procedures in IoT by means of BC-based solutions. To achieve such a goal, we are considering the simple scenario determined by activity detection in a room. A set of sensors is targeted to spot anomalous behaviour in a restricted access space, e.g. the data centre or any other critical premises in a company or organisation. A proper concretion of a security policy includes an audit plan to foster security forensics in case of the occurrence of a security incidence. This audit plan must preserve the chain of custody and thus the integrity of any piece of evidence. Therefore, event recording must be carried out by entities properly authenticated and authorised, which demands the selection of a BC architecture with adequate access and authorisation controls.

Correspondingly, in the next section we discuss the limitations of public BCs as those of Bitcoin and Ethereum, and we introduce the so-called permissioned BC. In specific, we centre our analysis in one of the most relevant proposals in the field of permissioned BC, Hyperledger Fabric. In Sect. 3, we describe our design and implementation of a protocol for evidence collection using Hyperledger Fabric. In the last section, we summarise the outcomes of the research associated with this paper and the main strands of future work.

2 The Hyperledger Initiative

Public BCs as Bitcoin and Ethereum have serious shortcomings to be considered as the main option to convey event recording and audit policies in IoT. First of all, any single entry in public BCs requires a fee, which is a clear drawback with respect to traditional on-premise procedures for event gathering and treatment. Scalability is another big problem when dealing with public BCs [6], which is even more critical if we take into account the throughput, i.e., the time necessary to actually consolidate a piece of data. Although the number of transactions per second in Ethereum is much larger than in Bitcoin, the security problems of Ethereum's smart contracts could lead to a degradation of the availability of the events stored in the BC [8,19]. As a matter of fact, the decentralised nature of Bitcoin and Ethereum is very dependent on the distribution of computational power among the entities that are collaborating to insert new transactions into the BC. Although Bitcoin and Ethereum were originally conceived to elude intermediate agents, their practical deployment shows a dangerous trend to centralise the acceptance of new transactions around the entities with higher computational power [10]. If we are considering BC solutions to store and exchange our logs and activity records, this re-centralisation dynamics actually implies an implicit trust on those entities with high computational power.

In addition, the identification of the evidence sources is a major concern in forensics, and any BC-based forensics procedure must guarantee that the entities in charge of storing evidence can be properly traced and linked. Certainly, the Locard's exchange principle establishes that the perpetrator of a non-authorised action always carries something into the action scene and leaves with something from it [16]. In other words, there is a trail that can lead to the origin of a security problem. This trail is based on items of evidence that can be connected to the perpetrator. Nonetheless, the admissibility of any sort of evidence requires to maintain the chain of custody [3]. That it is to say, it is necessary to deploy mechanisms for the authentication of entities with rights to write information onto the BC. Likewise, from the perspective of the security of a company, there is no reason to enable anyone to access information related to its inner structure and dynamics.

All in all, BC-based forensics procedures are more aligned with permissioned BCs that include explicit access control policy [15], and which have better performance in terms of scalability and throughput. One of the most relevant examples of permissioned BC is the Hyperledger initiative [1]. Hyperledger is a collaborative and open source consortium announced in December 2015 by the non-profit technological organisation Linux Foundation to investigate, evolve, transmit knowledge, and establish standards on DLT technology. Specifically, in the permissioned BC oriented to the economic and business environment where thanks to the ability to execute private transactions and the inclusion of novel features, the way in which business and commercial processes take place will be redirected. The objective of this initiative is possible thanks to the effort and dedication of numerous partners and members, whose number is increasing, including organisations focused on the BC (the Enterprise Ethereum Alliance, the R3 consortium, etc.), technology companies (IBM, Intel, Red Hat, VMware, Cisco, etc.), financial companies (JP Morgan, SWIFT, etc.) and banking entities (European Central Bank, Bank of Japan, Deutsche Bundesbank, BBVA, etc.), among others. Currently, this initiative provides a set of tools that includes frameworks that facilitate the deployment infrastructure of the BC and complementary tools for the maintenance and design of BC networks.[1]

Hyperledger Fabric is the most well-known project of this initiative and the first in obtaining the incubation phase [1]. Originated in the first place by the contribution of the companies Digital Asset Holdings and IBM, it is open source, oriented to business solutions that implement the permissioned BC technology, and in constant evolution[2] providing diverse features (modular and extensible architecture, high performance transactional network, scalability, privacy and identity management, chaincode or smart contract functionality, configuration of organisations, channels for private and confidential transactions, membership providers, and consensus algorithms) that tend to improve many aspects of productivity and reliability and that distinguish it from other alternatives. But where it really differs, apart from the absence of a cryptocurrency, it is in its BC of a private and discretionary character since it allows controlling the sensitivity about what and how much information participants can share in a business-to-business network. Indeed, Hyperledger Fabric allows defining who has access to the network, for what assets and with what privileges, limiting in such a way that the participants have known identities in front of a reliable Membership Service Provider (MSP).

3 Hyot

Hyot is an open source Proof of Concept for the traceability of a controlled IoT environment using Hyperledger Fabric technology. While it is true that this traceability can be done through traditional methodologies, such as databases

[1] See https://www.hyperledger.org/ for more details.
[2] Hyperledger Fabric v1.4.0 was released on January 9, 2019.

(DB) or log files, it is also evident that there is a lack of security and guarantee that the information considered relevant has remained immutable since its registration. Taking advantage of the support of all the advantages provided by the concept of BC, Hyot maintains part of its persistence in one of permissioned character to solve the problem of how to ensure the reliability of an irrefutable and reliable way of uncontrolled events that arise in IoT environments. As we have pointed above, the nature of Hyperledger Fabric favours security against public BCs in this case for several reasons. For instance, by default:

- It provides a much higher transaction rate than public BCs.
- It avoids the need to pay (high) fees for executing transactions.
- It allows a finer control on the authentication requirements of its participants.

Especially, through the existence of identities and levels of privacy and membership it can be controlled who can access what and with what privileges. In this way, the most important information is not publicly exposed to any other alien user and is always identified to the participant who registers it, which makes easier the tracking. But only a part of all the information tracked from the environment is recorded in the BC. In particular, the BC is applied to guarantee the integrity of the information that signals the occurrence of uncontrolled or anomalous events in a time frame. This type of event from which a piece of evidence is generated -video from the environment- is stored in a TTP, as it is a storage service in the cloud, due to its size. This external agent may not be fully secure and by assuming a zero trust model [17], the stored evidence must be protected by encrypting its content and subsequently signed. The remaining information, which is merely informative and allows broadening the context of the traceability carried out, is stored in DB.

This solution, which is highly configurable by the user, transparently manages a series of events -temperature, humidity and distance- of the environment that are constantly monitored with a default frequency of 3 s and in real time from input data sources such as the sensors, connected to a Single Board Computer (SBC) like Raspberry Pi, with the latest version available (v3) being used in the project. The information collected is analysed to determine if the read values are considered anomalous or not at a specific time and therefore determine if an unauthorised event is occurring in the environment.

At all times, the actions performed and the measurements made are notified to the user both through the terminal and through the output devices that the hardware prototype has. Each measurement, regardless of whether an anomalous event occurs or not, is composed of the following information: identifier, timestamp, value of each event (temperature, humidity and distance), indicator of whether an anomalous event occurs, sensor id, event and threshold of the event that originates an anomalous event, link to the file with the evidence, email of therecipient for notification purposes, and user that records the event in the BC.

If the current reading of values presents data that exceed one or more of the pre-established thresholds, then it is considered as a possible indication of an incidence on the environment. In this situation, the procedure to be followed is

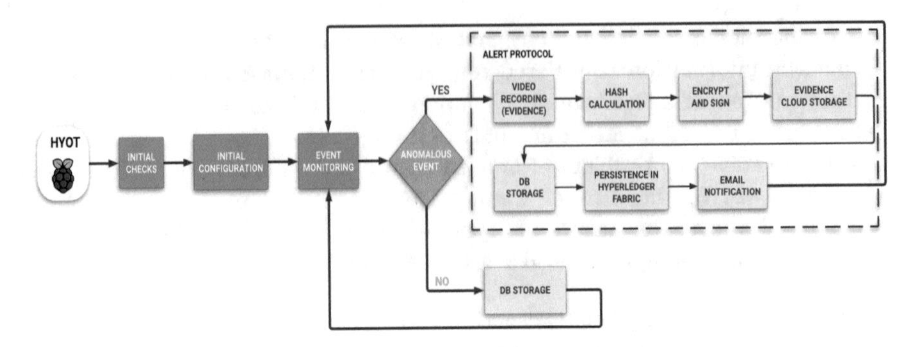

Fig. 1. Flow diagram - Event monitoring in the IoT environment.

more exhaustive in order to certify the originality and authenticity of recorded data. An incident is an uncontrolled event or action that takes place in the environment that is being monitored and originates the execution of an alert protocol that includes (see Fig. 1):

1. Recording a video through the PiCamera. This video represents the evidence.
2. Calculation of the hash value of the original content of the evidence, applying the hash function SHA-3, which acts as an integrity proof.
3. Encrypting and signing the evidence, since a zero trust model has been assumed in which all outsourced information must be protected in the client-side.
4. Cloud storage of evidence. This type of services are considered TTP that can put in risk the privacy so that the previous step is necessary.
5. Insert all the information of the measurement into the DB to ensure that the controlled environment presents one incidence at that time.
6. Registration of the incidence in the BC of Hyperledger Fabric by storing the hash value of the original content of the evidence.
7. Instant notification of the incident by sending an email to the destination email address configured.

The current architecture for Hyot consists of two environments and different services and platforms as shown in Fig. 2. It is important to underline that our design entails different approaches for data persistence, ranging from metadata insertion in a local database, integrity control by storing content hashes onto the BC to the outsourcing encrypted evidence through cloud storage. In short, the protocol for the management of metadata and evidence recording is organised around three main components:

- Component for monitoring events in the IoT environment through sensors located in a Raspberry Pi device.
- Incident registration protocol in the BC of Hyperledger Fabric and storage of protected evidence in the cloud.

- Web app that acts as a client and consumes in real time the information registered by the Raspberry Pi. This application contains an administration panel where the administrator manages the users and consumes all the information that has been recollected. On the other hand, the regular users can only consume the information that has been obtained by themselves.

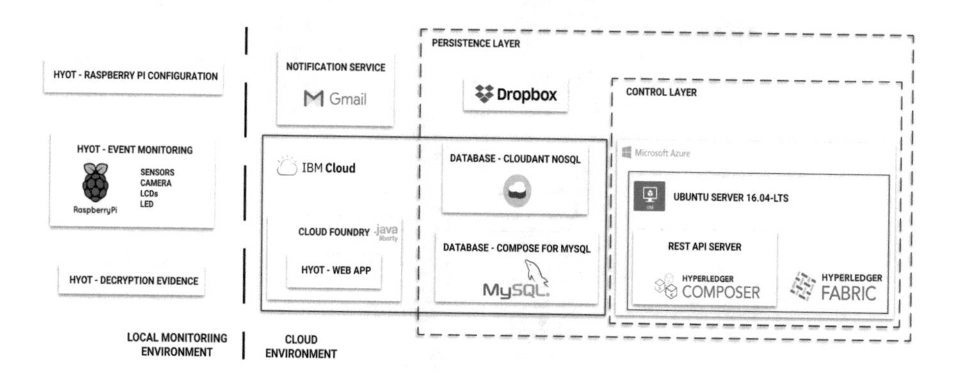

Fig. 2. Hyot architecture.

Other than that the main components, two additional elements have been developed in order to complete the project and facilitate the tasks of:

- Initial configuration of the Raspberry Pi device for the subsequent execution of the component for event monitoring in the IoT environment. This configuration includes the installation of packages and Python libraries, along with the activation of software interfaces.
- Decryption of the evidence previously encrypted and signed. In this component, in addition to the decryption process, the signature is shown and the integrity of the content of the evidence obtained is verified through the comparison of the hash value calculated on the content of the evidence after decrypting and the hash value that was initially stored in the BC when the incident occurred. At this point of verification, the confidence that is deposited in this technology is crucial since in case of hash mismatch, we can conclude that the evidence stored in the cloud was improperly modified.

The diagram of the prototype used in Hyot by the event monitoring component of the IoT environment is displayed below in Fig. 3. This prototype is composed of different electronic devices where it is important to ensure a correct connection for proper operation. In specific, the list of electronic components is given by:

- A Raspberry Pi 3 to implement the control system.
- A V2 camera module as video recording and main piece of digital evidence in our scheme.

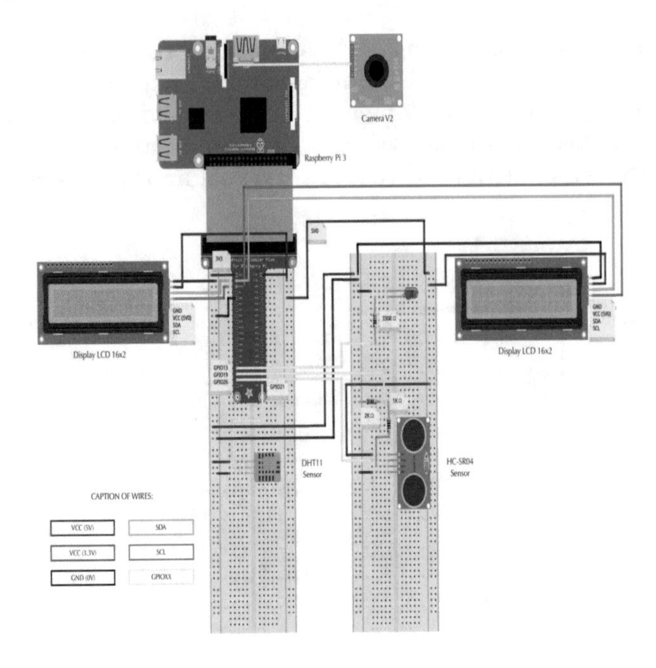

Fig. 3. Hyot prototype.

- A DHT-11 sensor to monitor the temperature and humidity.
- An HC-SR04 sensor to monitor the distance of any potential intruder with respect to the location of the control system (i.e., the Raspberry Pi).
- Two 16×2 LCD displays and a red LED as visual alert system.

On this point, we recall that video recording is initiated when one or more of the other two sensors register a value above a specific threshold.

4 Conclusions

Along this work we have used Hyperledger Fabric to build up a protocol for event recording in IoT. The implementation of such a protocol comprises low-level hardware components and high-level software elements. This being the case, we have implemented a hardware platform based on Raspberry Pi with a meaningful set of sensors. These sensors are targeted to monitoring the activity in a restricted access area in an organisation. The data obtained through the Raspberry Pi is stored in a cloud storage media, once it has been encrypted and signed to guarantee a zero trust model with respect to the cloud storage service provider. Hyperledger Fabric is on charge of guaranteeing the chain of custody of any piece of evidence, and thus a hash of the evidence is stored in the BC in case an anomalous behaviour is detected.

While we have provided a functional prototype, ensuring the most relevant security properties in the use case at hand, we also envision further improvements. Among them, for instance, we can point out a minor modification in the flow depicted in Fig. 1. Namely, we would invert the steps for DB storage and persistence in Hyperledger Fabric, with the purpose of also storing in the DB the transaction in which the specific event was persisted in the BC. While a minor change, this may help provide the needed pieces of evidence in a more natural way. Additionally, we can also consider possible combinations with public BC systems. While we have argued the suitability of permissioned BC systems for this case, it may also be a good approach to periodically submit bundles (e.g., as Merkle roots or equivalent types of cryptographic accumulators) of the events registered in Hyperledger Fabric. In this manner, we add yet one further integrity control based on a different set of trust assumptions, which may prove to be useful on concrete circumstances or for specific use cases.

Acknowledgements. This work has been supported by the Comunidad de Madrid (Spain) under the project CYNAMON (P2018/TCS-4566), co-financed with FSE and FEDER EU funds.

References

1. Androulaki, E., Barger, A., Bortnikov, V., Cachin, C., Christidis, K., De Caro, A., Enyeart, D., Ferris, C., Laventman, G., Manevich, Y., et al.: Hyperledger Fabric: a distributed operating system for permissioned blockchains. In: Proceedings of the Thirteenth EuroSys Conference, p. 30. ACM (2018)
2. Bartoletti, M., Pompianu, L.: An analysis of bitcoin OP_RETURN metadata. In: International Conference on Financial Cryptography and Data Security, pp. 218–230. Springer (2017)
3. Bernstein, D.E.: Expert witnesses, adversarial bias, and the (partial) failure of the Daubert revolution. Iowa Law Rev. **93**, 451 (2007)
4. Catalini, C., Gans, J.S.: Initial coin offerings and the value of crypto tokens. Technical report, National Bureau of Economic Research (2018)
5. Conoscenti, M., Vetro, A., De Martin, J.C.: Blockchain for the internet of things: a systematic literature review. In: 2016 IEEE/ACS 13th International Conference on Computer Systems and Applications (AICCSA), pp. 1–6. IEEE (2016)
6. Croman, K., Decker, C., Eyal, I., Gencer, A.E., Juels, A., Kosba, A., Miller, A., Saxena, P., Shi, E., Sirer, E.G., et al.: On scaling decentralized blockchains. In: International Conference on Financial Cryptography and Data Security, pp. 106–125. Springer (2016)
7. Danezis, G., Gürses, S.: A critical review of 10 years of privacy technology. In: Proceedings of Surveillance Cultures: A Global Surveillance Society, pp. 1–16 (2010)
8. De Domenico, M., Baronchelli, A.: The fragility of decentralised trustless socio-technical systems. EPJ Data Sci. **8**(1), 2 (2019)
9. Franco, P.: Understanding Bitcoin: Cryptography, Engineering and Economics. Wiley, Chichester (2014)
10. Gencer, A.E., Basu, S., Eyal, I., van Renesse, R., Sirer, E.G.: Decentralization in Bitcoin and Ethereum Networks (2018). http://arxiv.org/abs/1801.03998
11. Hassan, S., De Filippi, P.: The expansion of algorithmic governance: from code is law to law is code. Field Actions Sci. Rep. (17), 88–90 (2017). https://journals.openedition.org/factsreports/4518

12. Katz, J., Menezes, A.J., Van Oorschot, P.C., Vanstone, S.A.: Handbook of Applied Cryptography. CRC Press, Boca Raton (1996)
13. Lee, E.: The past, present and future of cyber-physical systems: a focus on models. Sensors **15**(3), 4837–4869 (2015)
14. Nakamoto, S.: Bitcoin: a peer-to-peer electronic cash system (2008)
15. Peters, G.W., Panayi, E.: Understanding Modern Banking Ledgers through Blockchain Technologies: Future of Transaction Processing and Smart Contracts on the Internet of Money. arXiv preprint arXiv:1511.05740, pp. 1–33 (2015)
16. Sammons, J.: The Basics of Digital Forensics: The Primer for Getting Started in Digital Forensics. Elsevier, Amsterdam (2012)
17. Sanchez-Gomez, A., Diaz, J., Hernandez-Encinas, L., Arroyo, D.: Review of the main security threats and challenges in free-access public cloud storage servers. In: Daimi, K. (ed.) Computer and Network Security Essentials, pp. 263–281. Springer International Publishing, Cham (2018). https://doi.org/10.1007/978-3-319-58424-9_15
18. Szabo, N.: Smart contracts: formalizing and securing relationships on public networks. First Monday **2**(9) (1997). https://ojphi.org/ojs/index.php/fm/article/view/548/469
19. Tsankov, P., Dan, A., Cohen, D.D., Gervais, A., Buenzli, F., Vechev, M.: Securify: practical security analysis of smart contracts, July 2018. http://arxiv.org/abs/1806.01143

Special Session: Looking for Camelot: New Approaches to Asses Competencies

Data Mining for Statistical Evaluation of Summative and Competency-Based Assessments in Mathematics

Snezhana Gocheva-Ilieva(✉), Marta Teofilova, Anton Iliev, Hristina Kulina, Desislava Voynikova, Atanas Ivanov, and Pavlina Atanasova

Faculty of Mathematics and Informatics, University of Plovdiv Paisii Hilendarski, 24 Tzar Asen Str., 4000 Plovdiv, Bulgaria
{snow,marta,aii,kulina,desi_voynikova,aivanov, atanasova}@uni-plovdiv.bg

Abstract. Measuring student achievement and competencies in mathematics is important for the teacher and the educational system, as well as in view of improving the motivation to learn among students. In this study we aim to develop an assessment methodology based on data mining approach. Two data mining techniques – cluster analysis and classification and regression trees (CART) are applied to investigate the influence of assessment elements on the final grade in two core mathematical subjects - linear algebra and analytical geometry. In addition, the specialty, academic year of education, and the sex of the students, as well as competency-based test results on content covered by secondary education curriculum are included. Using hierarchical cluster analysis the variables of scale type are classified into two clusters. CART models are built to regress and predict the summative assessment results in dependence of examined variables. The obtained models fit well over 90% of the data. It was established the relative importance of used variables in the model. The obtained results help to measure directly the student achievements and competencies in mathematics.

Keywords: Mathematical competencies · Summative assessment · Data mining · Cluster analysis · Classification and Regression Trees

1 Introduction

In a high-tech digital society, mathematics plays an important role in everyday activities of many professions – from engineers to accountants, sociologists and data scientists. In the field of engineering "A European comparison of competencies valued by employers and faculty showed that whereas teachers valued knowledge and research skills, employers valued planning, communication, flexibility, creativity, problem solving, and interpersonal skills. This raises questions of research design because it might be thought that the acquisition of research skills would necessarily involve problem solving", [1].

© Springer Nature Switzerland AG 2020
F. Martínez Álvarez et al. (Eds.): CISIS 2019/ICEUTE 2019, AISC 951, pp. 207–216, 2020.
https://doi.org/10.1007/978-3-030-20005-3_21

If acquired knowledge serves as the backbone of education, then its measurement in quantitative and qualitative terms with regard to an individual's ability to apply this knowledge is the basis of the notion of competency. This term was first introduced by the American psychologist R.W. White in 1959 as a concept for performance motivation [2]. In the field of mathematics, the concept is considered in varied forms. This study uses the concept of mathematical competency and its respective levels in the context of university engineering education in mathematics as developed by the Danish KOM project and the SEFI group, presented in [3, 4]. The main thesis is focused on the framework for Mathematics Curricula in Engineering Education and a competency-based approach to teaching and studying mathematics. The concept of mathematical competence includes eight distinguishable but overlapping mathematical competencies which are [4]: "C1: thinking mathematically, C2: reasoning mathematically, C3: posing and solving mathematical problems, C4: modeling mathematically, C5: representing mathematical entities, C6: handling mathematical symbols and formalism, C7: communicating in, with and about mathematics, C8: making use of aids and tools".

In the field of tools for assessment of mathematical competency, we outline papers [5, 6], as well as some publications on the application and role of computer-aided training in mathematics [7, 8]. It is also worth mentioning the achieved results within the framework of the RULES_MATH project [9], whose main objective is to develop assessment standards for a competency-based teaching-learning system for mathematics in engineering education in a European context. Papers [10, 11] present the project's strategies, objectives, and specific approaches to assessing mathematical competencies.

Another important aspect of scientific interest is the processing of results from formative, continuous, summative and other types of assessments of the knowledge, skills and attitudes which define the competencies of students. In recent years, new and powerful computer-based methods have been used for the statistical processing of various collected data. Special note is given to data mining (DM) methods, capable of extracting significant information, patterns and dependencies, which cannot be obtained through traditional statistical methods. In [12], decision trees are used to classify data from a survey about different skills and competencies possessed by the respondents. Based on the competencies, a prediction is made about the most suitable future jobs of students in IT companies. Authors of [13] report the results from a case study that uses neuro-fuzzy tool to classify and predict Electrical engineering students graduation achievement based on mathematics competency. An adaptive neuro fuzzy interference system (ANFIS) scheme is applied with back propagation and models are selected through cross-validation by two randomly selected data subsets. In [14], decision trees, random forest and naïve Bayes classification are applied in order to extract information about the most significant factors which influence the performance of students so as to reduce the number of drop-outs from the university. The conducted empirical study is based on an anonymous survey with data about the personal, social and cultural status of students, as well as their performance on tests during the previous semester. The recent review paper [15] provides numerous other examples of data mining, clear descriptions of this type of methods and their applications in education.

In this paper, we focus on student's achievements on just two mathematical subjects – in linear algebra (LA) and analytical geometry (AG), for which we used to develop an assessment methodology based on data mining approach. The objective is to study the results of math data assessment in the context of competencies for the Linear algebra and analytical geometry (LAAG) subject studied in the first year student's curriculum in informatics specialties in the Faculty of Mathematics and Informatics (FMI) at University of Plovdiv Paisii Hilendarski, Bulgaria. To this end, the DM methods of cluster analysis and decision trees (CART) are applied. The data are processed using the SPSS 25 package.

2 Empirical Data and Methodology

2.1 Data Description

Historical type data are used for the grades of LAAG subject and the solved exam problems, short 10-min competency tests in the same field and other information. The math assessment data were collected for 423 students from the first, second and third year of study at FMI in two specialties: Business Information Technologies (BIT) and Software Engineering (SE). At the time of the survey, the students were from the first, second and third year of SE and from the first and the second year of BIT. The main scale variable Final_grade is the summative assessment of the results from the final exam on LAAG during the first course of university studies. This assessment includes the points from 4 mathematical problems, where each of the problems is evaluated with a maximum of 20 points. The first two problems cover the LA subject and the other two – AG. An example of the final exam is shown in Fig. 1.

Problem 1. Given the matrices

$$A = \begin{pmatrix} -2 & 1 & 1 \\ 3 & -2 & -1 \\ 1 & 1 & -3 \end{pmatrix}, \qquad B = \begin{pmatrix} 2 & -1 & 3 \\ 1 & 1 & 1 \\ 0 & 1 & -1 \end{pmatrix}$$

find: a) the determinant of A; b) the inverse matrix A^{-1} of A; c) the matrix product $C = AB$.

Problem 2. Solve the system of linear equations

$$\left| \begin{array}{lcr} x_1 + 2x_2 + x_3 - 2x_4 & = & -2 \\ x_1 + x_2 - x_3 + 2x_4 & = & 1 \\ 2x_1 - 3x_2 - x_3 & = & 4 \\ 3x_1 + x_2 \qquad\quad - x_4 & = & 1. \end{array} \right.$$

Problem 3. In the Oxy-plane, there it is given a triangle ABC with $A(2,4)$, $B(-1,1)$, $C(-4,2)$. Find: a) the equation of the line passing through points A and B; b) the equation of the line containing the altitude passing through point A; c) the area of the triangle ABC.

Problem 4. Given the points $A(2,1,0)$, $B(1,-1,3)$, $C(1,2,4)$ and the plane $\alpha : 2x - y + 3z + 4 = 0$ find: a) the equation of the line p passing through point A and perpendicular to the plane α;
b) the equation of the plane β containing points A, B and C.

Fig. 1. Example of the LAAG exam.

Additionally, for 139 (or about 33%) of all involved students, a short 10-min competency test is conducted. This test includes 4 tasks from the secondary school mathematics curriculum from which 2 algebra problems and 2 geometry problems. An example of the 10-min competency test is shown in Fig. 2. Moreover, data on the current student academic year, specialty and sex from the educational arrays are also used.

Through the competency-based tests, as shown in the example in Fig. 2, our goal is to assess the mathematical competencies as follows. The first test question T1 evaluates the ability of the student with regard to competencies C2 and C6 (reasoning mathematically and handling mathematical symbols and formalism) since it requires recognition of the problem and understanding concepts such as the solution of equation, domain of definition, substitution in a function and verifying equality through elementary transformations. The second question T2 refers to competencies C2 and C3. T3 requires developed competencies for C3, and T4 – for C5. By applying this “pre-university” type assessment for students from the first, second and third year of their studies, we are in fact evaluating the persistence of competencies retained during the first, second and third years after attaining the school and university education.

TEST OF COMPETENCY
Name of the student:

Rules Math
NEW RULES FOR ASSESSING MATHEMATICAL COMPETENCIES

T1: Which one is the solution of the equation $\lg^2 x = 3 - 2.\lg x$:

a) $x_1 = 10,\ x_2 = 0.001$
b) $x_1 = -10,\ x_2 = 10$
c) $x = 10$
d) real solutions do not exist.

T2: Solve the inequality and give your answer $x^2 - 3|x| > 0$.

Your answer:

T3: Given the plot of the function $y = -3/4x + 3$ in rectangular coordinate system xOy find the value of the area of triangle ΔAOB.

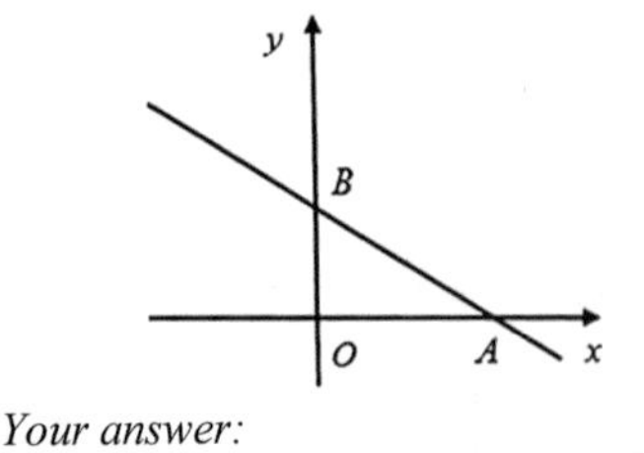

Your answer:

T4: Which one of the identities is true for any α:

a) $\sin(\alpha) = -\cos(-\alpha)$
b) $\sin(\alpha) = -\sin(\alpha + \pi/2)$
c) $\sin(\alpha) = -\sin(-\alpha)$
d) $\sin(\alpha) = \frac{1}{2}(\cos(2\alpha) + \sin(2\alpha))$

Fig. 2. Example of a 10-minute competency-based test.

The variables used in the analyses and descriptive statistics are presented in Table 1. This table shows that the mean values of P1, P2, P3, P4 decrease almost to the same extent as the corresponding mean values of T1, T2, T3, except for T4. This gives us an idea of the relatively good match of the exam problems with the selected competency tasks.

Table 1. Variables used in the analyses with descriptive statistics.

Variable	Description	Type	Values	Mean value	Mean value, %
Final_grade	Summative LAAG assessment	Scale	0, …, 80	51.16	64.0
P1	Score for exam Problem 1	Scale	0, …, 20	17.33	86.6
P2	– for Problem 2	Scale	0, …, 20	14.50	72.5
P3	– for Problem 3	Scale	0, …, 20	11.72	58.6
P4	– for Problem 4	Scale	0, …, 20	7.64	38.2
T1	Score for competency test, T1	Scale	0, 0.5, 1	0.63	63.0
T2	– for T2	Scale	0, 0.5, 1	0.43	43.0
T3	– for T3	Scale	0, 0.5, 1	0.28	28.0
T4	– for T4	Scale	0, 0.5, 1	0.56	56.0
Course_Year	Student's course year	Nominal	1, 2, 3	–	–
Specialty	BIT or SE, in number	Nominal	186/237	–	–
Sex	Male or Female, in number	Nominal	277/155	–	–

2.2 Data Mining Methods

The collected data are examined using well-known DM techniques – hierarchical cluster analysis (CA) and classification and regression trees (CART).

CA is a statistical DM technique with many applications in different areas such as medicine, biology, genetics, for data and text mining in computer science, climatology, education, etc. The CA method uses only quantitative variables which are rescaled in advance to avoid measurement unit differences. The objective is to group together closely spaced variables or cases in clusters. Small distance are interpreted as homogeneity or similarity and large distances – as difference. The distance between the clusters is defined using various clustering methods – Between-groups Linkage, Nearest Neighbor, Ward's Linkage, etc. A good cluster model includes well-differentiated clusters and clearly defined elements (variables or cases) within these. A detailed description of CA and its capabilities can be found for example in [16, 17].

The CART method is among the top 10 algorithms for DM to classify and discover dependencies [18]. It is used to process all data types – from nominal to interval without any special distribution requirements. This technique builds a tree-type diagram model, starting with a root containing a full data set. The objective is to separate and split down this set into subsets (classes) called nodes of the tree. In the case of regression with a dependent scale variable Y, at each step the CART algorithm selects the independent variable (predictor) X with the strongest influence and one of its threshold values x^*, according to which the dataset in current node is split into two child nodes. All cases in which x_k from X are smaller than x^* are assigned to the left child node and the rest – to the right. For each node, the value predicted by the CART model for Y is an arithmetic mean of y_k in that node. The selection of X and x^* depends on minimizing a selected error criterion in the regression for Y. Control parameters are used to stop the growth of the tree with the last leaves called terminal nodes, containing distributed initial data set. As a DM method, CART yields a large number of models. The standard method for model selection is based on cross-validation procedure by comparing the results from the prediction of learning and test random samples [18].

3 Results from Statistical Analyses

3.1 Initial Data Processing

By calculating the descriptive statistics and tests for the distribution of the quantitative variables it was found that these are not normally distributed. In order to investigate the presence and extent of correlation between these variables a nonparametric Spearmen's rho test was applied. Significant correlation coefficients for Final_grade and P1, P2, P3, P4 and lack or weak correlations with T1, T2, T3, and T4 were obtained as follows:

$$\begin{aligned} &(Final\_grade, P1, P2, P3, P4) = (0.426, 0.653, 0.822, 0.837), \\ &(Final\_grade, T1, T2, T3, T4) = (0.128, 0.108, 0.414, 0.105). \end{aligned} \quad (1)$$

From (1) we can conclude that P3 and P4 have a major influence on Final_grade.

3.2 Results from Cluster Analysis

In order to build a cluster model, all quantitative type variables are used with preliminary rescaling of their values in the interval [0, 1]. Two types of distance are applied in the examined 9-dimensional space, considered with reference to variables – standard Euclidean distance and squared Euclidean distance. Numerous models were built and compared. Figure 3 shows a dendrogram of the hierarchical grouping of the variables using Euclidean distance and Ward's cluster method. It is observed two well-differentiated clusters spaced at least by 5 units in terms of relative rescaled distance. This way the following 2-cluster solution is identified

$$K1 = \{Final\_grade, P3, P4, P1, P2, T1, T4\}, \quad K2 = \{T2, T3\}. \quad (2)$$

The dendrogram shows that all P1, P2, P3, P4 are in the same cluster as Final_grade and there is no problem in the exam which is isolated from this core group. In the first cluster are also classified T1 and T4. Despite having the same subject matter – algebra and geometry, variables T2 and T3 are significantly more distant and form a separate cluster. An almost analogous cluster model was obtained using squared Euclidean distance with Ward's method. Based on the examined data and the considered test, it can be concluded that problems P3 and P4 on AG have the strongest influence on the final grade in

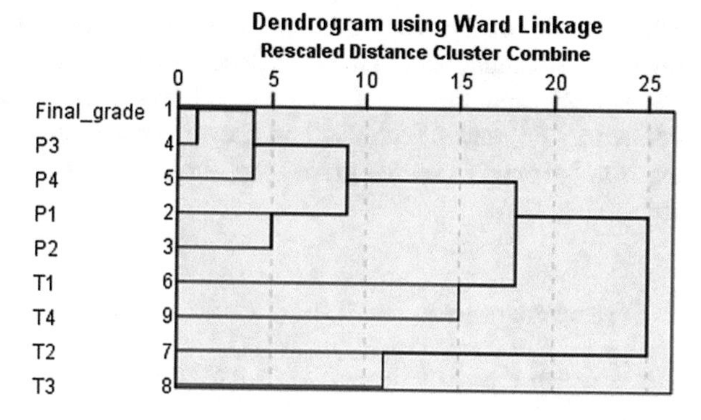

Fig. 3. Dendrogram with the two cluster solution (2) obtained using Euclidean distance and Ward's clustering method.

combination with competencies from T1 and T4 (reasoning mathematically, handling mathematical symbols and formalism, and representing mathematical entities).

3.3 Results from CART Modeling

Two statistical analyses were carried out using the CART method. The numbers (m1, m2) were set as control parameters to stop tree growth, where m1 is the minimum number of cases in a parent node and m2 is the minimum number of cases in a child node. The quality of the models for regressing the Y variable are assessed using goodness-of-fit criteria Root Mean Square Error (RMSE) and R^2 (coefficient of determination) according to the formulas

$$R^2 = \sum_{k=1}^{n} (p_k - \bar{Y})^2 / \sum_{k=1}^{n} (y_k - \bar{Y})^2, \; RMSE = \left(\sum_{k=1}^{n} (y_k - p_k)^2 / n \right)^{1/2}, \tag{3}$$

where p_k are predicted values for y_k, $k = 1, 2, \ldots, n$, $\bar{Y}$ is the mean value, n is the sample size. To avoid overfitting the model, we used the standard method of 10-fold cross-validation [18].

With the first analysis CART models of Final_grade were built with dependence on problems P1, P2, P3 and P4, the student's year of study, specialty and sex using the initial dataset for 423 students. When varying m1, m2, several close in performance models A1, A2, A3 were selected. The corresponding goodness-of-fit statistics are shown in the left part of Table 2. Model A3 is the one with the best performance achieving data fit $R^2 = 92.6\%$ and $RMSE = 5.587$.

Table 2. Summary statistics of the obtained best CART models of Final_grade.

Model (m1,m2)	R^2	RMSE	Model (m1,m2)	R^2	RMSE
A1 (80,40)	82.8%	8.519	B1 (40,20)	81.3%	7.732
A2 (40,20)	83.8%	8.256	B2 (20,10)	84.1%	7.129
A3 (20,10)	92.6%	5.587	B3 (10,5)	93.4%	4.601

Figure 4a presents a plot of the relative normalized importance of the variables in model A3, leading us to conclude that Final_grade depends to the greatest extent on the last problems P4 and P3, and then on P2 and P1. The specialty has about 12% significance compared to P4, and the course and the sex of the student are not significant. A comparison between observed grades and those predicted by model A3 is shown in Fig. 4b.

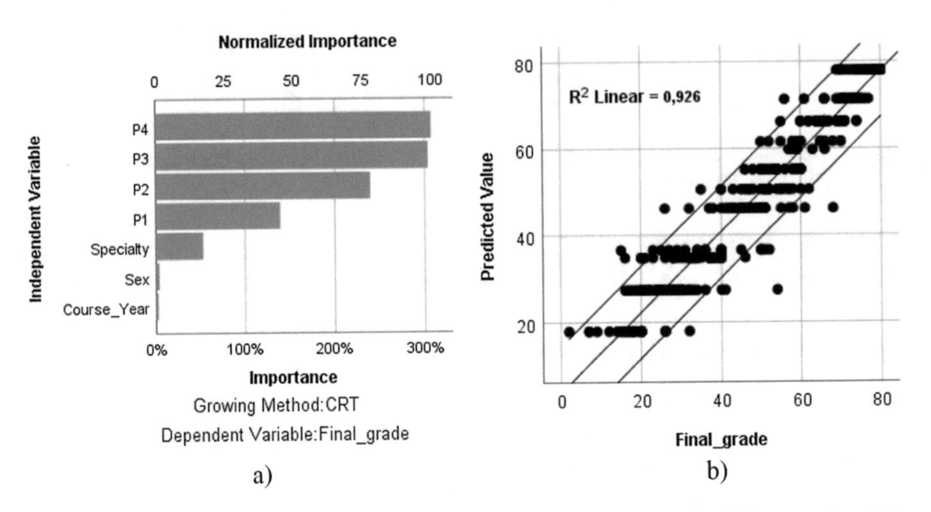

Fig. 4. (a) Importance of independent variables in CART model A3; (b) Values predicted by model A3 for Final_grade vs. measured ones.

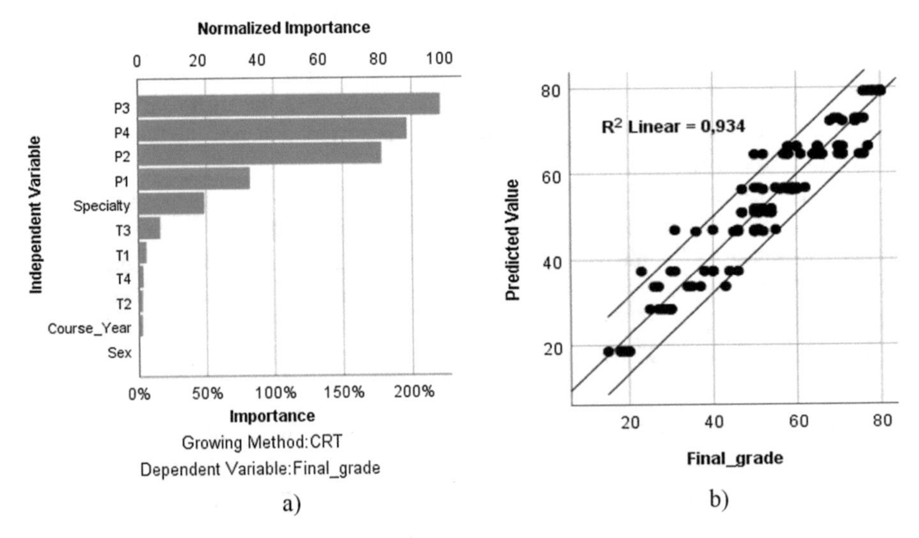

Fig. 5. (a) Importance of independent variables in CART model B3; (b) Values predicted by model B3 for Final_grade vs. measured ones.

The second CART analysis has been conducted using data from $n = 139$ students and inclusion of T1, T2, T3 and T4. The main models obtained are denoted by B1, B2 and B3. Model statistics are given in the right part of Table 2. We selected CART model B3 with $R^2 = 93.4\%$ and $RMSE = 4.601$. Relative plots are given in Fig. 5a, b.

Figure 5a shows the influence of the variables in model B3 very similar to that in model A3. Competency-based test variables do not have much influence. A more significant relative contribution to the model has T3 with 7.3%. This is explained by the indirect correspondence of T1, T2, T3 and T4 to the results of the LAAG exam. However, with a maintained ratio of control parameters, model B3 slightly improves A3 statistics while remaining very good in prediction, as illustrated in Fig. 5b.

4 Conclusion

In this study two DM approaches to study LAAG exam data are combined with data from short 10-min competency-based tests as well as with other type of information. The results of the several different analyzes made lead to close conclusions.

By cluster analysis a classification of quantitative variables is obtained in 2 clearly differentiated clusters. In particular, the cluster model established the grouping of competency tests with the specific test results, namely the proximity of the summative assessment for LAAG with the C2, C5 and C6 competencies.

A CART method was applied to two sets of data. Models were developed that reveal the specific weight of each of the participating variables in determining the final assessment grade for LAAG subject. The first type of models assess the impact of exam components and additional factors such as the student's specialty in attending, and others. The second type of models also assess the importance of 10-min competency-based tests. The best CART models describe over 90% of the data. A good agreement between the two types of CART models and the cluster model was obtained.

The achieved results from the empirical study show very promising capabilities of DM approaches. In particular, these demonstrate the extensive capabilities for flexible statistical modeling using CART method to investigate and predict results of different assessment approaches, as well as competency-based assessment in mathematics.

Acknowledgments. This work was co-funded by Erasmus+ program of the European Union under grant 2017-1-ES01-KA203-038491 (RULES_MATH). The first and third authors acknowledge partial support from the Grant No. BG05M2OP001-1.001-0003, financed by the Science and Education for Smart Growth Operational Program (2014–2020), co-financed by the European Union through the European structural and Investment funds.

References

1. Heywood, J.: The Assessment of Learning in Engineering Education, Practice and Policy. John Wiley & Sons Inc., Hoboken (2016)
2. White, R.W.: Motivation reconsidered: the concept of competence. Psychol. Rev. **66**(5), 297–333 (1959)

3. Niss, M.: Mathematical competencies and the learning of mathematics: the Danish KOM project. In: 3rd Mediterranean Conference on Mathematical Education, pp. 115–124. The Hellenic Mathematical Society, Athens (2003)
4. Alpers, B., Demlova, M., Fant, C.-H., Gustafsson, T., Lawson, D., Mustoe, L., Olsson-Lehtonen, B., Robinson, C., Velichova, D. (eds.): A Framework for Mathematics Curricula in Engineering Education: A Report of the Mathematics Working Group. European Society for Engineering Education (SEFI), Brussels (2013)
5. Carr, M., Brian, B., Fhloinn, E.N.: Core skills assessment to improve mathematical competency. Eur. J. Eng. Educ. **38**(6), 608–619 (2013)
6. Fhloinn, E.N., Carr, M.: Formative assessment in mathematics for engineering students. Eur. J. Eng. Educ. **42**(4), 1–13 (2017)
7. Sangwin, C.J., Köcher, N.: Automation of mathematics examinations. Comput. Educ. **94**, 215–227 (2016)
8. Albano G.: Knowledge, skills, competencies: a model for mathematics e-learning. In: Kwan, R., McNaught, C., Tsang, P., Wang, F.L., Li, K.C. (eds.) Enhancing Learning Trough Technology. Education Unplugged: Mobile Technologies and Web 2.0. ICT 2011. Communications in Computer and Information Science, vol. 177, pp. 214–225. Springer, Heidelberg (2011)
9. RULES_MATH Homepage. https://rules-math.com/. Accessed 25 Jan 2019
10. Rasteiro, D.M.L.D., Martinez, G.V., Caridade, C., Martin-Vaquero, J., Queiruga-Dios, A.: Changing teaching: competencies versus contents. In: Global Engineering Education Conference, EDUCON 2018, pp. 1761–1765. IEEE Press, New York (2018)
11. Queiruga-Dios, A., Sanchez, G.R., Del Rey, A.M., Demlova, M.: Teaching and assessing discrete mathematics. In: Global Engineering Education Conference, EDUCON 2018, pp. 1568–1571. IEEE Press, New York (2018)
12. Tapado, B.M., Acedo, G.G., Palaoag, T.D.: Evaluating information technology graduates employability using decision tree algorithm. In: The 9th International Conference on E-Education, E-Business, E-Management and E-Learning, IC4E 2018, pp. 88–93, ACM International Conference Proceedings Series, New York (2018)
13. Mat, U.B., Buniyamin, N.: Using neuro-fuzzy technique to classify and predict electrical engineering students' achievement upon graduation based on mathematics competency. Indonesian J. Electr. Eng. Comput. Sci. **5**(3), 684–690 (2017)
14. Jain, A., Choudhury, T., Mor, P., Sabitha, A.S.: Intellectual performance analysis of students by comparing various data mining techniques. In: 3rd International Conference on Applied and Theoretical Computing and Communication Technology (iCATccT), pp. 57–62. IEEE Press, Singapore (2017)
15. Dutt, A., Ismail, M.A., Herawan, T.: A systematic review on educational data mining. IEEE Access **5**, 15991–16005 (2017)
16. Rencher, A.C., Christensen, W.F.: Methods of Multivariate Analysis, 3rd edn. Wiley, New York (2012)
17. Rokach, L., Maimon, O.: Clustering methods. In: Maimon, O., Rokach, L. (eds.) Data Mining and Knowledge Discovery Handbook, pp. 321–352. Springer, Boston (2005)
18. Wu, X., Kumar, V.: The Top Ten Algorithms in Data Mining. Chapman & Hall/CRC, Boca Raton (2009)

Looking for the Antidote for Contaminated Water: Learning Through an Escape Game

María Jesús Santos[1(✉)], Mario Miguel[2], Araceli Queiruga-Dios[3], and Ascensión Hernández Encinas[1]

[1] Science Faculty, University of Salamanca, Salamanca, Spain
smjesus@usal.es
[2] Maristas School, Salamanca, Spain
[3] ETSII Béjar, University of Salamanca, Salamanca, Spain

Abstract. This article presents a Breakout played with students of the Master in Teaching of Compulsory Secondary Education and Higher Secondary School, at the University of Salamanca, as part of their training as future teachers. Both the difficulties and the advantages that this type of educational activities bring out are analyzed. Moreover, some possibilities of its possible utilization in the classroom, at different educational levels, are also discussed and presented.

Keywords: Escape Room · Breakout educative · Gamification · Competencies · Assessment

1 Looking for the Antidote for Contaminated Water

I am attending the master course with 14 fellows, in a "Physics and Chemistry Didactics" lesson. An unknown man arrives at the beginning of the class. He identifies himself as the person responsible for the Health Department of the Spanish Government. He asks for our help in solving a serious problem existing in the building of our Faculty: the water is absolutely contaminated. We have been selected to begin working as part of an interdisciplinary team of graduates in Chemistry, Chemical Engineering, Physics, and Energy Engineering.

On the classroom screen we see a video where the problem is set out: the polluted water of our building could be extended both to the University of Salamanca and to the rest of the city, endangering the life of the population. In addition it would tarnish the celebration of the VIIIth Centenary of our institution, it would project a negative image of our city. We have to find the chemical combination that could purify the water as soon as possible to avoid massive poisonings among the inhabitants of the city, students, and tourists who visit us.

The initial shock turned into an exciting experience. This was the introduction to a game. We have 60 min to solve some riddles and problems to find the antidote and to make the water drinkable.

© Springer Nature Switzerland AG 2020
F. Martínez Álvarez et al. (Eds.): CISIS 2019/ICEUTE 2019, AISC 951, pp. 217–226, 2020.
https://doi.org/10.1007/978-3-030-20005-3_22

We divide ourselves into two groups and each one receives a bag containing a roadmap (with a brief outline of the game), a meter, paper, scissors, and pens.

We enter a classroom in which there are a box with six digital locks (Fig. 1) and different objects both within sight and hidden. Interest, intrigue, and curiosity in knowing the next stage was increasing. Game starts!

Fig. 1. Box with six digital locks: view of the locks (left) and the QR code to access them (right). The QR codes were on the back of the padlock.

This is what students found in our Breakout implementation. Students had no idea about the activity that us, as supervisors, were proposing and they were anxious and full of expectation.

2 Solving the Riddles and Problems

The main part of a Escape Room or Breakout is the riddle solving. Nicholson *et al.* [8] defined them as: "Escape rooms are live-action team-based games where players discover clues, solve puzzles, and accomplish tasks in one or more rooms in order to accomplish a specific goal (usually escaping from the room) in a limited amount of time".

In our case, players must search and classify all the material they find in the room (Fig. 2). There were posters with Quick Response codes (QR), paper bows, water glasses, straws, pieces of a white puzzle, lab coats, balloons with messages, an invisible ink pen with a small flashlight on the lid, an augmented reality (AR) marker, etc. In the classroom there are also a computer and a screen in which a timer is displayed.

The students must verify that the QR codes corresponding to each lock give them access to a *Google form* questionnaire with one question. When they answer it properly they will open the corresponding padlock. Figure 1 shows the box with the six digital locks [7]. These are numeric, with letters and directional locks and each theme riddle/padlock has a different colour.

The game was designed for students of the Master in Teaching of Compulsory Secondary Education and Higher Secondary School. Therefore, it includes tests

Fig. 2. The players look for and classify all the material they find in the room.

of Physics, Chemistry and Education. In addition, an effort has been made to work different multiple intelligences carefully choosing the tests.

2.1 Parabolic Motion

Students need to find some hidden information throughout the room. Some drinking straws containing small papers are at their disposal, and one of these papers has a web address. When they access the site they see a scene from the film "In July" [10], which is about a parabolic throwing problem. Solving this problem they find four digits, i.e., the key that opens a numerical padlock.

They identify the padlock and open it. So, to get the password of the orange numerical padlock, students need to solve a parabolic throwing problem.

2.2 Types of Plastics

The *Google form* that is accessed with the QR code of the green padlock asks for three numbers. On the one hand, students found a crossword puzzle inside balloons. On the other hand, there is a stuck poster on the wall which gives them, through an AR App, the necessary information about the plastics with which the pieces of *Lego* are made. This is the key to solve the crossword puzzle. In addition there were selected, in a different colour three letters, which, using its corresponding place number in the alphabet, will be the key to the green padlock.

2.3 Refraction Index

In this case, the proposal is to solve an experimental test. Among the material found in the room there are an apparently white puzzle, an invisible ink pen with light on its lid (Fig. 3, left), a glass, a bottle of water and a stick. When the puzzle is illuminated with ultraviolet light, a formula can be observed (Fig. 3, right). This formula allows them to determine the refractive index of a liquid inside

Fig. 3. An apparently white puzzle and an invisible ink pen with light on its lid (left). When the puzzle is illuminated with ultraviolet light, a formula can be observed (right).

a glass, experimentally, after taking the corresponding measurements. When they find the refractive index of the water, comparing it with a key-code which associates number intervals with the names of different colours, they will find out the name of the colour that opens the correct padlock.

2.4 Musical Ladder with Glasses

The students found tall glasses in the classroom, a bottle of water and a spoon to perform this musical challenge (Fig. 4). In the paper made bow ties they found different *Twitter* accounts from authors working on Breakout issues. One of these tweets contains the instructions to carry out the experience. They had to tune a musical scale with a different volume of water in each glass, to record a video with the musical scale and upload it to a *Twitter* account. Then they got the key to the purple padlock.

Fig. 4. The students found in the classroom tall glasses, water, and a spoon to perform this musical challenge. They had to achieve a tuned musical scale with a different volume of water in each glass.

2.5 Metacognition Staircase

To open the blue lock, students should know the meaning of the *metacognition scale*. The players, when scanning a hidden QR code, could see a video that explains what the *metacognition scale* consists on. On the other hand, they have to find some incomplete words, as in the *hangman game* [1]. With the video information they should be able to complete the sentences of the *metacognition scale*. Once resolved, they should select the letters marked with red colour. It is completed in ascending order from top to bottom. This is the key with the correct combination of letters for the blue padlock.

2.6 Secondary Education Law

In this test the padlock's key is some information that should be known about the education law currently in force in Compulsory Secondary Education. To answer the questions, the players found the clues in an interactive image made with the web tool "Genially" [2].

2.7 Opening the Sixth Padlock

With all the tests passed and, therefore, with the locks open, the students were able to get the only real key that opens the box. The box contains the chemical compound that managed to turn poisoned turbid water into crystalline. Amazing! The big problem of the contaminated water in our building has been solved.

This is how the game ends. However, the analysis of the reasons that support the realization of this type of experiences in the classroom, with students of all age, begins here.

3 Innovation Through Breakout

A *Breakout* is an immersive game derived from the popular Escape Rooms that are being played as leisure spaces in cities around the world. Can it really contribute somehow to the teaching-learning process?

Innovation in the classroom is done for different reasons. First of all we are playing with the objective of learning. This is what is now known as *Gamification* or *Game-based learning (GBL)* [9]. It is also contextualized in a narrative, which is the motivation to solve the tests and the common thread of all the activity. As Tang *et al.* [11] says "Games-based learning takes advantage of gaming technologies to create a fun, motivating, and interactive virtual learning environment that promotes situated experiential learning".

It is also a *learning based on challenges* because students must overcome a great challenge and small challenges in an estimated time. Undoubtedly, it is a process of mobile learning or *M-learning* as technologies such as QR codes, augmented reality, searches in different databases through the Internet, *Google forms*, applications like *Genially*, etc. are usually used. If the Breakout has been

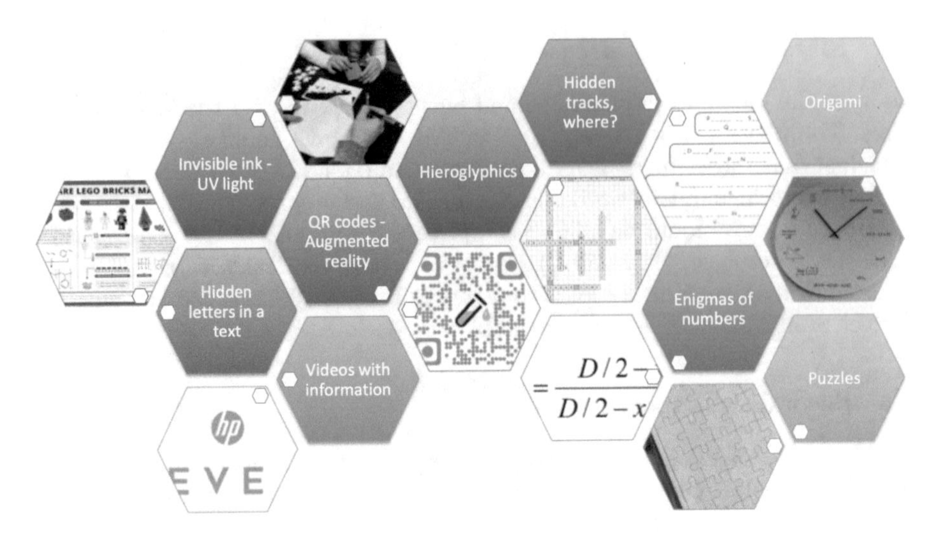

Fig. 5. Different type of tools used in the game. QR codes, augmented reality (AR) markers, puzzles, an invisible ink pen with a small UV flashlight on the lid, origami, hieroglyfics, etc.

properly prepared, *multiple intelligences* are also worked on, since the tests are of a diverse nature (Fig. 5), such as word games, riddles, maps, encrypted codes, hieroglyphics, puzzles, songs, experiments, origamis, etc.

This activity is designed so that the whole group of people involved collaborate, each one with their own skills, i.e., the most powerful Gardner intelligences [5]. It is played in teams made of 5 to 10 students, in which it is necessary to cooperate, to dialogue, to assume a role, and to reach a personal consensus to overcome the challenge. This is a fundamental aspect to take into account: the *collaborative work*.

With the use of a Breakout, different tests, challenges or problems are introduced in relationship to the *curricular content* of a subject.

There are many situations that allow themselves to introduce this tool in our teaching-learning methodology: initial motivation of a didactic unit, review of acquired learning, assessment of student learning, ask students to design tests for an activity of this type. For sure you, as a reader, also have some proposals.

The Escape Room experiences and Breakout in Education are growing, not only in Primary and Secondary Education, but also in Higher Education, as several studies show. In a recent study Clarke *et al.* [4] presents a framework for creating educational Escape Rooms for Higher/Further Education. The paper presents a pilot study that was used to assess the feasibility and acceptance of University teaching staff of embedding interactive GBL into a higher education environment. They concluded that there is a high level of interest from higher education staff on how to develop these experiences for their own teaching practice.

Vörös *et al.* [12] have developed an educational escape game for high-school students about physics of fluids. They concluded that gamification of educational process has multiple benefits: it is engaging (involves students towards active learning), stimulates curiosity, gives the "flow" experience, and gives real learning tasks. Even students with poor grades in physics had good results in the final quiz.

Hermanns *et al.* [6] presented a study to describe the use of a toolbox gaming strategy based on an Escape Room concept to help nursing students learn about cardiovascular medications in a pharmacology course. They ask for future studies to be conducted to quantify the students' academic success, i.e., test grades, self-confidence, and perceived competence after engaging in a non-traditional learning activity such as the toolbox puzzle exercise.

We think that Breakout is a really interesting activity to do with future high-school teachers, because it brings together all the aspects mentioned above. As noted by Castañeda Quintero and Adell Segura [3], an effective strategy for teacher professional development should incorporate exposure to innovations in teaching and technology, training to test new strategies and facilitate reflection and discussion aimed at a common purpose.

4 Discussion

Carrying out the evaluation with the group of students involved in the experience, the following reflection was achieved: it is an activity with numerous advantages but undoubtedly it also presents a series of difficulties, both are listed in Table 1.

Table 1. Breakout strengths and difficulties.

	For teachers	For students
Difficulties	Initial motivation	
	Calculate times adequately	Difficulty of the tests themselves
	Properly organize tests and clues	Teamwork
	Calculate the difficulty well	Working under pressure
	Excessive work	Frustration tolerance
	Teacher as manager of coaching	
Strengths		Learn playing
	Motivated/Motivator	Creativity
	Competencies-Based assessment	Multiple intelligences
	Multiple possibilities	Teamwork
	Learning	Student-centered learning

Among the difficulties that a teacher should take into account is, in the first place, the need of carrying out a good initial motivation of the activity to get all the students involved. In addition, it must be taken into account that the adequate preparation of an activity of this type takes a lot of time and dedication before the game is set up (compared this with the time used to solve it). It is necessary to organize tests and clues very well, thinking about all the phases of the process to be solved. Always putting yourself in player's shoes. It is also important to calculate the appropriate times for the resolution of each particular test and the Breakout in general. The teacher, as game master, will have a role of "coaching and scaffolding": he guides the students by providing the instructions to carry out the activity, and is aware of possible difficulties or doubts that may arise. If there is any difficulty arising in the group, the teacher will take care of giving clues or advices so that the activity can follow its course, without getting to solve it in any case.

From the student's point of view there are several clear difficulties. Of course, one of them is the intrinsic difficulty of the game's tests. But another fundamental challenge is the need to work as a team, coordinating the tasks of all the components. Otherwise, it is difficult for them to overcome the challenges, as they demand different skills, knowledge, and competencies. A third difficulty is the fact that they must solve the tests in a limited time, that means, working under pressure. It is a game and, it is not always winning. Some of the students do not have a good tolerance to frustration, they do not really accept to loose. In fact, this actually happened because none of the teams solved the six tests by themselves, they needed the teachers' help.

We think that these difficulties can become a wealth, if they are calmly analyzed. Reflection can contribute to improve in future occasions.

The strengths presented by carrying out this type of activity are greater than the difficulties. From the point of view of the teacher it is an activity that achieves a very high motivation of students to solve a problem properly. A competencies-based assessment is actually carried out, since students should be able to perform activities in a different way of the normal protocols used to solve problems. The possibilities that are open are multiple, both to include the contents of any topic and the type of tests that can be designed. But, without a doubt, the main advantage is that not only students learn, but also who prepares a Breakout learns both during the process and in conducting the test.

One of the Breakout main values for the student is the possibility of learning while playing. In addition to discover the power of teamwork, as they see how the skills of each individual are necessary for getting the outcome of the proposed challenge. At the same time, multiple intelligences are valued as something real, not just in a theoretical way. It also encourages creativity that is characterized by the ability to perceive the world in new ways, to find hidden patterns, to establish connections between apparently unrelated phenomena, and to generate solutions. Finally, it can be said that it is a learning process focused on the student. As previously mentioned, the teacher plays a supporting role to give clues only if the game does not progress.

5 Evaluation

This experience has been carried out with students of the Master in Teaching of Compulsory Secondary Education and Higher Secondary School, at the University of Salamanca. The studies of these students are: Degree in Chemical Engineering, Degree in Physics, Degree in Energy Engineering. Only 40% of the people who made up the group had played an Escape Room before, therefore it turned out to be a new experience for 60% of the students. Most of them were not familiar with the usual techniques of searching and solve the riddles and problems. Students with previous experience in Escape Rooms have more facility to look for clues.

This has been a first pilot experience on escape games and their application in teaching. We plan to carry out this activity with graduates who are being trained to be future teachers because their opinion is important once the experience has taken place. As young people, but at the same time mature, they are familiar with the innovative trends in education, and open to experimenting with new teaching-learning proposals and therefore their viewpoint are very important.

Some of the opinions of the participants in the game are gathered here, after making the evaluation:

"We did a Breakout in class. I loved it. I find it a great idea. I saw a lot of possibilities. I am motivated, I'm sure I'll make one. The balance between master classes and games is perfect to consolidate learning."

"And it is not that everything went well, neither of the two groups were able to solve the complete enigma, but we did not erase the smile from the face or the desire to get it."

"I thought it was a good experience especially to motivate and encourage teamwork, while providing different views of the subjects and work the multiple intelligences of each person."

"Probably one of the best lessons of my life."

"The experience from my point of view has been spectacular, because I have had a good time, I have learned many things, especially experimenting."

"Give thanks to those who made it possible. Motivated teachers create motivated students, and if the students are future teachers ..."

"When we repeat?"

Reading the comments of the students it is clear that it has been a really positive experience. Despite the fact that they were not able, on their own, to solve all the puzzles, it is an experience to be repeated.

6 Conclusion

A Breakout has been carried out with students of the Master in Teaching of Compulsory Secondary Education and Higher Secondary School, at the University of Salamanca. The motivation to start playing was that they had to solve the problem of non-potable water in the Faculty. Six riddles/problems with contents of Physics, Chemistry and Education have been proposed. The final prize was to discover the chemical compound that will allow to purify the water.

The evaluation of the activity, together with the group of players, future teachers of Secondary Education, allowed us to reflect on the difficulties and advantages of this type of activity.

The Breakout activity is conceived as a good practice of innovative teaching, which includes important elements for the teaching-learning process: motivation, work with multiple intelligences, evaluation by competencies, teamwork, promotion of creativity and the students-centered learning.

References

1. https://en.wikipedia.org/wiki/Hangman_(game)
2. https://www.genial.ly/en
3. Castañeda Quintero, L., Adell Segura, J.: El desarrollo profesional de los docentes en entornos personales de aprendizaje (PLE). In: La práctica educativa en la Sociedad de la Información: Innovación a través de la investigación, pp. 83–95. Editorial Marfil (2011)
4. Clarke, S., Peel, D.J., Arnab, S., Morini, L., Keegan, H., Wood, O.: escapED: a framework for creating educational escape rooms and interactive games for higher/further education. Int. J. Serious Game **4**(3), 73–86 (2017)
5. Gardner, H.: Frames of Mind: The Theory of Multiple Intelligences. Hachette, New York (2011)
6. Hermanns, M., Deal, B., Campbell, A.M., Hillhouse, S., Opella, J.B., Faigle, C., Campbell IV, R.H.: Using an "escape room" toolbox approach to enhance pharmacology education. J. Nurs. Educ. Pract. **8**(4), 89 (2017)
7. Martínez, A.M., Fernández, M., Poyatos, M.: Guía para diseñar un *breakout edu* y escape room, April 2018. http://www.blogsita.com/guia-para-disenar-un-breakout-edu-y-scape-room/
8. Nicholson, S.: Peeking behind the locked door: a survey of escape room facilities. White Paper (2015). http://scottnicholson.com/pubs/erfacwhite.pdf
9. Prensky, M.: Fun, play and games: what makes games engaging. Digit. Game-based Learn. **5**(1), 5–31 (2001)
10. Schubert, S., Schwingel, R.P., Akin, F.: In July (video film). Wueste Filmproduktion edn, Turkey, German (2000)
11. Tang, S., Hanneghan, M., El Rhalibi, A.: Introduction to games-based learning. In: Games-Based Learning Advancements for Multi-Sensory Human Computer Interfaces: Techniques and Effective Practices, pp. 1–17. IGI Global (2009)
12. Vörös, A.I.V., Sárközi, Z.: Physics escape room as an educational tool. In: AIP Conference Proceedings, vol. 1916, p. 050002. AIP Publishing (2017)

Enhancing Collaborative Learning Through Pedagogical Alignment

Júlia Justino and Silviano Rafael(✉)

Instituto Politécnico de Setúbal, Escola Superior de Tecnologia de Setúbal, Setúbal, Portugal
{julia.justino,silviano.rafael}@estsetubal.ips.pt

Abstract. This paper presents a study conducted on a mathematics' course unit where the learner-centred approach was implemented. The study follows a quantitative approach based on surveys, complemented by typical actions of qualitative approach. The pedagogical structure developed is based on the constructive alignment between the objectives, the learning activities and the assessment. A pedagogical strategy with emphasis on the application of the collaborative working group and a comparative study of the outcomes achieved over the academic years of 2017–18 and 2018–19, pointing out some relevant aspects experienced by students, are presented.

Keywords: Collaborative learning · Pedagogical strategy · Constructive alignment · Learner-centred approach · Active learning

1 Introduction

The use of collaborative learning in Portuguese tertiary education has increased in recent years for it is suitable for the students training into the real world of work, in particular the development of key competences needed by students for the modern economy (OECD 2018). However, the application of a pedagogical structure to help the teacher to establish the pedagogical technique to adopt is still not a standard procedure in Portuguese tertiary education.

This paper arises from a study conducted on a mathematics' course unit of a tertiary technological course taught in a Portuguese polytechnic institute where the collaborative working group, in addition to other active learning techniques, was the main generator of active learning space within the classroom environment where the learner-centred approach was applied. The pedagogical strategy chosen in the setting-up of this curricular unit, solidly based on the pedagogical alignment established, was amended according to the data obtained in the previous academic year 2017–18 and also to the initial survey results which allow to define the new students' profile. This amendment triggered a significant improvement on the outcomes produced by active learning implementation.

© Springer Nature Switzerland AG 2020
F. Martínez Álvarez et al. (Eds.): CISIS 2019/ICEUTE 2019, AISC 951, pp. 227–234, 2020.
https://doi.org/10.1007/978-3-030-20005-3_23

2 Research Methodology

Two research methods were applied in this study: a quantitative method and a qualitative method (Shaffer and Serlin 2004; Tashakkori and Teddlie 1998). In the quantitative method, based on students' investigations and assessments, three different surveys were carried out during the term time: initial, intermediate and final. The qualitative method, based on the interpretative approach given by the teacher's personal observation in classes upon the students' behavioural attitude along the learning process, provided useful information about how the learning process was being conducted.

3 Context of Application

In Portuguese polytechnics a new type of short-cycle tertiary educational programme (2 years – 180 ECTS) was added to the range of its course offerings, called Curso Técnico Superior Profissional (CTeSP). This programme, suitable for students from vocational education, has a strong practical and technological component in order to quickly provide the skills needed for students to join the labour market. The pedagogical method adopted in the mathematics' course unit presented in this paper was the learner-centered approach, in order to aggregate it with the other technical course units of the CTeSP. The pedagogical structure was designed taking into account the learning outcomes, the learning activities and the evaluation activities, being merged with the constructive pedagogical alignment (Biggs 1996) in all teaching and learning activities, consistently connected with the established learning outcomes. So the learning outcomes were related to the contents or the topics, the available resources, the active pedagogical techniques to be applied in each case and also to the time and type of evaluation carried out, forming a matrix. The implementation of the constructive alignment matrix requires reflection and continuous adaptation by executing a set of tasks such as: conceiving the pedagogical alignment between the learning outcomes and the learning activities, developing the learning activities and its necessary resources, implementing the pedagogical strategy that has to be in line with the learning activities, whether in classroom context or outside, continuously analysing the quality of training, which includes training and summative assessments, and adapting what is necessary according to previous results. To briefly summarise, all this procedure is developed to ensure that the pedagogical alignment exists and is respected during term time, allowing the articulation between the contents, the objectives, the students' training needs and the pedagogical techniques to be applied (Livingstone 2014). The pedagogical strategy implemented in the mathematics' course unit made use of several active learning techniques applied in the classroom environment. Qualitative and quantitative assessments were carried out to students during the classes in order to assess the progress of the application of the pedagogical structure. The development of the students' behavioural attitude (individually and in team), the specific objectives for each class within the mathematical contents and the need to optimize the students' working time, always providing them the necessary learning activities, were analysed throughout the term time. In this context, the collaborative working group (Barkley et al. 2014; Burke 2011; Chiriac 2014; Gokhale 1995; Laal and Ghodsi 2012) was the most applied active learning technique.

4 Pedagogical Strategy and Dynamics

The implementation of the pedagogical structure aforementioned within the learner-centred approach, undergoes continuous reflective interactions, allowing the correction of the learning paths during term time, the change of the contents' format, the adaptation of the learning techniques and the improvement of the learning activities. Despite the positive results obtained in the previous academic year (Justino and Rafael 2018), some aspects of the initial pedagogical alignment established were amended in order to improve the expected outcomes. In particular, a change in the order of the topics covered, the addition of one more assessment element, regarding the learning outcome considered more difficult by the students in the previous academic year, and the implementation of more pedagogical techniques. Thus, several pedagogical techniques were applied in the learning activities, being the collaborative working group the most applied one. In situations of failures regarding previous knowledge or concepts, a personalized Just-in-Time Teaching (Gavrin 2006) was the learning technique used. Some learning activities were based on the Problem-Based Learning (PBL) technique (Duch et al. 2001). Other learning activities were based on Jigsaw (Aronson and Patnoe 1997) and Gallery Walk (Chin et al. 2015). All these techniques form the pedagogical strategy established by the constructive alignment matrix applied in the mathematics' course unit.

5 Implementation of the Collaborative Working Group

In academic year of 2017–18, the course coordinator divided students by four teams of four members each, which remained the same in all course units. In the academic year of 2018–19 there were five teams of four or five members each. In the first class of mathematics' course unit an initial survey was carried out aimed to identify the students' study habits and the students' attitude in classroom.

Regarding the characterization of the differences in the profile between students of the academic year 2018–19 and those of the academic year 2017–18, there is an increase of 15% of students studying individually, by a total of 87% of students. Since greater interaction and team work had to be developed, including in study, the collaborative working group would contribute to bring the students closer in this context.

On students' study habits and preparation for evaluations, there is a reduction of 8% of students stated to often summarise the contents to study later and an increase of 10% of students that only prepare themselves in the run-up to the evaluation, since they have a good memory. This means that students were accustomed to achieve superficial learning. It can be noted that students memorized the contents by summaries, not showing another type of study mechanism leading to a deeper assimilation and consolidation of knowledge. Classroom observation reinforced that most students did not demonstrate ongoing study habits, either individually or in team. So the learning activity had to take into account the need to provide, based on the collaborative working group, moments of deep learning.

On students' attitude, there is a slight reduction on the ability to actively participate in classroom, with 87% of students stating to be passive. This passive attitude, from previous school years, is typical of a strongly teacher-centred education in which the participation of students in not encouraged. So the need for an active and participative attitude enhanced the need for applying collaborative group techniques as a mean of supporting effective learning.

In the mid-term an intermediate survey was conducted for assessing the implementation of the learning process applied. It can be noted a slight variation of 1% between the two academic years, with 82% of students welcoming the collaborative working group. The sharing of expertise, the mutual assistance in solving problems and the group dynamics were the most reported aspects pointed out in this survey. The most significant negative aspect, stated by 30% of the students in the academic year of 2017–18, was the difficulty in having a four hours activity within which 19% also stated that they would have liked more time for the activity. In the academic year of 2018–19 the most significant negative aspect, also stated by 30% of the students, was the limited amount of exercises performed before the training tests. This last information was useful to put more available exercises on the IT platform used by the curricular unit.

In short, the information given by surveys allowed to adapt the type and the form of the learning activities in order to work out students towards a more participative and collaborative work and a shared study. To that extent the method of evaluation implemented in the mathematics' course unit was defined following the constructive pedagogical alignment matrix which integrates the topics to be studied and the learning outcomes to be achieved, considering the depth of contents based on Bloom's taxonomy (Amer 2006). The students' assessment consisted in four training tests, one oral test, two summative tests and peer review, with students being encouraged to achieve 70% of the contents' skills on training tests. On the intermediate survey students pointed the method of evaluation as one of the major strengths of the learning process.

6 Outcomes of the Implementation

On the final survey, 78% of the students stated to have been encouraged to participate and discuss the topics over classes, representing a slight increase of 3% compared with the previous academic year. The remaining 22% stated not to always have an active and participative attitude during the learning activities. On collaborative work, 69% of the students stated to feel supported by the remaining team, facilitating their social integration through sharing of knowledge relating to the curricular unit, 25% stated that the support of the remaining team was not frequent and only 6% of the students showed social and interaction difficulties with the remaining team. On the working group dynamics, the mutual assistance during the learning activities was positively reported by 96% of the students that also considered the team to always have worked well, which stands for a significant increase of 27% compared with the previous academic year. This result satisfies the goal of greater student interaction with contents and knowledge sharing established in the pedagogical alignment matrix since only 4% of students considered that the team didn't work out so well. Also, on the group learning, 91% of the students stated to have learned more in team than individually, which stands for an increase of 10% compared with the previous academic year.

7 Outcomes of the Academic Performance

7.1 Group Performance

In Fig. 1 below, the average ranking of the teams' performance is presented in a bar graph (with teams A, B, C, D and E) Teams from different academic years were associated by profile's similarity and leadership characteristics. It can be noted a relevant similar evaluation of the competences achieved according to the teams' profile. This relationship is undergoing a deeper study in the next academic years. Some of the teams' characteristics were analysed in order to join them as follows.

Team A is characterised by shared leadership, good cohesion and knowledge consistency, presenting a fast learning capacity. Both teams A had one of the best pupils of the class. The collaborative working group provided a high dynamic in knowledge sharing and more commitment to competence development.

Team B is characterised by lack of leadership, weak cohesion and interpersonal relations, diversified knowledge and communication difficulties, presenting a differentiated academic performance among its members. Both teams B lost a member who dropped out by the end of term time. The collaborative working group fostered the communication and team work skills.

Team C is characterised by strong leadership with passiveness of the remaining members, little cohesion, diversified knowledge and some communication skills. Despite the positive outcomes presented, the excessive passivity hampered a wider development of the desired competences.

Teams D and E are characterised by shared and communicative leadership. Although there was a profound lack of background knowledge among its members, good commitment, great sharing capacity and good interpersonal relations could be observed. The collaborative working enabled students to share knowledge and provide mutual assistance, handling the lack of background knowledge.

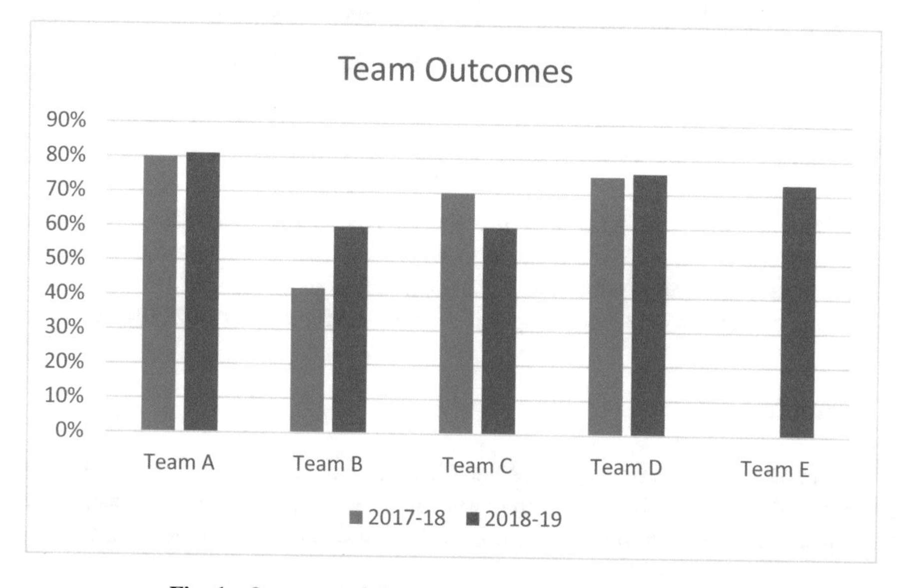

Fig. 1. Outcomes of the teams over the past academic year

In short, the collaborative working group provided the exercise of knowledge sharing and the metacognition application, highlighted in teams D, E and A, being the booster of motivation and commitment in problem-solving in all teams. The passivity and the lack of communication and sharing among the team members are the main factors that produce inertia into the development of the desired competences.

7.2 Individualized Performance

In Fig. 2 below, the performance outcomes of each student are presented by team. Overall, the success rate has increased from 80% to 93% in the academic year of 2018–19. Although the average score in 2018–19 is not higher than in 2017–18, more students achieved a positive score.

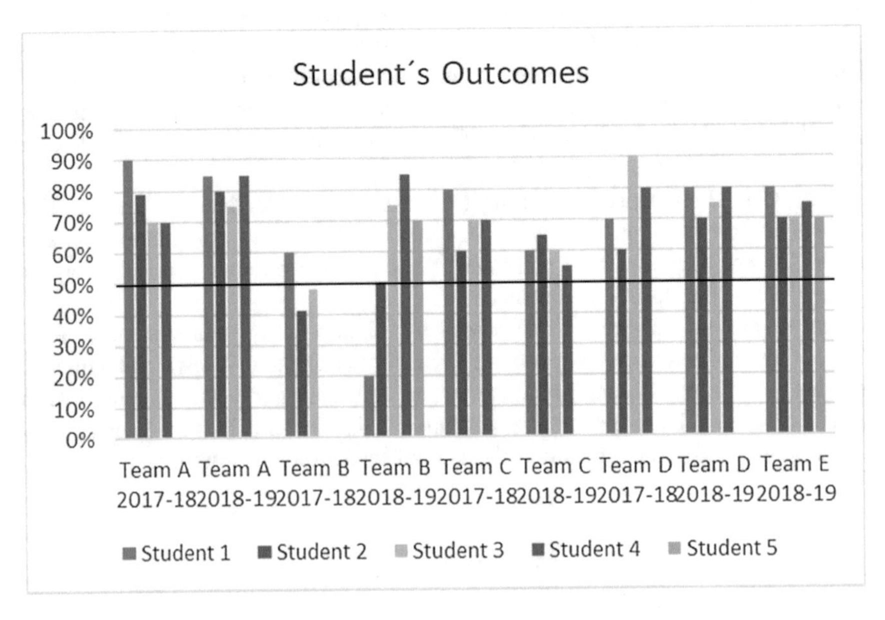

Fig. 2. Student's outcomes by team.

It should be noted that in the final survey carried out in 2018–19, 78% of the students classified the learning process applied to mathematics' course unit as good or excellent and the remaining 22% as satisfactory. This represents an increase of 18% concerning the degree of satisfaction when compared with the academic year of 2017–18.

8 Conclusions

Since Mathematics is a cross-cutting scientific area whose contents, although necessary for student's training, hardly fit into projects of strong technological nature, the application of active learning techniques allows its articulation with the other technical course units of a CTeSP. This contributes for the development of the students' soft skills, such as team work, autonomy, organisation, critical thinking, among others. Also, the teacher-student relationship that is established is closer, respectful and clear which is an important motivation factor for the student's academic performance, contributing to reduce the drop-outs rate in the course unit.

The collaborative working group, Jigsaw and Gallery Walk are pedagogical techniques that contribute positively for the students' interaction by developing problem-solving skills as well as motivation for the opportunity to confront their ideas with the other team members, to relate concepts and to apply critical thinking. It can also be noted that the collaborative working group, applied in a significant part of the educational context, contributes profoundly to improve self-esteem, mutual assistance and responsibility, changing the typical students' passive attitude in a classroom environment.

The learner-centred approach was implemented through the constructive alignment matrix. This instrument, which defines the entire pedagogical structure, is flexible, allowing an ongoing reflection based on qualitative and quantitative results that provides a continuous improvement of objectives, learning activities and assessment. The outcomes achieved reinforce the relevance of the application of the constructive alignment matrix. Furthermore, the amendment made on the constructive alignment matrix from one academic year to another triggered a significant improvement on the outcomes produced by active learning implementation.

References

Amer, A.: Reflections on Bloom's revised taxonomy. Electron. J. Res. Educ. Psychol. **8**(4), 213–230 (2006)

Aronson, E., Patnoe, S.: The Jigsaw Classroom: Building Cooperation in the Classroom, 2nd edn. Addison Wesley Longman, New York (1997)

Barkley, E.F., Cross, K.P., Major, C.H.: Collaborative Learning Techniques: A Handbook for College Faculty. Jossey-Bass Publishers, San Francisco (2014)

Biggs, J.: Enhancing teaching through constructive alignment. High. Educ. **32**, 1–18 (1996)

Burke, A.: Group work: how to use groups effectively. J. Effective Teach. **11**(2), 87–95 (2011)

Chin, C.K., Khor, K.H., Teh, T.K.: Is gallery walk an effective teaching and learning strategy for Biology? In: Biology Education and Research in a Changing Planet, pp. 55–59. Springer, Singapore (2015). https://doi.org/10.1007/978-981-287-524-2_6

Chiriac, E.H.: Group work as an incentive for learning – students' experiences of group work. Front. Psychol. Educ. Psychol. **5** (2014). Article 558. https://doi.org/10.3389/fpsyg.2014.00558

Duch, B.J., Groh, S.E., Allen, D.E.: The Power of Problem-Based Learning. Stylus Publishing, Virginia (2001). ISBN 1-57922-036-3

Gavrin, A.: Just-in-Time teaching. Metrop. Univ. **17**(4), 9–18 (2006)

Gokhale, A.A.: Collaborative learning enhances critical thinking. J. Technol. Educ. **7**(1), 22–30 (1995)

Justino, J., Rafael, S.: Teaching mathematics in tertiary education through collaborative work. In: 3rd International Conference of the Portuguese Society for Engineering Education (CISPEE 2018). IEEEXplore Digital Library (2018). https://doi.org/10.1109/cispee.2018.8593476

Laal, M., Ghodsi, S.M.: Benefits of collaborative learning. Procedia Soc. Behav. Sci. **31**, 486–490 (2012). https://doi.org/10.1016/j.sbspro.2011.12.091

Livingstone, K.A.: Constructive alignment and the curriculum: a call for improved pedagogical practices in higher education. J. Bus. Manag. Soc. Sci. Res. **3**(12) (2014)

OECD: Review of the tertiary education, research and innovation system in Portugal (2018). https://www.santamariasaude.pt/sgc/Assets/Plugins/CKEditor/kcfinder/Uploads/files/Review%20of%20TERI%20in%20Portugal%206%20February%20DRAFT.pdf. Accessed 12 Mar 2019

Shaffer, D.W., Serlin, R.C.: What good are statistics that don't generalize? Educ. Res. **33**(9), 14–25 (2004). https://doi.org/10.3102/0013189X033009014

Tashakkori, A., Teddlie, C.: Mixed Methodology: Combining Qualitative and Quantitative Approaches. SAGE Publications, Thousand Oaks (1998)

Rules_Math: Establishing Assessment Standards

Araceli Queiruga-Dios[1(✉)], Ascensión Hernández Encinas[2], Marie Demlova[3], Deolinda Dias Rasteiro[4], Gerardo Rodríguez Sánchez[5], and María Jesús Sánchez Santos[2]

[1] ETSII University of Salamanca, Béjar, Salamanca, Spain
queirugadios@usal.es
[2] Science Faculty, University of Salamanca, Salamanca, Spain
[3] The Czech Technical University in Prague, Prague, Czech Republic
[4] Instituto Superior de Engenharia de Coimbra, Coimbra, Portugal
[5] EPS University of Salamanca, Zamora, Spain

Abstract. In 2017, we proposed a European project: Rules_Math, in order to find new rules to assess mathematical competencies. This proposal was a consequence of what we were facing separately in our daily classes. We teach mathematics in several engineering schools, and we want to change the way of teaching and learning for engineering students. Some university teachers from different departments usually teach mathematics as we have learned mathematics. They only give master classes but students usually want to write numbers and formulas to learn and practice mathematical reasoning and to distinguish mathematical symbols. To communicate in, with, and about Mathematics, they need mathematical thinking, and to use aids and tools for mathematical activity. We have included in this papers our proposal to make this possible, and to assess mathematical competencies.

Keywords: Engineering · Mathematics · Competencies · Assessment

1 What Happens the First Day of a Math Course?

In some departments and some universities, lecturers do not choose their courses and they change subjects every year. This makes them to try to adapt to their students. To teach Mathematics to mathematicians, or physicists is (should be) different than to teach future engineers. When you get this, you have half of your work done.

When a trainer first looks at a new course, he usually has an academic guide at his disposal, where all the information about the course is included: credits, previous recommendations, objectives, contents, methodology, assessment criteria, bibliography, etc. So, what a beginning lecturer usually do is to take the contents and prepare some presentations on the course topics.

The thoughts about a new course should start with establishing the goals that the teacher wants students to acquire. Usually, the goals are listed as list

© Springer Nature Switzerland AG 2020
F. Martínez Álvarez et al. (Eds.): CISIS 2019/ICEUTE 2019, AISC 951, pp. 235–244, 2020.
https://doi.org/10.1007/978-3-030-20005-3_24

of learning outcomes rather content oriented. But there should be other goals emphasized, namely mathematical competencies that should be supported.

What would happen if the trainer had competencies instead of contents? Or what would happen if he had both things: competencies and contents?

To use a competencies-based methodology, it is necessary to change the usual way of teaching and learning. Students should really improve their knowledge taking an active role in classes [11].

Project based learning (PBL) [5], design thinking [3], flipped classroom [7] or gamification [6] are methodologies that are coming to university studies. They are use, to a greater or lesser extent, in several primary and secondary schools, but not so much at university level. On the other hand, some of these "new" methodologies have always been used as part of the activities developed during a course. If we propose our engineering students to work in small groups to simulate the population growth in a specific ecosystem, this could be worked as a PBL [9]. Similar situation could be the proposal of team or individual work to develop on their own some course topics, such as solving systems of linear equations using Jacobi or Gauss–Seidel methods.

2 The New Engineering Paradigm: Competencies–Based Assessment

In recent years several university teachers have changed from a content–based methodology to a competencies–based one. Since students often learn what is necessary to successfully pass the course, the assessment process must also change to a system stressing competencies acquirement. Generally speaking, mathematical competence is understood as the ability to develop and apply mathematical reasoning in order to solve different problems in daily situations.

First, let us be more precise in what competencies we do have in mind; they are 8 mathematical competencies introduced by Niss [8], and developed for engineering students [1]:

1. Thinking mathematically.
2. Reasoning mathematically.
3. Posing and solving mathematical problems.
4. Modelling mathematically.
5. Representing mathematical entities.
6. Handling mathematical symbols and formalism.
7. Communicating in, with, and about mathematics.
8. Making use of aids and tools.

Several papers, from different educational levels, have been published that focus on these competencies; even though e.g. OECD/PISA report (assessment of 15-year-old student in reading, mathematics and science) is also based in these competencies, but they use 7 competencies instead of 8:

1. Think and reason.
2. Argue.

3. Communicate.
4. Model.
5. Pose and solve problems.
6. Represent.
7. Use symbolic, formal and technical language and operations.

Lot of research has been done concerning assessment methods. Let us point out what Biggs [2] suggests:

- The criteria for the course and to describe the assessment tasks (clearly outlined as rubrics that the students fully understand).
- One assessment task may address several learning outcomes (LO) (they do not speak about contents [4], but learning outcomes). One appropriate assessment per LO can easily lead to an overload of assessment for the student.
- One LO may be addressed by more than one assessment task.
- In selecting assessment tasks, the time spent by students performing them and by staff assessing students' performances, should reflect the relative importance of the LOs. This is frequently breached when they have compulsory final examinations.
- An important practical point is that the assessment tasks have to be manageable, both by students in terms of both time and resources in performing them and by staff in assessing students' performances. For example, a portfolio would be difficult to manage in a large class.

What we usually find while planning our courses is that the assessment is established by the study programme in the faculty, and it already includes the percentage of the final mark for each assessment task. A common distribution of those percentages is: from 20 to 30% for problems, software practices, or team work, and the remaining 80 to 70% is dedicated to the final written or written and oral exam.

The current educational system uses several and different tools for assessing competencies and contents. Some activities and tools that are being used for all education level are team work, problem solving (including in some cases the use of mathematical software), and written exams (a final written exam or/and partial written exams).

Rules_Math project was presented and accepted for funding in 2017, and it will be finished in 2020. As part of this project, 9 higher education institutions from 8 different EU countries are working together trying to define the rules or standards that allow us the competencies assessment in engineering degrees [10].

Of course, the rules and standards must be content sensitive; different parts of mathematics curriculum support different mathematical competencies. Here we present a case study of the Calculus course.

3 Proposal: Definition of Assessment Standards

Procedures and methods for students' assessment and evaluation are fundamental activities of teacher's work. As it is well known, assessment is a core component for effective learning. The competencies–based learning makes the assessment methods even more important. As teaching and learning are based on acquiring competencies, the assessment should determine the acquisition of such competencies. There is a strong relationship between learning and assessment. On the other hand, the change from a teacher–centred instruction towards a student–centred one led us to the necessity of developing activities to enable the students an autonomous evaluation.

The proposed competencies–based assessment starting point is the set of learning outcomes, i.e., the knowledge and skills that should be acquired by students by the end of the course.

Furthermore, in outcomes–based teaching and learning, it is important that students clearly understand the LO they are meant to achieve, and accordingly they are written from a student perspective [2].

The starting point of this study is the document developed by the mathematical working group from the European Society for Engineering Education (SEFI) [1]. This framework for Mathematics Curriculum in Engineering Education has detailed the LO that students have to do in order to achieve the 8 mathematical competencies, not what teachers have to do.

Trying to define the best way to assess the acquisition of the 8 mathematical competencies by engineering students we found that:

1. The assessment tasks must be aligned to the learning outcomes we intend to address.
2. The assessment activities depend on the number of students attending the course.
3. Team work (PBL, flipped classroom, etc.) and the use of a mathematical software are methods recommended for a comprehensive assessment.
4. A written exam is needed to assess the acquisition of mathematical competencies.
5. The written exam usually includes "classical" problems, multiple choice tests and small projects (engineering applications).

We propose a comprehensive assessment for engineering students that will include those items.

4 Case Study: Written Exam for a Calculus Course

We started with the learning outcomes from the framework curriculum [1] aimed to engineering freshmen, i.e., core level 1. This document is separated in 4 levels (from core zero to level 3) according (more or less) to the mathematical levels in engineering degrees.

In the case of Analysis and Calculus, we have defined a written exam, including some LO about hyperbolic functions, functions, differentiation, methods of integration, and applications of integration. For all of these, the proposed exam will allow students the acquisition of the 8 competencies (called C1 to C8) in more or less degree (from green, totally covered to red, not so much). This scheme is shown in Fig. 1.

		C1	C2	C3	C4	C5	C6	C7	C8
	Analysis and Calculus								
AC1	1. Hyperbolic functions								
AC11	Define and sketch the functions sinh, cosh, tanh								
AC12	Sketch the reciprocal functions cosech, sech and coth								
AC13	State the domain and range of the inverse hyperbolic functions								
AC14	Recognise and use basic hyperbolic identities								
AC15	Apply the functions to a practical problem (for example, a suspended cable)								
AC16	Understand how the functions are used in simplifying certain standard integrals								
AC4	4. Functions								
AC41	Define and recognise an odd function and an even function								
AC42	Understand the properties 'concave' and 'convex'								
AC43	Identify, from its graph, where a function is concave and where it is convex								
AC44	Define and locate points of inflection on the graph of a function								
AC5	5. Differentiation								
AC51	Understand the concepts of continuity and smoothness								
AC7	7. Methods of integration								
AC71	Obtain definite and indefinite integrals of rational functions in partial fraction form								
AC72	Apply the method of integration by parts to indefinite and definite integrals								
AC73	Use the method of substitution on indefinite and definite integrals								
AC74	Solve practical problems which require the evaluation of an integral								
AC75	Recognise simple examples of improper integrals								
AC76	Use the formula for the maximum error in a trapezoidal rule estimate								
AC77	Use the formula for the maximun error in a Simpson's rule estimate								
AC8	8. Applications of integration								
AC81	Find the length of part of a plane curve								
AC82	Find the curved surface area of a solid of revolution								

Fig. 1. Learning outcomes with degree of coverage of competencies involved in this assessment activity.

We have separated this section in: Objectives, contents, and the proposed examination, which contains problems, multiple–choice tests and a final project. With this scheme, students would acquire the 8 competencies regarding the learning outcomes detailed in Fig. 1.

The time required to complete this exam will be 3 h (aprox.).

4.1 Objectives

The objectives, related to the learning outcomes detailed in Fig. 1, to be achieved with the development of this activity are the following:

1. To define and to sketch the functions sinh, cosh, tanh and their reciprocal functions.
2. To state the domain and range of the inverse hyperbolic functions.
3. To recognize and use basic hyperbolic identities.
4. To apply the functions to a practical problem (for example, a suspended cable).
5. To understand how the functions are used in simplifying certain standard integrals.
6. To define and to recognize an odd function and an even function.
7. To understand the properties 'concave' and 'convex' and to identify, from its graph, where a function is concave and where it is convex.
8. To define and locate points of inflection on the graph of a function.
9. To understand the concepts of continuity and smoothness.

10. To obtain definite and indefinite integrals of rational functions in partial fraction form.
11. To apply the method of integration by parts to indefinite and definite integrals.
12. To use the method of substitution on indefinite and definite integrals.
13. To solve practical problems which require the evaluation of an integral.
14. To recognize simple examples of improper integrals.
15. To use the formula for the maximum error in a Trapezoidal rule estimate and in a Simpson's rule estimate.
16. To find the length of part of a plane curve and the curved surface area of a solid of revolution.

4.2 Contents

1. Hyperbolic functions.
2. Points of inflection, continuity, smoothness, concavity and convexity on the graph of a function.
3. Different integration methods to indefinite and definite integrals.
4. Simple examples of improper integrals.
5. Formula for the maximum error in a Trapezoidal rule estimate and in a Simpson's rule estimate.
6. Length of part of a plane curve and the curved surface area of a solid of revolution.

4.3 Problem

1. Calculate the next integral:

$$\int \left(\sinh(x+a) - x\log(x) + \frac{x^3+3x}{x^2+4x+10} \right) dx$$

2. Obtain the path, domain, relative maxima and minima and inflection points, the possible symmetries, the concavity and continuity of function shows in Fig. 2.

   LO: AC1, AC4, AC5.

4.4 Multiple Choice Tests

1. The value of $\int_0^1 |x^2+x-6| dx$ is
   (i) $-31/6$
   (ii) $31/6$
   (iii) None of the above.
2. If $f(x) = \int_{-x}^{2x} 3t^2 \sin(t) dt$ then $f'(\pi/2)$ is
   (i) $3/4\pi^2$
   (ii) $-3/4\pi^2$
   (iii) None of the above.

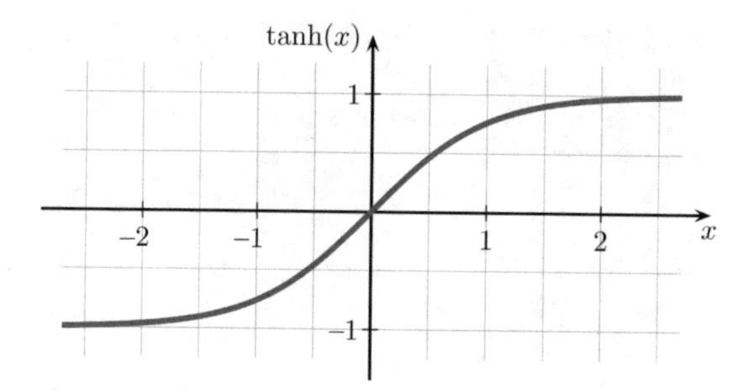

Fig. 2. Graph of tanh(x).

3. If $f(x)$ is a positive function such that $\lim_{x\to\infty} \frac{f(x)}{\sqrt[4]{x^3}} = 1$ then $\int_1^\infty f(x)dx$ is
   (i) Convergent
   (ii) Divergent
   (iii) The criterion doesn't decide.
4. Applying Simpson's formula to the function $f(x) = sin(x)$ in the interval $[-\pi, \pi]$ then $\int_{-\pi}^{\pi} \sin(x)dx$ is:
   (i) -1
   (ii) 4
   (iii) 0

LO: AC7.

4.5 Project

Reading some newspapers we found several news about a broken catenary, New Haven Catenary Replacement Project, or trains that were not running because a catenary problem. Taking into account the information that appears in online newspapers or at the internet, answer the following questions:

1. Do you know what a catenary is? The dictionary of the Royal Spanish Academy (RAE) defines the catenary with 3 meanings:
   (a) Pertaining or relative to the chain.
   (b) Curve formed by a chain, rope or similar thing suspended between two points not located in the same vertical.
   (c) Suspension system in the air of an electric cable, maintained at a fixed height of the ground, from which by means of a trolley or pantograph, some vehicles take over, such as trams, trains, etc.

   Which of the meanings is the one applied in the news?
2. It is a very important function and its rupture causes great losses to some companies, despite the discomfort of the passengers from some public transports like metro or train (Fig. 3).
   From the mathematical point of view, the catenary is the curve taken in the second meaning of the dictionary of the RAE. Their equation is $y = a \cosh \frac{x}{a}$,

Fig. 3. Some catenaries along the railroad tracks.

where $a = \left(\frac{\tau_h}{\lambda}\right)$ is a positive real number, considering the weight per unit length, λ, constant and where τ_h is the horizontal tension that appears at the ends of the cable.
You can see some catenaries (Fig. 4) in the following representation for different values of parameter a.

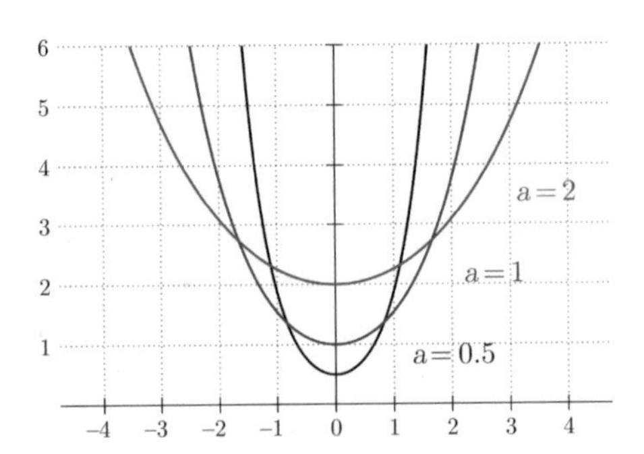

Fig. 4. Catenaries graphs.

3. Calculate the area bounded by the curve and the OX axis in the interval $[0, 1]$ for $a = 1$. Also calculate the arc length of the catenary between 0 and a, knowing that the length of a curve is calculated through the formula

$$L = \int_A^B \sqrt{1 + (f'(x))^2} dx,$$

for the parameter values that appear in the previous graphs.

Finally, calculate the lateral surface obtained by rotating around the axis OX of the previous catenary, knowing that said area is given by

$$A_L = 2\pi \int_A^B |f(x)|\sqrt{1+(f'(x))^2}dx$$

LO: AC1, AC7, AC8.

5 Conclusions and Future Work

As part of the European project Rules_Math ("New rules for assessing mathematical competencies") we perform some analysis to find assessment standards to assess mathematics in engineering degrees. We understood standards as rules or regulations that allow us to assess students. This assessment should include the acquisition of mathematical competencies taking into account what a learner knows and what is able to do, and indeed what he will be able to do after successfully completing the course, i.e., the learning outcomes will be achieved.

The initial result of these analysis showed us that almost all students were submitted to a type of written exam that includes "classical" problems, multiple–choice tests, and a small project. From the performed analysis we were also able to concluded that the mathematical background of the European students involved at the project was, in general, very similar. The mathematical subjects taught at high school were similar.

Our next proposal is to test this assessment method with students from different universities and different countries and try to find a feedback from them. Furthermore, we plan to extend the assessment activities and define some projects including guided exercises, real applications, and the use of mathematical software.

The results obtained by the different students of each country will be then analyzed and compared. It is also our aim to have a Moodle virtual discipline where students from different countries may interact and learn. This platform will also be evaluated and analyzed.

References

1. Alpers, B.A., Demlova, M., Fant, C.H., Gustafsson, T., Lawson, D., Mustoe, L., Olsen-Lehtonen, B., Robinson, C.L., Velichova, D.: A Framework for Mathematics Curricula in Engineering Education: A Report of the Mathematics Working Group. European Society for Engineering Education (SEFI), Brussels (2013)
2. Biggs, J.B.: Teaching for Quality Learning at University: What the Student Does. McGraw-Hill Education, New York (2011)
3. Brown, T., et al.: Design thinking. Harvard Bus. Rev. **86**(6), 84 (2008)
4. Caridade, C.M., Encinas, A.H., Martín-Vaquero, J., Queiruga-Dios, A., Rasteiro, D.M.: Project-based teaching in calculus courses: estimation of the surface and perimeter of the Iberian Peninsula. Comput. Appl. Eng. Educ. **26**(5), 1350–1361 (2018)

5. Jones, B.F., Rasmussen, C.M., Moffitt, M.C.: Real-Life Problem Solving: A Collaborative Approach to Interdisciplinary Learning. American Psychological Association, Washington, D.C. (1997)
6. Méndez, M.D.C.L., Arrieta, A.G., Dios, M.Q., Encinas, A.H., Queiruga-Dios, A.: Minecraft as a tool in the teaching-learning process of the fundamental elements of circulation in architecture. In: International Joint Conference SOCO16-CISIS16, pp. 728–735. Springer (2016)
7. Merchán, M.D., Canedo, M., López-Gil, F.J., Usero, J.L.: Motivating students of chemical engineering through a cooperative work recording educational videos. In: Proceedings of the Fourth International Conference on Technological Ecosystems for Enhancing Multiculturality, pp. 805–809. ACM (2016)
8. Niss, M.: Mathematical competencies and the learning of mathematics: the Danish KOM project. In: 3rd Mediterranean Conference on Mathematical Education, pp. 115–124 (2003)
9. Queiruga-Dios, A., Hernández Encinas, A., Martín Vaquero, J., Martín del Rey, Á., Bullón Pérez, J.J., Rodríguez Sánchez, S.: How engineers deal with mathematics solving differential equation. Procedia Comput. Sci. **51**, 1977–1985 (2015)
10. Queiruga-Dios, A., Sánchez, M.J.S., Pérez, J.J.B., Martín-Vaquero, J., Encinas, A.H., Gocheva-Ilieva, S., Demlova, M., Rasteiro, D.D., Caridade, C., Gayoso-Martínez, V.: Evaluating engineering competencies: a new paradigm. In: Global Engineering Education Conference (EDUCON) 2018 IEEE, pp. 2052–2055. IEEE (2018)
11. Rasteiro, D.M., Caridade, C.M.: Assessing statistical methods competencies and knowledge in engineering. In: The 19th SEFI Mathematics Working Group Seminar on Mathematics in Engineering Education, pp. 92–99 (2018)

Evaluate Mathematical Competencies in Engineering Using Video-Lessons

Cristina M. R. Caridade(✉) and Deolinda M. L. D. Rasteiro(✉)

Coimbra Institute of Engineering, Coimbra, Portugal
{caridade,dml}@isec.pt

Abstract. There are eight Mathematical competencies proposed by B. Alpers *et al.* in the Framework for Mathematical Curricula in Engineering Education, previously proposed in the Danish KOM. However, some of the competencies identified are difficult to evaluate in higher education since there is a large number of students in the classrooms and therefore the personal contact between the teacher and the students is limited. When we want to evaluate the competency "communicating in, with, and about mathematics" or "make use of aids and tools for mathematical activity", we need other moments of evaluation besides written exams. With this intention we developed a practical project during the semester for all the students who wanted a continuous evaluation. The project consisted in the development of a video lesson about one of the contents studied in the curricular unit. The students had to create an initial script and then produce the video where we were able to evaluate the referred mathematical competencies. At the end of the semester, we identified if the content developed in the video lesson by the student was acquired and is a part of their knowledge. For this, in the written evaluations made by each student the exercises that contained the video lesson content were analysed and verified if they were correctly solved. Students also presented their videos to their peers and to the teachers and their presentation was also evaluated.

Keywords: Interactive classes · Video-lessons · Significant learning · Competencies · Mathematics · Engineering

1 Introduction

Competencies are combinations of attitudes, skills and knowledge that students develop and apply for successful learning, living and working. Competencies help students draw and build upon what they know, how they think and what they can do. Mathematical competencies allow students to develop the ability to understand the role of mathematics in the world; studying issues, situations

Project Erasmus+ 2017-1-ES01-KA203-038491 "New Rules for Assessing Mathematical Competencies".

© Springer Nature Switzerland AG 2020
F. Martínez Álvarez et al. (Eds.): CISIS 2019/ICEUTE 2019, AISC 951, pp. 245–251, 2020.
https://doi.org/10.1007/978-3-030-20005-3_25

or events of global significance that require a mathematical approach or solution; and use mathematics to support conclusions, arguments, and decisions that lead them to act as reflective, constructive and concerned citizens of the world. The goal of these Mathematics competencies, which are expected to be acquired by engineering students, is to provide a clear and coherent idea about the mathematics that students need to know and to be able to do, to be successful in their courses. According to B. Alpers *et al.* in the Framework for Mathematical Curricula in Engineering Education (Alpers et al. 2013), previously proposed in the Danish KOM project (Niss 2003) the mathematical competencies are: thinking mathematically; reasoning mathematically; posing and solving mathematical problems; modelling mathematically; representing mathematical entities; handling mathematical symbols and formalism; communicating in, with, and about mathematics and make use of aids and tools for mathematical activity. Recent research indicates that the more students possess and can activate these competencies, the better able they will be to make effective use of their mathematical knowledge to solve contextualised problems. In other words, the possession of these competencies relates strongly to increased levels of mathematical literacy (Turner 2010). The development of efficient methods and measuring procedures of competence assessment is very useful to be able to assess whether a given competence has been acquired or not by the student. However, some of the competencies identified by (Alpers et al. 2013) are difficult to assess in higher education, since there is a large number of students in the classrooms and the personal contact between the teacher and the students is limited. When we want to evaluate "communicating in, with, and about mathematics" or "making use of aids and tools" we need other moments of evaluation besides written exams. One way to enrich mathematical learning experiences is through the use of different types of activities. Although the use of these experiences does not determine learning by itself, it is important to provide several opportunities for contact with different activities to arouse interest and involve the student in mathematical learning situations. Videos have assumed a growing and promising role in teaching, particularly as a different and motivating activity at the service of education (Stefanova 2014). A number of initiatives and studies have already been carried out in this area, such as (Moran 1995), who discusses different possibilities for the use of this material in the classroom, as "simulation" (to simulate an experiment) or as "teaching content" (to show a certain subject) or (Willmot et al. 2012) who describes the design and development of an attractive new resource to encourage academics to incorporate video reporting into their student-centered learning activities, among others. Within this context, video as a medium continues to have an on-going impact on higher education, on the role of the student, challenging the (traditional) role of the lecturer and the format of delivering course contents via lectures. Currently the videos available on the internet of small parts of classes are the most viewed by the students, who watch them when they have some conceptual questions. Videos of this nature have also gained space in distance education (Koppelman 2016). In the scope of mathematics education (Clarke et al. 2013) pointed out that video can be used in an interactive environment in order to enhance expression and communication, as well as a pedagogical action that motivates learning. With this intention, during this academic year (2018/2019), in the curricular unit (CU) of Mathematical Analysis I of the Electrotechnical Engineering course at Coimbra Institute of Engineering, we proposed to the students the development of a video lesson about the contents taught in this curricular unit. This experience follows the work already developed in the previous year with the students of this course (CMRCaridade, DMLDRasteiro 2018). The experience consisted in the development a video lesson about one of the contents studied in the curricular unit. From the students enrolled on the CU, 58 of them, in groups of two, developed and presented their video lessons. The students had to create an initial script where we intend to evaluate the above mentioned competencies: thinking mathematically (C1), reasoning mathematically (C2), posing and solving mathematical problems (C3), modelling mathematically (C4), representing mathematical entities (C5), handling mathematical symbols and formalism (C6) and then produce the video where you can also evaluate the competencies: communicating in, with, and about mathematics (C7) and making use of aids and tools (C8). At the end of the semester we identified, in some cases (some students did not take the exam), if the content developed in the video lesson by the student was assimilated in his knowledge. For this, in the written evaluations made by each student, the exercises that contained this content were analysed and verified if they were correctly solved.

2 Description of the Study

In this study, we analysed the possibilities of video lessons development in order to evaluate some of the mathematical competencies suggested by (Alpers et al. 2013). The CU of Mathematical Analysis I had 156 teaching hours, in the classroom, of which 14 h tutorials and another 20 h attributed to group work. We dedicated 4 tutorial hours to let students know some materials for video production, studying mathematical concepts, understanding how to present and discuss them, knowing the potentialities and limitations of their use in video lessons. For some students, this was the first time they had contact with a video production, especially videos to teach mathematical content. The idea was to present various types of videos to inspire their own productions. During these classes, various educational videos that were available on YouTube were presented to students in order to exchange ideas and experiences between groups and teachers. Some of the cases presented were video lessons produced only with producer narration, slides with animations or computer screen capture and others were more elaborated. The activity proposed to the students consisted of the development, creation and edition of a video lesson on one of the contents taught in the CU of Mathematical Analysis I. The video lesson was developed by a group of two students and who were assigned to a specific subject (chosen by the teacher). The experience was divided into 4 stages as described below:

- Study the content and prepare a video lesson script.
- Choose one or more explanatory examples and prepare it.

- Preparation (editing) of the video lesson.
- A written report of their work to be delivered to teachers and a presentation.

At the end of the course, the student should present the video lesson to teachers and other classmates during the last week of classes. At this stage, the teacher will assess the activity and the students also do a self-assessment.

3 Student Experiences

In the previous work presented in (CMRCaridade, DMLDRasteiro 2018) it was analysed and investigated the possibilities of using video-lesson in the classroom. The results were very interesting, enriching and attractive as teaching and learning tools. Many different types of video lesson were presented by the students. Video lessons with different forms were presented (animated slides, film white-board, or white paper); using different materials (pens, pointers); highlight or explanation of some details in different ways; presenting the auxiliary calculations of different forms and using various mathematic applications (GeoGebra, Matlab Mathematica and calculator) (CMRCaridade, DMLDRasteiro 2018). In this academic year, we proposed the same experience with an extra challenge, students had to write also a video script and register on it some required issues. The script allowed teachers to evaluate the mathematical competencies that the student acquired when developing the video lesson. The realization of a script was a new idea that came to us to evaluate other mathematical competencies that we could not assess only with video lesson edition and presentation. In Fig. 1 an example of a video lesson script is represented.

4 Experience Evaluation and Learning Outcomes

Fifty eight students in groups of two presented their video lesson. The evaluations, focused on the form (argument, sound, aesthetics and, editing), language (including mathematical language) and content (clarity, narration, creativity, research and exploration) of video lessons. Each of the topics evaluated is related to the competencies acquired by the students. In Table 1 it is represented the competencies associated with each evaluated topic. For example, the presentation topic is very related to "communicating in, with, and about mathematics" (C7) and "making use of aids and tools" (C8). In the same table, it is possible to see the mean and standard deviation of the partial grades obtained by the students in this video lesson experience.

The grades' mean obtained by students was 2.58, out of 4.0, with a standard deviation of 0.59 which may lead to admit that the activity was relatively appropriated and worked out by the students.

In order to be able to evaluate if competencies and their acquisition, were assimilated in the development of the experience by the students it was analysed, during the written test, if students learned the content they developed in their video lesson. When comparing the students' performance at the exam with their grades obtained with the video-lesson activity, see Fig. 2, we may conclude that our objective was not totally achieved although the Pearson correlation value of the exam and the video-lesson grades is 0.422 which allows us to admit that there is a moderate correlation between both variables.

At the exam when we observe only the grades obtained on the related video lessons contents, we notice that approximately 45% of the students did not take the exam which is a different problem that teachers have to worry about - students absence to the exams.

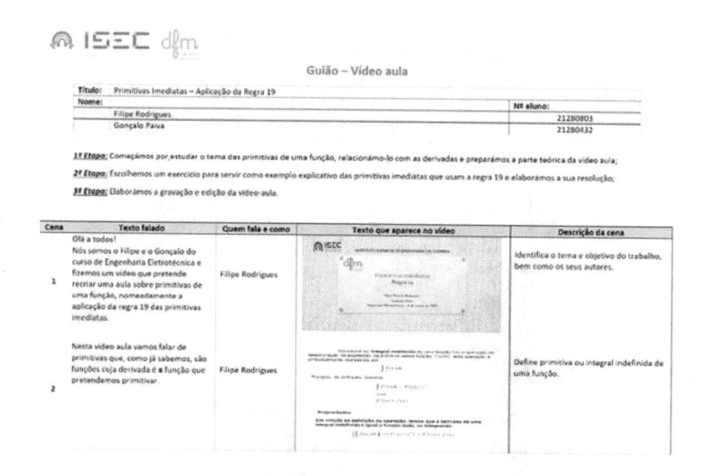

ISEC dfm

Guião – Vídeo aula

Título:	Primitivas Imediatas – Aplicação da Regra 19	
Nome:		Nº aluno:
	Filipe Rodrigues	21280803
	Gonçalo Paiva	21280432

1º Etapa: Começámos por estudar o tema das primitivas de uma função, relacionámo-lo com as derivadas e preparámos a parte teórica da vídeo aula;

2º Etapa: Escolhemos um exercício para servir como exemplo explicativo das primitivas imediatas que usam a regra 19 e elaborámos a sua resolução;

3º Etapa: Elaborámos a gravação e edição da vídeo-aula.

Cena	Texto falado	Quem fala e como	Texto que aparece no vídeo	Descrição da cena
1	Olá a todos! Nós somos o Filipe e o Gonçalo do curso de Engenharia Eletrotécnica e fizemos um vídeo que pretende recriar uma aula sobre primitivas de uma função, nomeadamente a aplicação da regra 19 das primitivas imediatas.	Filipe Rodrigues	[illegible]	Identifica o tema e objetivo do trabalho, bem como os seus autores.
2	Nesta vídeo aula vamos falar de primitivas que, como já sabemos, são funções cuja derivada é a função que pretendemos primitivar.	Filipe Rodrigues	[illegible]	Define primitiva ou integral indefinida de uma função.

Fig. 1. Example of a video lesson script.

Table 1. Competencies associated, partial grade, mean and standard deviation to each evaluated topic.

	Competencies	Partial note	Mean	Standard deviation
Presentation	C7; C8	0.25	0.18	0.06
Argument		0.5	0.28	0.12
Sound		0.20	0.17	0.06
Aesthetics and editing		0.20	0.16	0.05
Language		0.50	0.32	0.10
Clarity		0.30	0.23	0.07
Narration		0.30	0.23	0.07
Creativity		0.30	0.12	0.12
Research		0.30	0.10	0.08
Exploration		0.40	0.07	0.11
Video-lesson guide		0.75	0.64	0.14
Final grade		**4.0**	**2.58**	**0.59**

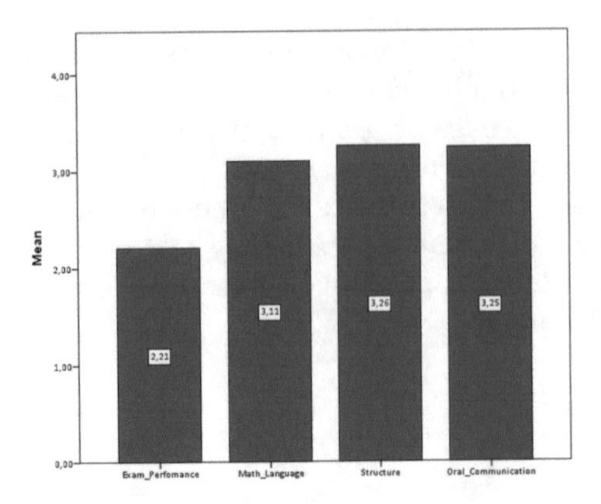

Fig. 2. Exam and Video-Lesson performance.

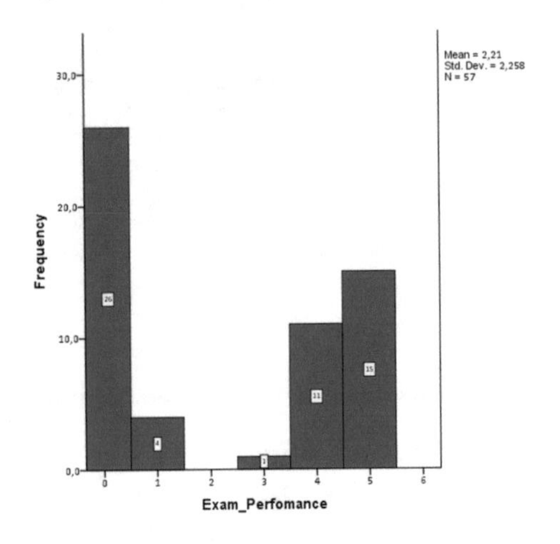

Fig. 3. Exam and Video-Lesson performance.

From Fig. 3 above we may notice that all students that solved the proposed exercise were able to do it almost right.

5 Conclusions and Future Work

The influence of digital videos in our daily lives is undeniable. Online video sites have a very large audience. Hence it seems natural to extend the videos to the educational environment. Today's students use educational videos as a tool to learn everything. The visual and auditory nature of the videos attracts and allows each user to process the information in a natural way. The use of videos in teaching and learning serves not only to benefit students, but also teachers and their institutions. With the development of video lessons by students of Mathematical Analysis I, we verified that abstract topics that once seemed difficult to teach and learn are now more accessible and understandable to students. The motivation of students in the accomplishment of this experience has been enormous and their involvement and great dedication as well. Students were very fond of doing the videos and some were extremely creative in using additional mathematical tools to present a topic in a different and engaging way. The analysis of the mathematical competencies in the accomplishment of this experience gave us an idea of the competencies that traditionally were evaluated in the written exams of the CU of mathematics and those that we could evaluate. We have verified that some of the competencies cannot be evaluated only with written exams. It is necessary to create other forms of assessment so that students' evaluation is more accurate and learning becomes effective.

Acknowledgements. The authors would like to acknowledge the financial support of Project Erasmus+ 2017-1-ES01-KA203-038491 "New Rules for Assessing Mathematical Competencies".

References

Turner, R.: Exploring mathematical competencies, No. 24, Article 5 (2010). https://research.acer.edu.au/resdev/vol24/iss24/5

Alpers, B., et al.: A framework for mathematics curricula in engineering education, SEFI (2013). http://sefi.htw-aalen.de/

Niss, M.: Mathematical competencies and the learning of mathematics: the Danish KOM project. In: Proceedings of the 3rd Mediterranean Conference on Mathematical Education. Hellenic Mathematical Society, Athens (2003)

Caridade, C.M.R., Rasteiro, D.M.L.D.: Involve Me and I Learn – video-lessons to teach math to engineers. In: 19th SEFI-MWG, pp. 107–114 (2018)

Project Erasmus+ 2017-1-ES01-KA203-038491 "New Rules for Assessing Mathematical Competencies". https://www.researchgate.net/project/New-Rules-for-assessing-Mathematical-Competencies

Moran, J.M.: O vídeo na sala de aula. Comun. Educ. São Paulo **1**(2), 27–35 (1995)

Stefanova, T.A.: Using of training video films in the engineering education. Procedia Soc. Behav. Sci. **116**, 1181–1186 (2014)

Clarke, D., Hollingsworth, H., Gorur, R.: Facilitating reflection and action: the possible contribution of video to mathematics teacher education. J. Educ. **1**(3), 94–121 (2013)

Koppelman, H.: Experiences with using videos in distance education. A pilot study: a course on human-computer interaction. Issues Inf. Sci. Inf. Technol. **13**, 269–277 (2016). http://www.informingscience.org/Publications/3472

Willmot, P., Bramhall, M., Radley, K.: Using digital video reporting to inspire and engage students (2012). http://www.raeng.org.uk/publications/other/using-digital-video-reporting

Statistical Methods – What Is the Problem Here?

Deolinda M. L. D. Rasteiro(✉) and Cristina M. R. Caridade(✉)

Coimbra Institute of Engineering, Coimbra, Portugal
{dml,caridade}@isec.pt

Abstract. The concepts taught during a Statistical Methods course make use of different mathematical skills and competencies. The idea of presenting a real problem to students and expect them to solve it from beginning to end is, for them, a harder task then just obtain the value of a probability given a known distribution. Much has been said about teaching mathematics related to daily life problems. In fact, we all seem to agree that this is the way for students to get acquainted of the importance of the contents that are taught and how they may be applied in the real world. The definition of mathematical competence as was given by Niss (Niss 2003) means the ability to understand, judge, do, and use mathematics in a variety of intra– and extra – mathematical contexts and situations in which mathematics plays or could play a role. Necessary, but certainly not sufficient, prerequisites for mathematical competence are lots of factual knowledge and technical skills, in the same way as vocabulary, orthography, and grammar are necessary but not sufficient prerequisites for literacy. In the OEDC PISA document (OECD, 2009), it can be found other possibility of understanding competency which is: reproduction, i.e, the ability to reproduce activities that were trained before; connections, i.e, to combine known knowledge from different contexts and apply them to different situations; and reflection, i.e, to be able to look at a problem in all sorts of fields and relate it to known theories that will help to solve it. The competencies that were identified in the KOM project (Niss 2003; Niss and Højgaard 2011) together with the three "clusters" described in the OECD document referred above were considered and adopted with slightly modifications by the SEFI MWG (European Society for Engineering Education), in the Report of the Mathematics Working Group (Alpers 2013). At Statistical Methods courses often students say that assessment questions or exercises performed during classes have a major difficulty that is to understand what is asked, i.e, the ability to read and understand the problem and to translate it into mathematical language and to model it. The study presented in this paper reflects an experience performed with second year students of Mechanical Engineering graduation of Coimbra Institute of Engineering, where the authors assessed Statistical Methods contents taught during the first semester of 2017/2018 and 2018/2019 academic years. The questions in the assessment tests were separated

Project Erasmus+ 2017-1-ES01-KA203-038491 "New Rules for Assessing Mathematical Competencies".

© Springer Nature Switzerland AG 2020
F. Martínez Álvarez et al. (Eds.): CISIS 2019/ICEUTE 2019, AISC 951, pp. 252–262, 2020.
https://doi.org/10.1007/978-3-030-20005-3_26

into two types: ones that referred only to problem comprehension and its translation into what needed to be modelled and calculated and others where students needed only to apply mathematical techniques or deductions in order to obtain the required results. The research questions that authors want to answer are:

- What are the competencies that students found, in a Statistical Methods course, more difficult to obtain?
- Having the idea that learning concepts by applying them to reality is much more fun and worthy for students, is it really what we should assessed them for? If not, how can knowledge be transmitted to students and be transformed into significant learning?
- What are the most frequent mistakes performed by students at Statistical Methods exams?

Keywords: Mathematical competencies · Higher education · Statistical methods

1 Introduction

In higher education, mathematics has an important role in engineering courses (OECD (1996)). From the curriculum of the first and second years there are Curricular Units (CU) in the area of Mathematics that are fundamental for students to acquire the necessary basic knowledge. One of those CU is Statistical Methods. The concepts taught during a Statistical Methods course make use of different mathematical skills and competencies. The idea of presenting a real problem to students and expect them to solve it from beginning to end is, for them, a harder task then just obtain the value of a probability given a known distribution. At least, from our perception, this is what students believe. Often the concept of mastering a subject does not have the same definition for students and math teachers. Regarding students we, as teachers, also should make a difference to which students we are teaching. Mathematics is of course the same but the usage that will be given to their math knowledge is different if they are going to be mathematicians or engineers or else. The authors are math teachers at Coimbra Engineering Institute and for them to teach math is much more than to transmit concepts and resolution methods. It also involves the ability of looking at a real life problem and to be able of selecting, among all the variety of mathematical tools and concepts, the ones that may be applied to solve the problem in hand. (Niss *et al.* 2017) formulated the questions "What does it mean to possess knowledge of mathematics? To know mathematics? To have insight in mathematics? To be able to do mathematics? To possess competence (or proficiency)? To be well versed in mathematical practices?" and gave a big insight to this discussion. They attempted to present significant, yet necessarily selected, aspects of and challenges to what some call "the competency turn" in mathematics education, research and practice. During an Engineering course, students learn and consolidate the basic principles of mathematics to solve practical problems, reinforcing their conceptual mathematical knowledge. However, although mathematics is a basic discipline regarding the admission to any Engineering degree, difficulties related to mathematics' basic core are identified by almost all engineering students at each CU. In this context, it seems relevant to identify the mathematics competencies attained by engineering students so that they can use these skills in their professional activities. Mathematics competencies is the ability to apply mathematical concepts and procedures in relevant contexts which is the essential goal of mathematics in engineering education. Thus, the fundamental aim is to help students to work with engineering models and solve engineering problems (SEFI (2011)). According to (Niss 2003) eight clear and distinct mathematics competencies are: thinking mathematically, reasoning mathematically, posing and solving mathematical problems, modelling mathematically, representing mathematical entities, handling mathematical symbols and formalism, communicating in, with, and about mathematics and making use of aids and tools. Gaps were detected between engineers' required mathematics competencies and acquired mathematics competencies of engineering students under the current engineering mathematics curriculum. There is a need to revise the mathematics curriculum of engineering education making the achievement of the mathematics competencies more explicit in order to bridge this gap and prepare students to acquire enough mathematical competencies (Rules_Math Project). Hence an important aspect in mathematics education for engineers is to identify mathematical competencies explicitly and to recognize them as an essential aspect in teaching and learning in higher education. It is the fundamental that all mathematics teaching must aim at promoting the development of pupils' and students' mathematical competencies and (different forms of) overview and judgement (Niss and Højgaard 2011; Alpers et al. 2013; DMLDRasteiro *et al.* 2018).

This research pretends to evaluate and recognize what are the competencies that engineering students can have or, acquire, when Statistical Methods contents are taught to them and also to identify the most frequent errors that are performed by students at this course exams. In the past few years the CU teacher used to assess students with questions where the recognition of the probability distribution models were necessary to perform further calculus. Students usually complained that they were not able to solve the problem if they failed the first part of the resolution (identification process) and therefore their success was conditioned by the ability to full understand the problem. Even though we consider that surely the complete success must involve the full understanding of the problem, maybe we can accept that not all engineers need to be modelers and some of them will not work directly with the theoretical part of the questions. Once, in the last two years we have performed and experience regarding assessment. The questions were, as much as possible, separated into calculus items and models identification and deduction items. Then, and according with the competencies defined by Niss, we analysed the perception of students, in the acquisition of the taught competencies regarding mathematics. Two tests were performed during the first year semesters of 2017/2018 and 2018/2019 and final exams for students that preferred a global evaluation. Students were from Mechanical Engineering second year degree.

2 Description of the Study

At the second year of Mechanical Engineering degree, 108 and 112 students engaged in the Statistical Methods curricular unit (CU) of 2017/2018 respectively 2018/2019 academic year. This CU belongs to the second year of their course curricula and has the duration of one semester. At the beginning of the course it was discussed with the students the evaluation possible methods. They had the opportunity to choose between regular final exam and distributed tests along the semester. The opinion of students that were engaged for the second time was very important for the students that were attending the course for the first time. Since the former ones were properly empowered to have an opinion about the subject specially which were the main difficulties that they found in the previous year their opinion was most valued. From the discussion it was agreed that students could choose between distributed assessment tests (two tests along the semester) and final exam. It was also agreed that questions of pure calculus were to be separated, as much as possible, from question where students should recognise which was the probability model/distribution and also to recognise what should be determined. One example of the test questions is given in Figs. 1 and 2.

11. A company producing television cameras has three production sectors: A, B and C. It is known that:
    - 50% of the television cameras is produce at sector A;
    - 10% of the television cameras are defective;
    - at sector C no defective televison cameras are produced;
    - 2% of the television cameras are produced at sector B and are defective.

    A television camera is randomly chosen from the company production.

    (a) Define in understanding the events referred to in the statement and extract from it all the data provided.
    (b) Indicate, without calculating, what probability value allows you to:
        i. determine the probability that the television is defective, knowing that it was produced in sector A.
        ii. knowing the probability that the television camera does was not produced at sector B knowing that it is defective.
        iii. knowing that, from the non-defective television cameras 40% were produced at sector C, to determine the probability of being produced at sector C.
    (c) Considering a sample of 10 television cameras from the company production, explain the whole process of defining the variable as well as its law of probability, which will allow you to determine the probability that, in these ten television cameras, there is at most one that is defective and comes from sector A? What probability value would you calculate?

Fig. 1. Comprehension/identification/model recognising type of questions

6. Let Ω be the space results associated with a certain random experience and A and B two random events contained in Ω. We know that $P(A) = 0.4$, $P(B) = 0.3$ and $P(\overline{A}/B) = 0.25$. Determine the value of $P(\overline{B}/\overline{A})$.
7. Let X and Y be two real random independent variables such that $X \sim \mathcal{P}(1)$ and Y has its probability function given by the following table

y	1	2	3	4
$P(Y = y)$	$\frac{1}{3}$	$\frac{1}{4}$	$\frac{1}{6}$	$\frac{1}{4}$

Determine $P(X^2 + Y = 4)$.

8. Let $X \sim \mathcal{P}(3)$.
   (a) Determine $P(X \geq 3/0 < X \leq 10)$.
   (b) Obtain $E((X-4)^2)$ and $V(4(X-2))$.
9. Let X_i, $i = 1, \ldots, 20$, be real random independent variables with $X \sim \mathcal{P}(1)$.
   (a) Determine $P\left(\frac{1}{20}\sum_{i=1}^{20} X_i \geq 0.8\right)$.
   (b) Obtain the mean value and the variance of the real random variable $\sum_{i=1}^{20} \frac{X_i}{10}$.

Fig. 2. Calculus type of questions

3 Findings and Discussion

In this section we outline the essential findings concerning the type of assessment chosen by Statistical Methods of Mechanical Engineering degree students and enumerate some of the possible reasons for those findings. We start by mentioning that, in 2017/2018, more that 73% of the engaged students submitted themselves to the distributed assessment. The results obtained were transformed into relative frequencies in order to be able to compare them.

From Table 1 and Fig. 3 we may see that, although the mean values of both types of questions are very similar, in fact calculus questions have higher classification. We did a t-student test and at significance level of 5% we reject the equality means hypothesis. The mean belongs, with 95% confidence, to the interval $]0.4122, 0.4894[$ in case of modelling/comprehension questions and to the interval $]0.4695, 0.5753[$ in the calculus questions.

Table 1. Test 1 results

Test 1 - results				95% confidence	*Bootstrap*[a] interval
		Statistic	Std. error	Lower	Upper
Modelling_1	N	79	0	79	79
	Mean	0.4541	0.0197	0.4122	0.4894
	Std. Deviation	0.18013	0.01301	0.15160	0.20375
Calculus_1	N	79	0	79	79
	Mean	0.5353	0.0204	0.4952	0.5753
	Std. Deviation	0.18488	0.01368	0.15715	0.21082
Test1_n	N	79	0	79	79
	Mean	0.5060	0.0185	0.4695	0.5416
	Std. Deviation	0.16823	0.01317	0.14022	0.19240

[a]Unless otherwise noted, bootstrap results are based on 1000 bootstrap samples

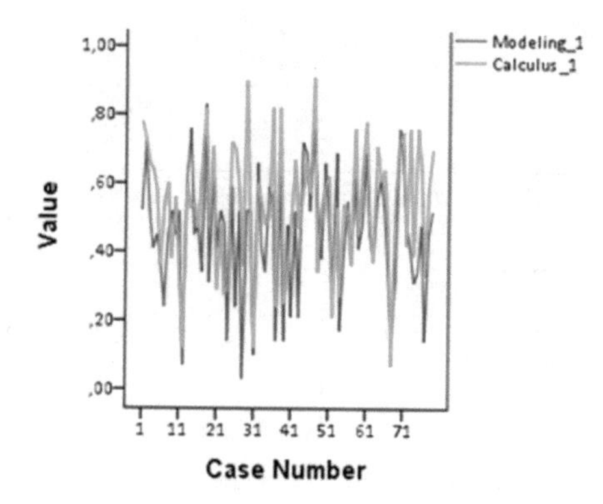

Fig. 3. Line graph of test 1 results

Only 44,4% of students engaged in the CU submitted themselves to the second test. From Table 2 and Fig. 4 we may see that the mean values of both types of questions are very similar. Performing the t-student test at 5% level we do not reject, in this case, the equality of means. The mean belongs, with 95% confidence, to the interval $]0.2706, 0.4242[$ in case of modelling/comprehension questions and to the interval $]0.2898, 0.3969[$ in the calculus questions.

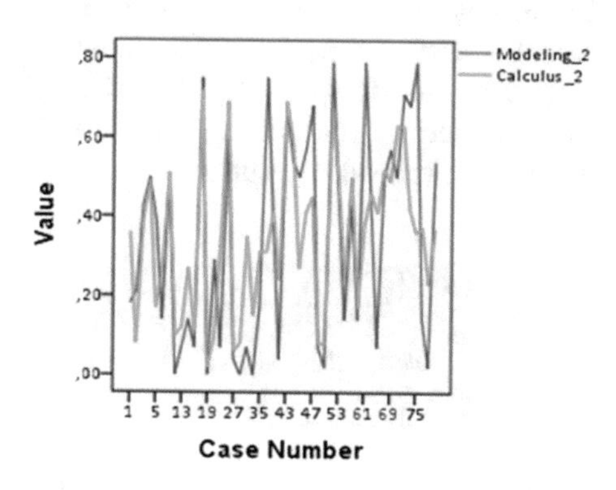

Fig. 4. Line graph of test 2 results

In terms of final results, the ones obtained at 2018/2019 academic year were analogous and therefore were omitted. Since the results were similar within both years, and better than in 2016/2017 academic year, we looked for the most frequent errors in order to be able to identify what students found more difficult to apprehend and acquire as significant learning. The most common errors done

Table 2. Test 2 results

Test 2 - results				95% confidence	*Bootstrap*[a] interval
		Statistic	Std. error	Lower	Upper
Modelling_2	N	48	0	48	48
	Mean	0.3454	0.0391	0.2706	0.4242
	Std. Deviation	0.26942	0.01493	0.23664	0.29412
Calculus_2	N	48	0	48	48
	Mean	0.3446	0.0278	0.2898	0.3969
	Std. Deviation	0.19292	0.01492	0.15935	0.21888
Test2_n	N	48	0	48	48
	Mean	0.3448	0.0300	0.2867	0.4045
	Std. Deviation	0.20663	0.01361	0.17723	0.23098

[a]Unless otherwise noted, bootstrap results are based on 1000 bootstrap samples

by students at Statistical Methods course exams include the reunion of probability of events as if they were sets, Fig. 5, even though the students know a lot of formulas they are enable to reasoning and obtain the desired value, Fig. 6, the inability of relating theoretical properties, Fig. 7, to extract data from the exercise statement, Fig. 8, not being able to translate into mathematic symbols expressions that represent statistic variables making confusion between the variable and the parameter of its distribution, Fig. 9, and perform approximations that make no sense in the exercise resolution, Fig. 10. Below we present some photos taken from the exams perform by students at 2018/2019 academic year.

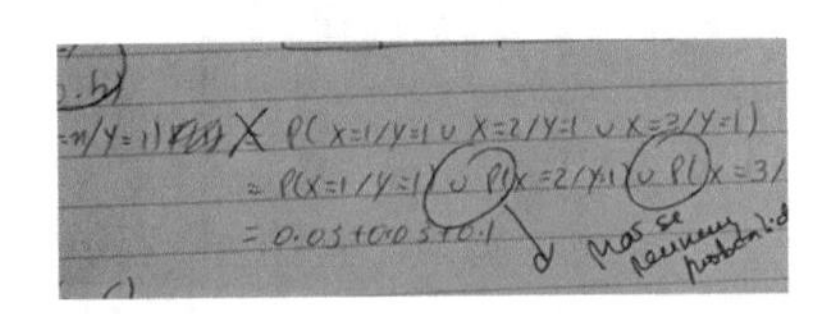

Fig. 5. The reunion of probability of events as if they were sets

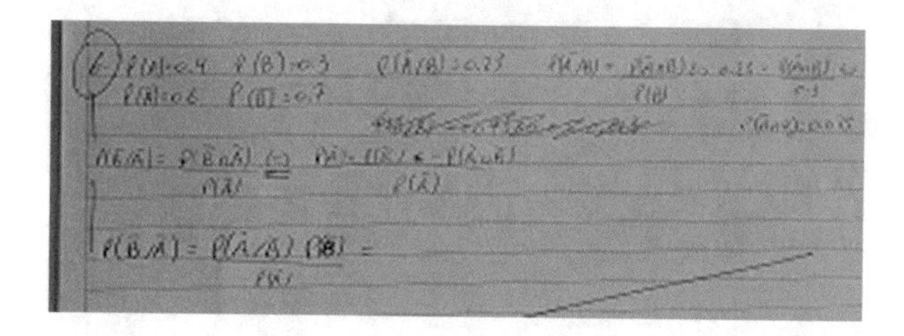

Fig. 6. Students know a lot of formulas but are enable to reasoning to obtain the desired value

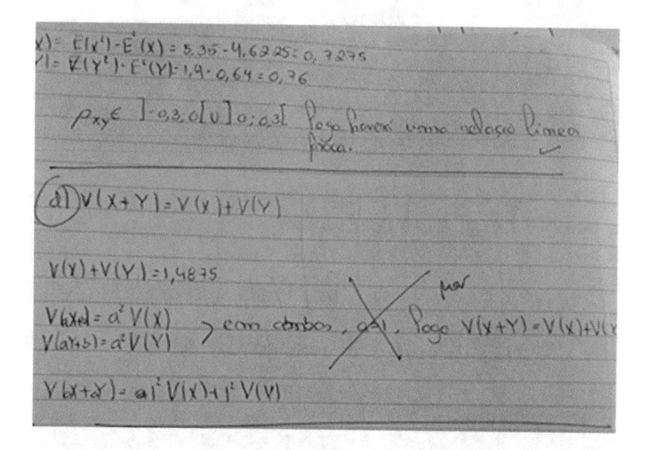

Fig. 7. Inability of relating theoretical properties

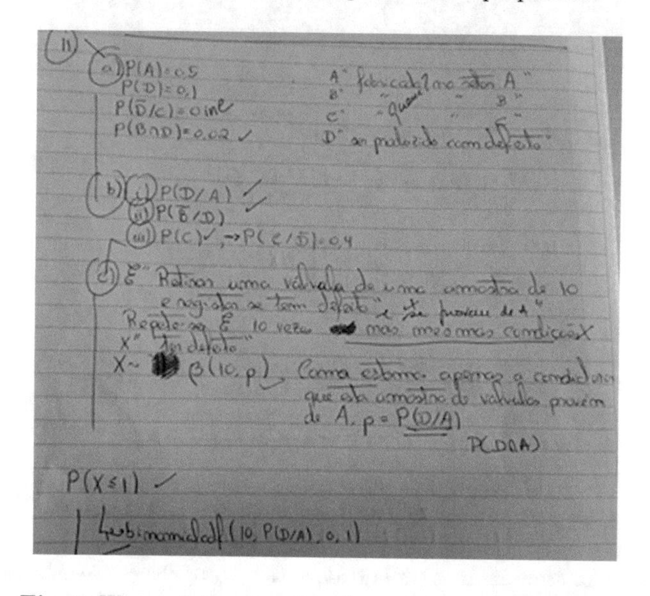

Fig. 8. Wrong data extraction from the exercise statement

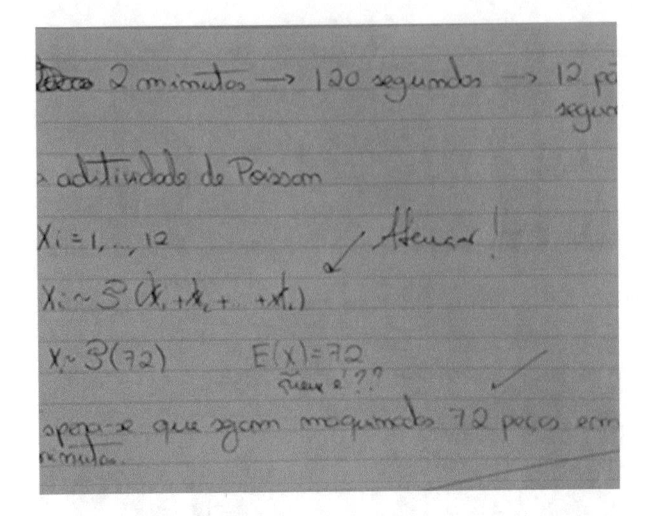

Fig. 9. Confusion between the variable and the parameter of its distribution, inability to represent the summation of a set of variables

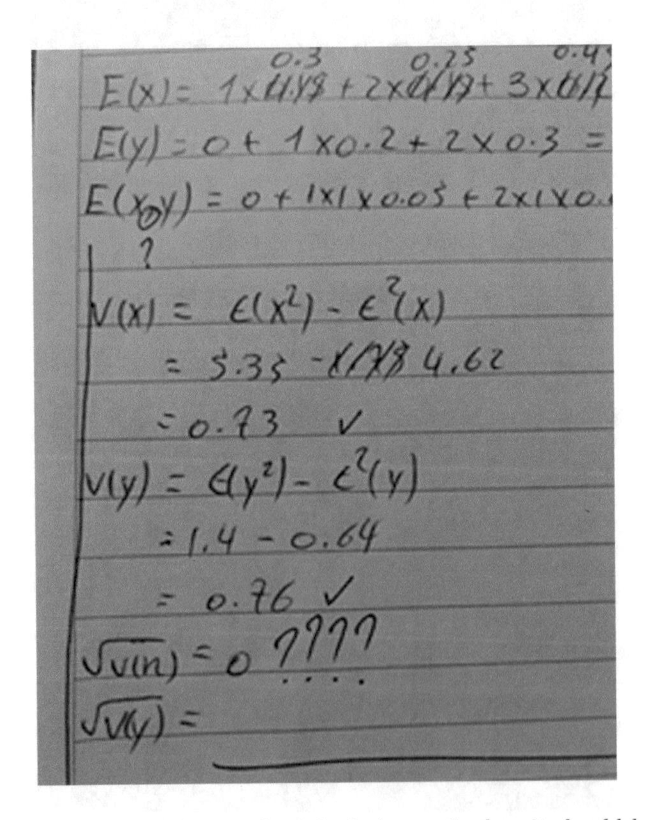

Fig. 10. Approximation of the standard deviation to 0 when it should be the square root of 0.73

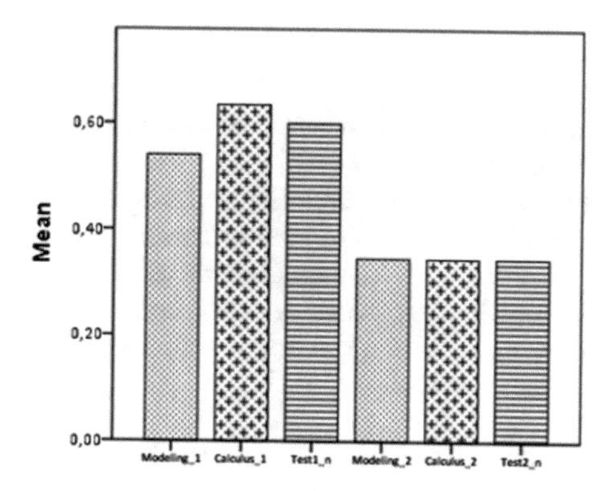

Fig. 11. Test 1 and Test 2 results

4 Conclusions for Education and Future Work

Analysing the results obtained namely Table 1 and Fig. 3 we may conclude that although the mean values of both types of questions are very similar in fact, calculus questions have higher classification. The same does not happened on the second test. The contents of the second part of syllabus are basically statistical inference. From results obtained on previous years and also this year (Fig. 5), we notice that the grades are not that high as the ones obtained on the first test and the difficulties felt on modelling and handling mathematical symbols are smaller. Although we may notice that, despite the efforts perform, the existence of non-explicable errors is still a fact. In the near future some activities to prevent this type of errors are meant to be designed together with the assessment tests refinement (Fig. 11).

Still, as teachers and with these results we are convinced that our students did acquire the competencies of modelling mathematically, representing mathematical entities, handling mathematical symbols and formalism yet they sure need to work more on them and dedicate time exploring real life problems in order to become real "doers" and apply their knowledge. We also defend, as Alpers et al. (2013), that mathematical education aims to provide mathematical expertise needed in the students future but also has to provide the mathematical concepts and procedures needed in application subjects and more theoretically considered contents need to be assessed end the knowledge acquisition measured. Modelling and working with models plays an important role for efficient work. Thus, setting up models and solving problems with models should be an essential part of engineering education without disregarding calculus, solution analysis and communication.

References

Alpers, B., et al.: A framework for mathematics curricula in engineering education, SEFI (2013). http://sefi.htw-aalen.de/

Niss, M.: Mathematical competencies and the learning of mathematics: the danish KOM project. In: Proceedings of the 3rd Mediterranean Conference on Mathematical education. Hellenic Mathematical Society, Athens (2003)

Niss, M., Højgaard, T. (eds.): Competencies and Mathematical Learning. Ideas and inspiration for the development of mathematics teaching and learning in Denmark, English Edition, Roskilde University (2011)

Niss M., Bruder R., Planas N., Turner R., Villa-Ochoa J.A.: Conceptualisation of the role of competencies, knowing and knowledge in mathematics education research. In: Kaiser G. (eds.) Proceedings of the 13th International Congress on Mathematical Education. ICME-13 Monographs. Springer, Cham (2017)

Rasteiro, D.D., Martinez, V.G., Caridade, C., Martin-Vaquero, J., Queiruga-Dios, A.: Changing teaching: competencies versus contents. In: EDUCON, 2018 - "Emerging Trends and Challenges of Engineering Education". Tenerife, Spain (2018)

Project Erasmus+ 2017-1-ES01-KA203-038491 "New Rules for Assessing Mathematical Competencies". https://www.researchgate.net/project/New-Rules-for-assessing-Mathematical-Competencies

Enhancing the Learning of Cryptography Through Concept Maps

Amalia Beatriz Orúe[1(✉)], Alina Maria Orúe[2], Liliana Montoya[3], Victor Gayoso[1], and Agustin Martín[1]

[1] Institute of Physical and Information Technologies (ITEFI), CSIC, Madrid, Spain
{amalia.orue,victor.gayoso,agustin}@iec.csic.es
[2] Ortocrip S.L., Madrid, Spain
alina@ortocrip.es
[3] Universidad Complutense de Madrid, Madrid, Spain
lmorue@ucm.es

Abstract. This paper proposes an approach of teaching centered on students which emphasizes the creative and effective use of technology. This approach meets students learning objectives using concept maps and shows how they can be used as an instructional tool to enhance the learning of Cryptography. The use of concept maps allows students to acquire a set of key mathematical competencies, such as representing mathematical entities, manipulating symbols and formalisms, and communicate in, with, and about mathematics.

Keywords: Meaningful learning · Concept map · Teaching cryptography · Education · Competencies

1 Introduction

The unstoppable technological progress of recent decades and the changes it causes in the most demanded work profiles, makes it essential that education systems take firm steps towards their continuous adaptation to this reality. Certainly, it is no longer enough to equip college students with a specific set of skills and knowledge; it is necessary to base their learning on a set of competencies, which allows them to adapt to changes since they should assume the need to continue learning and the acquisition of new relevant skills throughout their lives [1].

The Bologna Accord draws a clear path towards achieving this goal, being the improvement of the quality and relevance of teaching and learning in Higher Education one of its priorities. Higher Education in Spain continues taking steps to comply with this Accord. Traditional teaching methods are no longer sufficient to achieve the objectives of this competencies-based educational model that places the students as the protagonists of their learning. Therefore, in order to adapt to the learning needs of students, teaching strategies must also be changed [2].

© Springer Nature Switzerland AG 2020
F. Martínez Álvarez et al. (Eds.): CISIS 2019/ICEUTE 2019, AISC 951, pp. 263–272, 2020.
https://doi.org/10.1007/978-3-030-20005-3_27

This new paradigm can be possible through strategies that involve the active and constructive learning of students, allowing them to develop relationships between new, and previously assimilated knowledge, with the goal to accomplish a meaningful learning.

1.1 Cryptography as a Discipline Within Technical Careers

The need for providing security through efficient cryptographic algorithms has always been a concern in almost all areas of society. Formerly it was limited to military and diplomatic environments but nowadays it affects both the business world and the personal scope. The extraordinary rise of new information technologies, the development of the Internet, the electronic commerce, the e-government, mobile telephony and the pervasiveness of Internet of Things, have provided great benefits to the whole society. However, as it might be expected, this brings new challenges for the protection of information, such as the loss of confidentiality, privacy, and integrity in digital communications. Modern Cryptography plays an important role by providing the necessary tools to ensure the security of these new media [3].

Teaching Cryptography as part of the curriculum of technical careers is, therefore, a necessity, bearing in mind that it currently serves as a basis for the security of information and data communications networks. Note that, specifying and implementing security mechanisms and communications security have become key competencies for Information Technology (IT) security professionals [4].

The quality of the learning depends to a large extent on the quality of the teaching strategies. Traditional methods for teaching Cryptography have involved the transmission of the content of the course in one direction, from teachers to students. These methods can be effective on some topics if performed correctly but, in most cases, students become passive recipients of information who do not get involved in the subject or move away from the learning experience.

For students to develop skills that can best meet their personal aspirations and social needs, many teachers have introduced new teaching approaches that encourage interaction, participation, and collaborative activities [5].

Additionally, the theory and application of Cryptography encompass several disciplines that offer different degrees of complexity, such as Mathematics (number theory, abstract algebra, probability), Information Theory, Computer Science, and Telecommunications, which are difficult to understand and to interrelate using conventional teaching methods. Learning Cryptography requires the development of cognitive skills that include general mathematics competencies [1], analysis and synthesis, innovative solutions, working with priorities, and clear and precise communications [6,7].

This paper describes how the use of concept maps in the teaching of Cryptography has contributed to a certain extent to students achieving and developing key competencies that include creativity, communication skills, and continuous learning. The rest of the paper is organized as follows: Sect. 2 gives an overview of concept maps and their applications as instructional tools and describes the

CmapTools software (available at http://cmap.ihmc.us), a tool that helps to construct the concepts maps. Section 3 deals with examples of applications of the methodology that we propose. Finally, Sect. 4 summarizes the results reached and the viability of this methodology to enhance the teaching and learning of Cryptography.

2 Concept Maps

Developed by Prof. Joseph D. Novak and his colleagues at Cornell University in the early '70s, concept maps are a pedagogical tool that facilitates meaningful learning. Their work was based on a constructivist model of human cognitive processes, in particular, the assimilation theory of David Ausubel, who emphasized the importance of prior knowledge in being able to learn new concepts that would lead to meaningful learning [8]. Knowledge models, organized as concept map structures linked together, enable to organize complex fields of knowledge dynamically. When concept maps are implemented correctly, the information is organized and clearly structured, being much easier to remember and understand.

Concept maps are more than just a graphic tool to make a schematic and easy to understand representation about the assimilated knowledge; they act as pedagogical mediators and as visual organizers to manage and structure thoughts and ideas efficiently, allowing the capture of individual or group cognitive processes. In addition, the use of software for their elaboration provides them the added value of increasing the motivation in learning, making possible to carry out this process in a collaborative way.

There are many different definitions of concept maps [9]. In this work, we have adopted the definition given by their creators, Novak and Gowin, who stated that concept maps are *tools for organizing and representing knowledge* [8]. The main elements of knowledge are concepts, generally enclosed by circles or boxes. Relationships between concepts are indicated with lines and linking words, where concepts linked by these words form propositions. By convention, links run top-down unless annotated with an arrowhead. The concepts are defined as *perceived regularities or patterns in events or objects, or records of events or objects, designated by a label* [10]. Concepts are represented hierarchically; the most general and inclusive concept is positioned at the top of the hierarchy and the most specific and least general concept at the bottom.

Although concept maps have generally been used to evaluate an individual's specific knowledge or cognitive structure, they have also been used to present information to a scientific modern organizer [11].

2.1 Applications of Concept Maps

As aforementioned, concept maps are an effective means of representing and communicating knowledge. When concepts and linking words are carefully chosen, these maps can be useful tools for helping users to organize their thinking and summarizing subjects of study. The technique of concept maps has been broadly applied in many disciplines and various educational levels, eventually expanding beyond its original intent [12]. There are numerous applications for concept maps that include their use as an educational tool [13,14], as an interactive learning tool for Mathematics Education [15,16], as a tool to create pedagogical curricular designs, as a knowledge acquisition tool during the construction of expert systems, as a tool to improve scientific communications [17], as a means for capturing and sharing the knowledge of experts [18], etc.

2.2 The CmapTools Software to Construct Concept Maps

The CmapTools software was created at the Institute for Human and Machine Cognition. It brings together the strengths of concept mapping and the power of technology, the Internet and the Cloud. It allows users to construct their maps, navigate, share, and improve their maps in a collaboratively environment. According to [13], the CmapTools software facilitates the construction of concept maps in the same way that a word processor supports the task of writing text. CmapTools offers a wide collection of additional features that help users in the reorganization of knowledge models in a collaborative way. Users can create a hierarchy of folders in their computers, on a server or on the cloud, to organize their concept maps, images, videos, etc., all those resources associated to a project.

3 Case of Use

The increasing application of cryptographic techniques in nowadays society has meant that universities devote more attention to the discipline of Cryptography. The contents of the Cryptography courses are deep and complex, and require a consistent mathematical training, which makes it quite difficult for some students [6]. This situation easily originates demotivation and negative feelings towards learning. On the other hand, the traditional methods of teaching used by many academics, the insufficient number of class hours devoted to the subject and the shortage of practical activities, make students feel overwhelmed and they fail to achieve the association of concepts in an interdisciplinary matter as Cryptography is, becoming passive recipients of information.

The experience described in this contribution aims to offer an alternative and flexible approach in the teaching of this discipline, focused on the student, using concept maps as a tool to achieve the interrelation of concepts and fields included in the subject, and helping all students to advance in the development of analytical reasoning, criticism, collaborative working and communication skills.

The activities that we propose have already been applied in a Cryptography course in the context of an on-line Master Program by the UCAM (Catholic University of Murcia), and the following Cmaps that we explain have been elaborated by the students of this program.

The content of the discipline is structured through the following topics:

1. Introduction and classic algorithms.
2. Symmetric key algorithms and hash functions.
3. Asymmetric key algorithms.
4. Digital certificates and Public Key Infrastructure (PKI).
5. Cryptographic applications.

To illustrate the methodology followed, we have used the activities carried out to construct the concept maps in topic 3, Asymmetric key algorithms, specifically those related to the Diffie-Hellman key exchange protocol. In the first lecture on this topic, the problem of the discrete logarithm is presented. To understand it, we need to review some definitions, theories, and propositions belonging to the field of abstract algebra. Then the problem itself is explained, and finally, we study some asymmetric ciphers whose security is based on this mathematical problem. In the second lecture, the Diffie-Hellman key exchange is dealt with in greater depth, referencing some basic concepts necessary to understand it and then explaining its operation and applications.

Note that the procedure to elaborate a concept map includes a process of brainstorming, so the Cmap is elaborated in an interactive way. Therefore, the students need to rebuild the map to improve the first one that they had made. In this learning process, students will acknowledge themselves the concepts and theories by reviewing the contents several times and adding to the Cmap the new knowledge they have acquired. To guide the student in the aforementioned process we proposed two sorts of activities, one set of individual tasks and another one of team building.

3.1 Individual Activities

We proposed the following activities using CmapTools software to create the concept maps:

1. Read [3, §8.1], create a parking lot[1] to select the concepts that, in your opinion, are relevant, link them and then, using CMapTools, create a concept map and save it.
2. Create a concept map in which the definitions, theories, and propositions of Abstract Algebra that serve for the Discrete Logarithm Problem are included.
3. Build a concept map based on the definition of the Discrete Logarithm Problem (DLP) in $\mathbb{Z}_p^*$.
4. Build a concept map from the following list of concepts: Cryptography, symmetric key algorithms, asymmetric key algorithms, the discrete logarithm problem, the Diffie-Hellman key exchange protocol, and encryption systems based on the discrete logarithm. Explain each concept adding resources to the concept map.

[1] This is a brainstorming process where a list of concepts is made using a single word or short phrase associated with the main idea to identify the key concepts that apply to this domain.

5. Create a concept map of the second lecture, summarizing and relating the most important aspects.
6. Add resources to the concept map created in the previous exercise. Students are encouraged to delve into the topics covered, adding the revised bibliography as a resource within the concept map.
7. Construct a concept map through which the operation of the Diffie-Hellman key exchange protocol is explained.

3.2 Team Building Activities

The followings activities were oriented:

1. Comment through a Knowledge soup[2] [19]. Add the map made in activity 2 (concept map based on the definition of the Discrete Logarithm Problem) to a knowledge soup called "Discrete Logarithm". Comment on the maps of the rest of the students.
2. Create a concept map about the theorems and mathematical definitions that underlie the Diffie-Hellman key exchange protocol. To do that you should follow these steps:
   - In CmapTools, activate the button "Allow synchronous collaboration".
   - Use CmapTools to record the participation in the elaboration of the map of the different members of the group.
   - Add a Discussion list to a concept that has been of particular complexity for you. First, carry out the communication tests with the rest of the students of the group. Note: Discussion lists can be added to a single map where Cmaps are published on the concept maps server.
   - Compare Cmaps.
3. Given the following focus question: What is the Diffie-Hellman key exchange protocol?
   (a) Build a concept map that answers this focus question.
   (b) Add resources to the fundamental concepts of your map, based on a search on the Internet, which allows you to expand your knowledge about the concept. You can add images, documents, web pages, etc.
   (c) Publish the concept map.
   (d) Compare your map with the one made by another student. To do so, you should request the corresponding permission.
   (e) Add a discussion list to a concept on your map, in order to share or argue criteria with the author of the map you compared yours.
4. From the previous discussion, build a new map using the same focus question, in conjunction with the author of the map you compared yours.

Figure 1 represents one of the first concept maps drawn up individually by a student in the team building task activity item 3(a). It can be seen that by means of the inspection of the concept map by the working team, certain changes are proposed that are subsequently reflected in the final concept map, (see Fig. 2),

[2] This concept is explained throughout a concept map in [19].

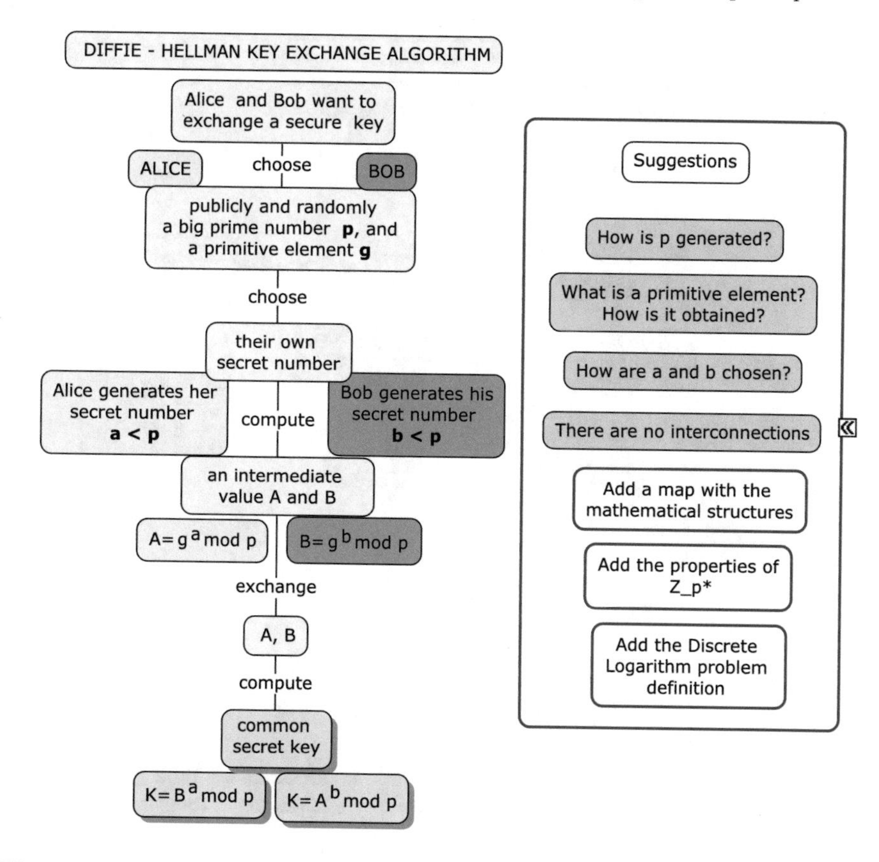

Fig. 1. Example of one the first concept maps drawn individually by a student in the Subsect. 3.2, item 3(a) and the suggestions to improve it.

elaborated by the student team as part of the team building task item 4. As we can see the final Cmap shows how the group of students consensually chose and interrelated the main concepts that are the basis of this protocol, interrelating it with the mathematical foundations, and adding the necessary resources so that the subject would be self-contained. The final Cmap can be seen in the following link:
https://cmapspublic3.ihmc.us/rid=1TDKSJ9N9-11817GL-50X/Diffie-Hellman %20Protocol.cmap.

With these activities, students were able to acquire skills to work on the synthesis of information, ordering ideas, and understanding topics of some level of complexity, as well as encouraging group work and collaborative learning through the concept maps.

The use of them as a tool in the teaching and learning process has also promoted collective intelligence and has provided us as teachers the necessary feedback to know which subjects and concepts to reinforce, and the aspects of the topic that have not been sufficiently clarified.

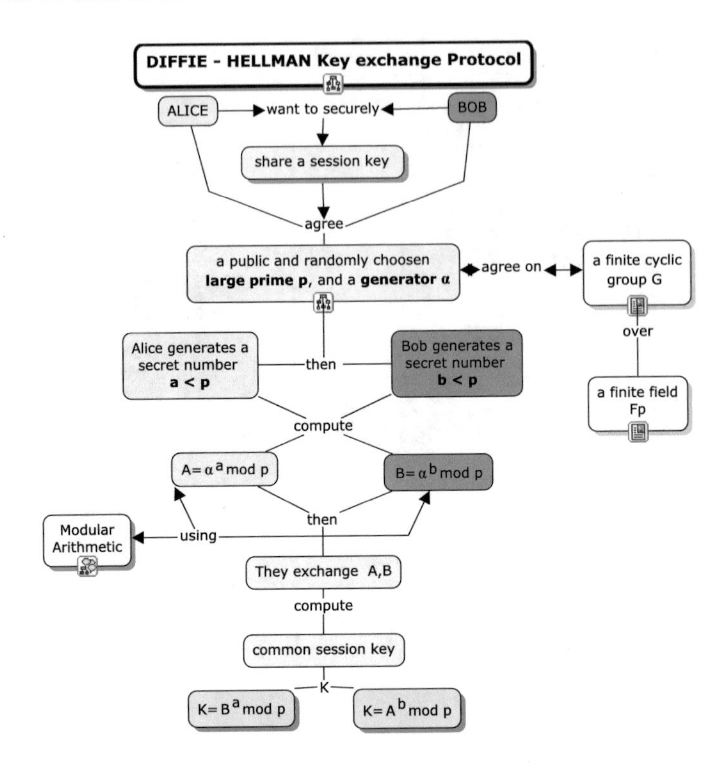

Fig. 2. Example of a final concept map elaborated collaboratively by a group of Master's students that describes the Diffie-Hellman key exchange protocol and illustrates some mathematical concepts mentioned in Subsect. 3.1, item 1.

4 Conclusion

Using for the first time concept maps tools in a Cryptography course was an amazingly enriching experience as well as a challenge. The use of concept maps allowed us to achieve much better results, evidenced by a better understanding of the algorithms and their mathematical bases. Students demonstrated the development of key competencies, such as representing mathematical entities, manipulating mathematical symbols and formalisms, communicating in, with and about mathematics and making use of aids and tools provided, in this case, by concept maps.

Interacting with the students and analysing the evolution of their concept maps, which showed their cognitive interests and their difficulties in understanding some topics, caused us to alter the content of the course on several occasions throughout the semester, which not only improved our way of teaching, but also allowed us to visualize the way to do it in future courses.

As it can be seen from the results of this experience, concept maps are not only an effective means to represent and communicate knowledge, but have also proven to be effective tools for teaching complex subjects, as they help students to discover by themselves the relationship between concepts and their link with previous knowledge, promoting their individual and collective learning, and transversally, fostering the development of key competencies such as communication and work in groups skills.

Acknowledgments. This research has been partially supported by Ministerio de Economía, Industria y Competitividad (MINECO), Agencia Estatal de Investigación (AEI), and Fondo Europeo de Desarrollo Regional (FEDER, EU) under project COPCIS, reference TIN2017-84844-C2-1-R, and by the Comunidad de Madrid (Spain) under the project CYNAMON (P2018/TCS-4566), co-financed with FSE and FEDER EU funds, and by the European Union program ERASMUS+ under the project 2017-1-ESO1-KA203-038491 (Rules_Math).

References

1. Alpers, B., et al. (eds.): A Framework for Mathematics Curricula in Engineering Education. A Report of the Mathematics Working Group. European Society for Engineering Education (2013)
2. Gómez, M., Aranda, E., Santos, J.: A competency model for higher education: an assessment based on placements. Stud. High. Educ. **42**(12), 2195–2215 (2017)
3. Paar, C., Pelzl, J.: Understanding Cryptography: A Textbook for Students and Practitioners. Springer, Heidelberg (2010)
4. Patel, A., Benslimane, Y., Bahli, B., Yang, Z.: Addressing IT security in practice: key responsibilities, competencies and implications on related bodies of knowledge. In: IEEE International Conference on Industrial Engineering and Engineering Management (2012)
5. Karst, N., Slegers, R.: Cryptography in context: co-teaching ethics and mathematics. PRIMUS **1**, 1–28 (2019)
6. Garera, S., Vasconcelos, G.: Challenges in teaching a graduate course in applied cryptography. ACM SIGCSE Bullet. **41**(2), 103–107 (2009)
7. Moon, B.M., Hoffman, R.R., Novak, J.D., Cañas, A.J.: Applied Concept Mapping Capturing, Analyzing, and Organizing Knowledge. CRC Press, Boca Raton (2011)
8. Gowin, D.B., Novak, J.D.: Learning How to Learn. Cambridge University Press, New York (1984)
9. Milam, J.H., Santo, S.A., Heaton, L.A.: Concept maps for web-based applications. ERIC technical report, National Library of education, WDC (2000)
10. Novak, J.D., Cañas, A.: The origins of the concept mapping tool and the continuing evolution of the tool. Inf. Visual. **5**(3), 175–184 (2006)
11. Gómez, J.L., Castro, J.L., Gómez, I.M: Organisers, concept maps, and text summarisation in the CLIL classroom. In: Cañas, A., Zea, C., Reiska, P., Novak, J. (eds.) Concept Mapping: Renewing Learning and Thinking, Proceedings of the Eighth International Conference on Concept Mapping, pp. 181–191 (2018)
12. Roessger, K.M., Daley, B.J., Hafez, D.A.: Effects of teaching concept mapping using practice, feedback, and relational framing. Learn. Instruct. **54**, 11–21 (2018)
13. Novak, J.D., Cañas, A.J.: Theory underlying concept maps and how to construct and use them. Technical report, IHMC CmapTools (2008)
14. Ozdemir, A.: Analyzing concept maps as an assessment (evaluation) tool in teaching mathematics. J. Soc. Sci. **1**(3), 141–149 (2005). ISSN 1549-3652

15. Novak, J.D., Cañas, A.J.: The Development and Evolution of the Concept Mapping. Tool Leading to a New Model for Mathematics Education, pp. 3–16. Springer, Boston (2009)
16. Flores, R.P.: Concept Mapping: An Important Guide for the Mathematics Teaching Process, pp. 259–277. Springer, Boston (2009)
17. Orúe, A.B., Alvarez, G., Montoya, F.: Using concept maps to improve scientific communications. In: ENMA Education International Conference, pp. 88–95 (2008)
18. Coffey, J.W., Hoffman, R.R., Cañas, A.J., Ford, K.M.: A Concept Map-Based Knowledge Modeling Approach to Expert Knowledge Sharing. Institute for Human and Machine Cognition. Report (2002)
19. CmapTools Knowledge Soup. Description of Knowledge Soups in CmapTools (2014). https://cmapskm.ihmc.us/viewer/cmap/1064009710027_604311754_27108

Special Session: Innovation in Computer Science Education

Analysis of Student Achievement Scores: A Machine Learning Approach

Miguel García-Torres[1(✉)], David Becerra-Alonso[2], Francisco A. Gómez-Vela[1], Federico Divina[1], Isabel López Cobo[2], and Francisco Martínez-Álvarez[1]

[1] Division of Computer Science, Universidad Pablo de Olavide, 41013 Seville, Spain
{mgarciat,fgomez,fdivina,fmaralv}@upo.es
[2] Universidad Loyola Andalucía, Seville, Spain
{dbecerra,ilopez}@uloyola.es

Abstract. Educational Data Mining (EDM) is an emerging discipline of increasing interest due to several factors, such as the adoption of learning management systems in education environment. In this work we analyze the predictive power of continuous evaluation activities with respect the overall student performance in physics course at Universidad Loyola Andalucíia, in Seville, Spain. Such data was collected during the fall semester of 2018 and we applied several classification algorithms, as well as feature selection strategies. Results suggest that several activities are not really relevant and, so, machine learning techniques may be helpful to design new relevant and non-redundant activities for enhancing student knowledge acquisition in physics course. These results may be extrapolated to other courses.

Keywords: Educational Data Mining · Classification · Feature selection

1 Introduction

The increasing usage of Information Technologies for educational systems (ES) has led to the current availability of a large amount of data in this field [3]. The analysis of these data may well be seen as an opportunity to improve ES, e.g., students failing advanced subjects could have not acquired basic pieces of information from these [1,4]. Therefore, analysing students' academic records could potentially identify causes for academic failure, and in case, take action to fetch these [1].

Moreover, due to the large number of students registered in the education systems each year these data cannot be handled manually. In this context, Data Mining techniques arise as a groundbreaking tool for the comprehensive understanding of the data, providing valuable insights from such data. The application of data mining in the educational field is called Educational Data Mining (EDM) and it deals with pattern identification and prediction-making for the characterization of students' behaviour and merits, including subject knowledge, student assessments educational functions and applications [9].

© Springer Nature Switzerland AG 2020
F. Martínez Álvarez et al. (Eds.): CISIS 2019/ICEUTE 2019, AISC 951, pp. 275–284, 2020.
https://doi.org/10.1007/978-3-030-20005-3_28

A major goal of classification and predictions task is to estimate student's performance in order to assist educators whilst providing an effective teaching approach [17]. These predictions could help in the implementation of special programs, like modified versions of the school curriculum for students showing significant difficulties and, thus, improving ES [13]. To do so, some works have used techniques such as Decision Trees or Neural networks [12,13].

In addition, classification algorithms are commonly applied after a feature selection process on the input data set [15]. In the EDM field, FS techniques are applied in the search for the most relevant subset of features for achieving highly-predictive student performance [11]. Feature selection (FS) algorithms remove irrelevant and/or redundant features from educational data sets and hence increasing the performance of classifiers used in EDM techniques [19].

In this paper we present a study of 41 students from physics course at Universidad Loyola Andalucíia, in Seville, Spain. The aims of this study can be summarized as follow:

- To analyze how informative are the performance of the continuous evaluation activities with respect the overall student performance in the course.
- To asses if, by using the machine learning techniques can be useful to identify relevant activities.

On the one hand, the first objective allows to get insight about the suitability of the continuous activities and, on the other hand, the second objective facilitates the work on developing an adequate set of activities.

The rest of the paper is organized as follows. Section 2 introduces the methodology applied and the data collected. Then, in Sect. 3, the different machine learning techniques are described. Finally, the experiments and the conclusions and future work are in Sects. 4 and 5 respectively.

2 Methodology

In this section we describe the tools used to collect the data and how it works, as well as the characteristic of the data.

2.1 Data Collection Methodology

In this work the data was collected using EduZinc [2], an automatic grading platform that allows to design the complete process of creating and evaluating personalized activities of a course.

This platform differentiates three roles: editors, teachers and students. Editors are those that create the activities while the teachers follow the performance of the students through a report, that they receive daily, with statistics about the overall progress of the students. Finally, the students are those who have to complete the different activities.

The workflow of EduZinc is shown in Fig. 1. As we can see, the framework contains several templates to cover the types of exercises proposed by teachers.

Such exercise templates are parametrized so that even in the case that two students have to solve the same activity, the results will be different. This is because the values depend on some input parameters that are created randomly.

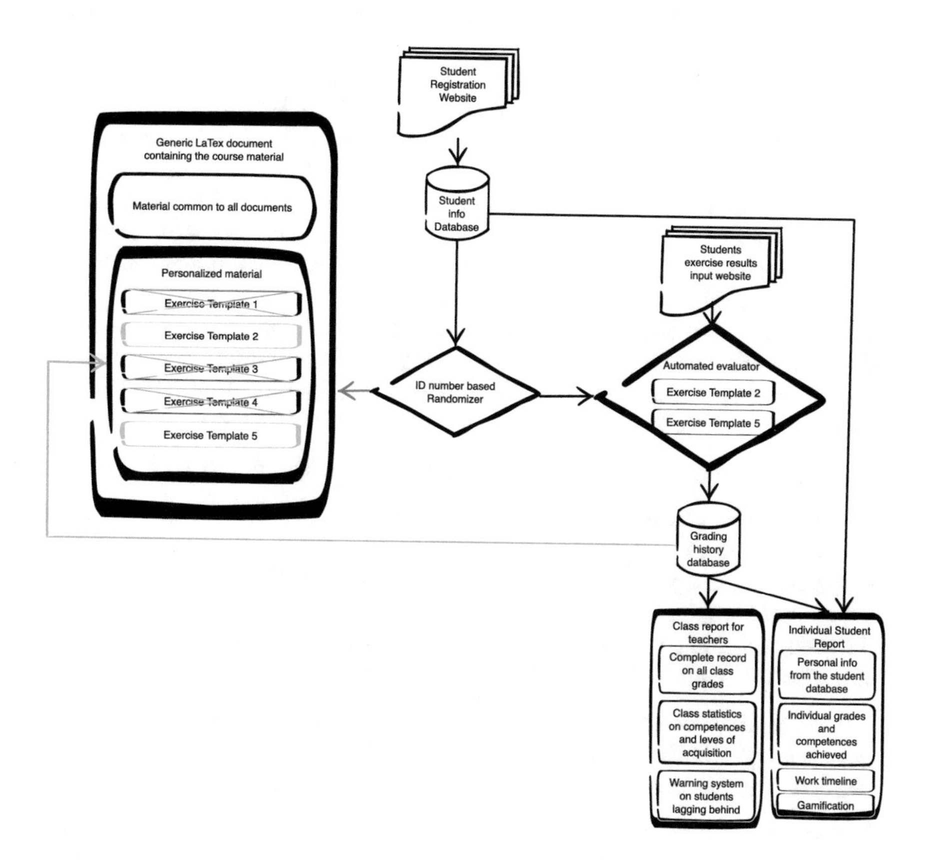

Fig. 1. Diagram of EduZinc platform.

Each student has to register to log in the platform and the system associated, to the student, a random ID. Such ID will be used by the system to assign a subset of activities. Then, the student solves such exercises and the results are stored in a database. Finally this information is reported to the teacher and student.

2.2 Data

The data used for this manuscript was collected during the fall semester of 2018, during Physics course at Universidad Loyola Andalucía, in Sevilla, Spain. This is a first year course for students in the School of Engineering. The course is taught in 3 months and covers 9 units.

During the course, a series of activities are proposed for many of this units. Such in-class activities, that correspond to features in th data, include:

- **eInstruction** [16]. The features are *EinstructionT12*, *EinstructionT7* and *EinstructionT8*, where the number indicates that this data was obtained for units 1, 2, 7 and 8.
- **Lab reports.** These features refer to lab experiments related to the contents in the course: *LabP1*, *LabP2*, *LabP3*, *LabP4* and *LabP5*.
- **EduZinc Questionnaires.** The variables *TestT1b*, *TestT4*, *TestT5* and *TestT6* attest to the units where this activity took place.
- **Mid-term exam.** Includes questions from units 2 to 5. The variables collected are *ExT2*, *ExT3*, *ExT4* and *ExT5*. Variable *ExRP* is a simple indicator of how much students present their exercises correctly, include measures, and use proper mathematical notation.
- **Final exam.** It includes questions from units 6 to 9. There is an extra emphasis on unit 9, where one question is about Newton's laws and another is on the conservation of angular momentum. The variables are *ExT6*, *ExT7*, *ExT8*, *ExT9* and *ExT9b*. Variable *ExRPb* is just like *ExRP* but for the final exam.

At home, the students are expected to participate in the following activities:

- **EduZinc Homework.** The variables collected are *HwT1*, *HwT2*, *HwT3*, *HwT4* and *HwT7*.
- **Flipped Classroom Questionnaires.** These features include questions about some topics that student have to answer after reading some material and watching videos. The names of these features are *TestT1*, *TestT2* and *TestT3*, again referring to the units where this activity took place.

All these variables are to be compared against a class variable that will simply be *yes* or *no* to whether students passed the course.

Students in this course are offered to be evaluated in one of two ways:

- **Evaluation focused on daily work:** This is the default form of evaluation. It is meant for those students who have worked during the semester and want to be acknowledged for it. Activities are graded in the following way: *Einstruction* 10%, *Lab* 15%, *Test* 10%, *Hw* 20%, *ExT2* to *ExT5* 15% and *ExT6* to *ExT9b* 30%. Exams don't take up the majority of the grading. However, in order to pass under this system, students must complete 70% of their work, where the final exam has to be above 40% of its value. The dataset created for this system is called $\mathcal{D}$-7.
- **Evaluation focused on knowledge:** This system is meant for two kinds of students. First, those who know the subject really well, and are good on exams. Second, those who didn't work enough but want to have an opportunity on the final exam. Activities are graded in the following way: *Einstruction* 5%, *Lab* 10%, *Test* 10%, *Hw* 15%, *ExT2* to *ExT5* 10% and *ExT6* to *ExT9b* 50%. Exams take up the majority of the grading. In order to pass under this system, students must complete 50% of their work, where the final exam has to be above 40% of its value. The dataset created for this system is called $\mathcal{D}$-5.

3 Machine Learning Algorithms

In this section we provide the concepts of the Machine Learning algorithms. First we introduce the classification task followed by the classifiers. Then, we describe formally the feature selection problem and the strategies studied in this work.

3.1 Classification

The aim of classification is to learn a function that maps an input to an output based on example input-output pairs. Formally, let $\mathcal{T}$ be a set of n instances described by pairs $(\mathbf{x}_i, y_i)$, where each $\mathbf{x}_i$ is a vector described by d features, and its corresponding y_i is a qualitative attribute that stands for the associated class to the vector. Then, the classification problem can be defined as the map $\mathcal{C} : \mathbf{X} \to \mathcal{Y}$, such that $\mathcal{C}$, which is called classifier, maps from a vector $\mathbf{X}$ to class labels $\mathcal{Y}$. In this work we use the following classifiers: Naive Bayes (NB), C5.0 and Support Vector Machine (SVM).

Naive Bayes (NB) [8] classifier applies the Bayes rule to compute the probability of a given class label given an instance. It assumes that all features are conditionally independent given the values of the class. Besides this unrealistic assumption, the performance of this classifier is surprisingly good. On the other hand, C5.0 [10] is a decision tree based classifier that uses the concepts of information theory to build the classifier. The model consists of an inverted tree where each node represents a feature, the branch a decision rule and each leaf specifies the expected value of the class. The attribute with the highest normalized information gain is chosen to make the decision. Finally, Support Vector Machine (SVM) [14] is based on the idea of finding a hyperplane that optimally separates the data into two categories by solving an optimization problem. Such hyperplane is the one that maximizes the margin between classes. For such purpose it uses a kernel function for mapping the original feature space to some higher-dimensional feature space where the data set is separable.

3.2 Feature Selection

Feature selection is the process of selecting the most relevant or useful features of the problem under study. Therefore, it seeks to reduce the number of attributes in the dataset, while keeping (or improving) the predictive power of the classifier induced. The goodness of a particular feature subset is evaluated using an objective function, $J(S)$, where S is a feature subset of size $|S|$. The algorithms considered in this work are: Fast Correlation Based Filter (FCBF), Scatter Search (SS) and Genetic Algorithm (GA).

Fast Correlation Based Filter (FCBF) [18] is basically an efficient strategy that performs a backward search strategy that uses Symmetrical Uncertainty (SU) as objective function $J(S)$ to estimate non-linear dependencies between features. Scatter Search (SS) [5] is a population-based metaheuristic that uses intensification and diversification mechanisms to generate new solutions. Solutions evolve as a result of combination and local search mechanisms. Genetic

Algorithm (GA) [7] is another population based metaheuristic that uses bio-inspired mechanisms to evolve solutions. In this work we use, as $J(S)$ for SS and GA, the Correlation Feature Selection [6] (CFS) measure; which considers, as good subsets of features, those that include features highly correlated with the class but uncorrelated to each other.

4 Experiments

The following section describes the experiments performed and presents the results achieved. We divide the experimentation in three parts:

(a) First, in Sect. 4.1, we analyze the predictive power of the features involved in the study.
(b) Then, we apply, in Sect. 4.2, feature selection techniques to analyze the relevance of the different features and identify a subset of highly informative features.
(c) Finally, Sect. 4.3 is dedicated to discuss the results to get insight in the area of education.

All the experiments were conducted using cross-validation to assess the predictive performance of the classification models. The number of folds k considered is important since lower values of k lead to higher bias in the error estimates and less variance. Conversely increasing k decreases the bias and increases the variance. In our experiments we use Leave-one-out (LOOCV) cross validation which is a special case of k-fold cross validation with $k = n$, where n is the sample size.

The predictive models are evaluated by means of the classification error averaged over the folds. In the feature selection analysis, in addition to the error we also report the average number of features selected by each strategy and the robustness of each algorithm. The robustness, also called stability, assesses the variations of a feature selection algorithm when it is applied to different samples of a dataset. Therefore, this analysis improves the confidence in the analysis of the feature selection strategies. In this work we asess the robustness $\Sigma(S)$ of all feature subsets found from LOOCV using the Jaccard index.

4.1 Baseline

In this section we compare the different classification models trained. Table 1 presents the baseline performance on both datasets with the classifiers described in this work: naive Bayes, Bayesian Network Classifier, Support Vector Machine and C5.0.

In general, all models (except C5.0 for $\mathcal{D}$-7) achieve models with good performance. However, on $\mathcal{D}$-5 dataset the classifiers induced show higher predictive power than those induced on $\mathcal{D}$-7 data. This applies in all cases, specially with C5.0, in which the difference is approximately of 26 points.

Table 1. Classification results on $\mathcal{D}$-5 and $\mathcal{D}$-7 datasets.

$\mathcal{A}$	$\mathcal{D}$-5	$\mathcal{D}$-7
NB	87.81 ± 33.13	85.37 ± 35.78
SVM	85.37 ± 35.78	80.49 ± 40.12
C5.0	82.93 ± 38.10	56.10 ± 50.24
Average	85.37	77.22

4.2 Feature Selection

In this section we apply feature selection techniques to study the relevance of the different features used in this study and how they relate with the class value. The results of the dimensionality reduction can be found in Table 2. First, we present for each dataset and feature selection strategy, the average number of features selected over all folds with the standard deviation and the robustness. Then, the accuracy of the different predictive models induced on $\mathcal{D}$-5 and $\mathcal{D}$-7 respectively.

We can see that on $\mathcal{D}$-5 dataset, SS is the strategy that, on average, reduces the most followed by FCBF and GA. When focusing on robustness, FCBF is the one that exhibits a higher robustness followed by SS. The stability of GA is, however, far below the other two strategies. Results on $\mathcal{D}$-5 data are similar although we can find that, in general, the subsets of features are smaller. In this case FCBF is the algorithm that finds the smallest subsets of features while now SS is the most stable strategy. In both measures GA is the algorithm that exhibits the worst results.

When paying attention to accuracy on $\mathcal{D}$-5, we can find that SS is, on average, the strategy which achieves the highest accuracy followed closely by GA and FCBF. SVM and NB achieve the best predictive power for SS and GA respectively. In contrast to the previous dataset, on $\mathcal{D}$-7 the classifiers degrade, especially with C5.0. On average, models learned using the features found by SS continues to be the classifiers with the highest rates of accuracy followed by FCBF and GA based models respectively. The best models found were SVM with SS and GA and NB with GA. In general, the accuracy on $\mathcal{D}$-5 dataset tends to be higher than on $\mathcal{D}$-7 data.

4.3 Discussion

In this section we analyze the features selected in order to get insights about the problem addressed. Figure 2 presents, for both datasets, the ranking of each feature according to the SU measure. Such Figure only displays the features with $SU > 0$. The features of $\mathcal{D}$-5 are shown in the lower horizontal axis while those associated to $\mathcal{D}$-5 can be found at the upper horizontal axis. This Figure also presents the top features most selected for each feature selection algorithm. Filled symbols refer to those features selected in all folds while empty ones refer to those features selected with an occurrence higher or equal to 90%.

Table 2. Classification results after feature selection on $\mathcal{D}$-5 and $\mathcal{D}$-7 datasets.

	$\mathcal{A}$	SS		FCBF		GA	
	$\mathcal{D}$	#features	$\Sigma(S)$	#features	$\Sigma(S)$	#features	$\Sigma(S)$
	$\mathcal{D}$-5	9.93 ± 1.31	0.81	10.68 ± 0.88	0.86	14.24 ± 2.36	0.57
	$\mathcal{D}$-7	9.44 ± 1.16	0.75	8.63 ± 1.04	0.68	13.88 ± 2.36	0.57
	$\mathcal{C}$	Accuracy		Accuracy		Accuracy	
$\mathcal{D}$-5	NB	90.24 ± 30.04		87.81 ± 33.13		92.68 ± 26.37	
	SVM	92.68 ± 26.37		87.81 ± 33.13		87.81 ± 33.13	
	C5.0	85.37 ± 35.78		85.37 ± 35.78		85.37 ± 35.78	
	Avg	89.43		87.00		88.62	
$\mathcal{D}$-7	NB	82.93 ± 38.10		82.93 ± 38.10		85.37 ± 35.78	
	SVM	85.37 ± 35.78		85.37 ± 35.78		85.37 ± 35.78	
	C5.0	63.42 ± 48.77		60.98 ± 49.40		53.66 ± 50.50	
	Avg	72.24		76.43		74.80	

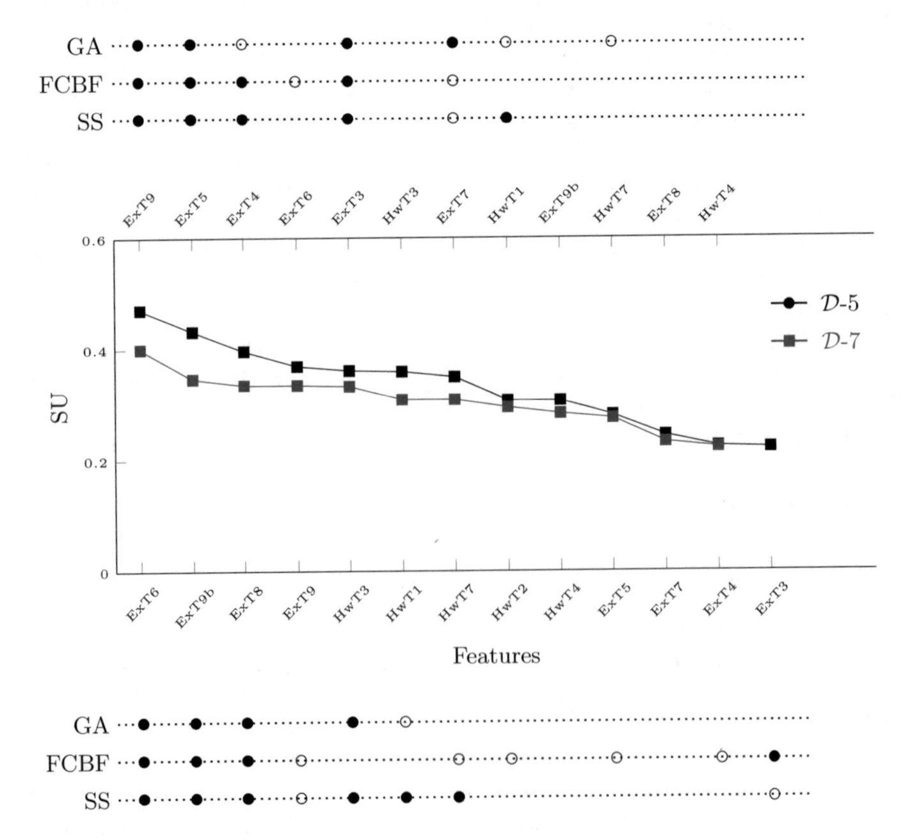

Fig. 2. Ranking of features of $\mathcal{D}$-5 and $\mathcal{D}$-7 datasets according to SU and the most frequently selected features from each feature selection algorithm.

The similar SU found for $\mathcal{D}$-5 and $\mathcal{D}$-7 speaks of them both being comparable systems of evaluation. However, since the final exam is more important under $\mathcal{D}$-5, many exam questions have a SU slightly higher than that of $\mathcal{D}$-7.

The most revealing fact derived from Fig. 2, however, is the list of features selected. Exam exercises take up the top of the list since, as it was mentioned, they are defining the final grade under both $\mathcal{D}$-7 and $\mathcal{D}$-5. This doesn't mean however that they are critical when it comes to defining how good a passing grade is. Remember the output all these features are being compared with is Pass or No-pass label. Exams, although not the most desired form of evaluation, remain the only opportunity the students have to show that they can tackle physics exercises in solitude.

The fact that *ExT9*, *ExT9b* and *ExT8* rank high in this feature selection is a satisfactory outcome. It confirms that some of the most important units are relevant to the final grade. This was the initial intent of the teacher, but he had no initial means to make it happen. *ExT7* is also very relevant to the course, however, it is often a long exercise that students tend to struggle with. That would explain it ranking within the mid-to-low part of both lists: it is important and has been selected, yet not alongside the most relevant exam exercises.

5 Conclusions and Future Work

In this work we have analyzed educational data collected during the fall semester of 2018 in physics at Universidad Loyola Andalucía, in Seville, Spain.

The results of the analysis, performed using machine learning approach suggest that this type of techniques are suitable for getting insight about the information contained in the data. Furthermore, feature selection has shown to be suitable to identify relevant activities with respect the performance of the students in the course.

Therefore, these results encourages us to refine and expand the use of EduZinc in the future, where it will be applied to subjects other than physics, and contexts other than higher education A systematic data collection will be used to reinforce or update the initial conclusions presented in this manuscript.

References

1. Algarni, A.: Data mining in education. Int. J. Adv. Comput. Sci. Appl. **7**(6), 456–461 (2016)
2. Alonso, D.B., Lopez-Cobo, I., Gomez-Rey, P., Fernandez-Navarro, F., Barbera, E.: A new tool to create personalized student material for automated grading, Working paper 2019
3. Asif, R., Merceron, A., Ali, S.A., Haider, N.G.: Analyzing undergraduate students' performance using educational data mining. Comput. Educ. **113**, 177–194 (2017)
4. Baker, R.S., Yacef, K.: The state of educational data mining in 2009: a review and future visions. JEDM— J. Educ. Data Min. **1**(1), 3–17 (2009)

5. García-López, F.C., García-Torres, M., Melián-Batista, B., Pérez, J.A.M., Moreno-Vega, J.M.: Solving the feature selection problem by a parallel scatter search. Eur. J. Oper. Res. **169**(2), 477–489 (2006)
6. Hall, M.A.: Correlation-based feature subset selection for machine learning. Ph.D. thesis, University of Waikato, Hamilton, New Zealand (1998)
7. Holland, J.H.: Adaptation in Natural and Artificial Systems: An Introductory Analysis with Applications to Biology, Control and Artificial Intelligence. University of Michigan Press, Ann Arbor (1975)
8. Pearl, J.: Probabilistic Reasoning in Intelligent Systems: Networks of Plausible Inference. Morgan Kaufmann Publishers Inc., San Francisco (1988)
9. Peña-Ayala, A.: Educational data mining: a survey and a data mining-based analysis of recent works. Expert Syst. Appl. **41**(4), 1432–1462 (2014)
10. Quinlan, J.R.: C4.5: Programs for Machine Learning. Morgan Kaufmann Publishers Inc., San Francisco (1993)
11. Ramaswami, M., Bhaskaran, R.: A study on feature selection techniques in educational data mining. J. Comput. **1**(1), 7–11 (2009)
12. Saa, A.A.: Educational data mining & students performance prediction. Int. J. Adv. Comput. Sci. Appl. **7**(5), 212–220 (2016)
13. Shahiri, A.M., Husain, W., et al.: A review on predicting student's performance using data mining techniques. Procedia Comput. Sci. **72**, 414–422 (2015)
14. Vapnik, V.N.: Statistical Learning Theory. Wiley, New York (1998)
15. Velmurugan, T., Anuradha, C.: Performance evaluation of feature selection algorithms in educational data mining. Int. J. Data Min. Tech. Appl. **5**(02), 131–139 (2016)
16. Ward, D.: einstruction: Classroom performance system (computer software). EInstruction Corporation, Denton, TX (2007)
17. Xing, W., Guo, R., Petakovic, E., Goggins, S.: Participation-based student final performance prediction model through interpretable genetic programming: integrating learning analytics, educational data mining and theory. Comput. Hum. Behav. **47**, 168–181 (2015)
18. Yu, L., Liu, H.: Efficient feature selection via analysis of relevance and redundancy. J. Mach. Learn. Res. **5**, 1205–1224 (2004)
19. Zaffar, M., Hashmani, M.A., Savita, K.: Performance analysis of feature selection algorithm for educational data mining. In: 2017 IEEE Conference on Big Data and Analytics (ICBDA), pp. 7–12. IEEE (2017)

Analysis of Teacher Training in Mathematics in Paraguay's Elementary Education System Using Machine Learning Techniques

Viviana Elizabeth Jiménez Chaves[1], Miguel García-Torres[2(✉)], José Luis Vázquez Noguera[1], César D. Cabrera Oviedo[3], Andrea Paola Riego Esteche[1], Federico Divina[2], and Manuel Marrufo-Vázquez[4]

[1] Universidad Americana, Asunción, Paraguay
{viviana.jimenez,jose.vazquez,andrea.riego}@americana.edu.py
[2] Division of Computer Science, Universidad Pablo de Olavide, 41013 Seville, Spain
{mgarciat,fdivina}@upo.es
[3] Dpto. de Matemáticas, Facultad de Ciencias Exactas y Naturales, Universidad Nacional de Asunción, San Lorenzo, Paraguay
ccabrera@facen.una.py
[4] Dpto. de Psicología Experimental, Facultad de Psicología, Universidad de Sevilla, 41018 Seville, Spain
marrufo@us.es

Abstract. In Paraguay, despite the fact that Elementary Education is one of the cornerstones of the educational system, it has not always received the recognition it deserves. Recently, the Paraguayan government has started to focus its effort on evaluating the quality of its education system through the analysis of some factors of the teachers. In this work, which falls into the context of such project, we study the ability to understand the different evaluation types structures in mathematics. The data, collected from elementary mathematics teachers from all over the country, is analyzed by applying an education data mining (EDM) approach. Results show that not all questions are equally important and it is necessary to continue through different lines of action to get insight about the action policy to improve the educational system quality.

Keywords: Educational data mining · Feature selection · Classification

1 Introduction

Elementary education system is one of the main axes of the Paraguayan education since it plays an essential role in acquisition of basic competences such as communication, mathematics, personal development, socialization, etc. These competences are focused on establishing the improvements of people's capacity

© Springer Nature Switzerland AG 2020
F. Martínez Álvarez et al. (Eds.): CISIS 2019/ICEUTE 2019, AISC 951, pp. 285–294, 2020.
https://doi.org/10.1007/978-3-030-20005-3_29

that make students integrate in their learning process. Despite its importance, it has not received the resources and attention necessary to develop a high-quality education system.

In Paraguay, elementary mathematics education has a number of problems that make challenging succeed in reaching high levels of student performance. Two main problems are the lack of resources and poor teaching skills. The economical improvement that has experienced in recent years has led the government to develop a project to identify needs in the education system in general, and in mathematics in particular, to establish action policies. In this context, the Universidad Americana, in conjunction with the Ministry of Education and Science conducted a survey, based on the Third Regional Comparative and Explanatory (TERCE) [11] survey, for mathematics teachers from all over Paraguay to assess their preparation in mathematics and in the didactics and pedagogy of mathematics.

The large number of teachers and the complexity of the data make data mining tools suitable for the analysis and get insight to identify potentially needs of the teachers. This interdisciplinary study, which applies data mining techniques to educational data, is referred as Educational Data Mining (EDM) [6] and deals with common predictive and descriptive tasks from machine learning area.

In EDM, classification is a major task that has been widely applied in previous works. In [10], the prediction is used to implement special programs focused on students with difficulties. Other works [1,7] have focused their effort on the student performance predictions. These techniques have also been successively applied in online courses. In [2] the analysis is used to provide a feedback to teachers and students from the development of the course.

In this work we study the ability of mathematics teacher, from all over Paraguay, to understand the different evaluation types structures in mathematics. For such purpose a survey was designed to identify possible background and pedagogical skill needs from the teachers. This analysis can serve as a basis for the development of policies and pedagogical interventions to improve the performance of teachers.

Despite the good results achieved on these works, it is worth noting that not all features contribute equally to the class concept. Therefore some recent works have started applying feature selection techniques [13,15] showing the improvement of the performance of the classifier after removing irrelevant features.

The rest of the paper is organized as follows. Section 2 introduces the project conducted and the main characteristics of the data collected. Then, in Sect. 3, the different machine learning techniques are described. Finally, the experiments and the conclusions and future work are in Sects. 4 and 6 respectively.

2 Data Collection Methodology

The Latin American Laboratory for the Assessment of the Quality of Education (LLECE) is a key instrument from UNESCO to monitor a follow-up the Education 2030 global agenda, with an integral focus on the quality of education and on ways to evaluate it.

LLECE developed, in 2012, the Third Regional Comparative and Explanatory Study (TERCE), which was a large scale study on the performance of mathematics that is focused on 4 major areas of mathematics: (i) evaluation of mathematical competences, (ii) didactics of arithmetics, (ii) didactics of geometry, and, (iv) didactics of statistics. The aim of TERCE is to get insight for the discussion on education quality in the region and to orient decision making in educational public policies. Results of TERCE showed that students from Paraguay have low skill in mathematics.

Based on TERCE results, the Universidad Americana has developed a new study, which falls within a Paraguayan Education Ministry public tender focused on pedagogy of teaching in mathematics, language and natural sciences. However, in contrast to TERCE, this new study is focused on teachers to identify the main gaps in the teaching of mathematics.

2.1 Data

Data was collected among Elementary schools from 216 towns that covered the 17 departments of Paraguay. The collection was performed in three steps in the months of December 2017, February and July 2018. Figure 1 shows a map of Paraguay with departments highlighted with different colors according to the step in which the survey was conducted. In the first step, that took place between December 11 and 15 of 2017, the survey was conducted in the east departments. Next, it was carried out between February 12 and 16 of 2018 in the departments located in the south. Finally, the survey was performed in the central and west departments between July 13 and 15 of 2018.

Each sample, denoted as $\mathcal{D}_i, i = \{1, 2, 3\}$, corresponds to the data collected in the departments located in the east, south and central and west, respectively.

The questions of the survey were designed to cover the 4 major areas of mathematics that TERCE considered:

1. **Didactics of arithmetic problem solving:** Applies problem solving to transform its pedagogical practice from the specific approach of arithmetic.
2. **Didactics of geometric problem solving:** Analyzes information about the Geometry approach focusing on problem solving to transform your pedagogical practice.
3. **Didactics of statistics and their applications:** Analyzes, interprets, and represents data through frequency tables and statistical graphs.
4. **Evaluation of mathematical competence:** It includes the different structures that must have the types of evaluation for the improvement of the teaching task in the area of Mathematics.

Based on this survey we want to study if we can assess the suitability of the mathematics teacher to identify possible background and pedagogical skill needs. The characteristics of the different datasets used in this work are summarized in Table 1.

Fig. 1. Map of the departments of Paraguay.

Table 1. Characteristics of the samples used in this study.

Dataset	#instances	#features	Locations of the departments
$\mathcal{D}_1$	71	17	East
$\mathcal{D}_2$	68	17	South
$\mathcal{D}_3$	77	17	Central and West
$\mathcal{D}$	216	17	All

3 Machine Learning Techniques

In this section we introduce the classification task and describe the algorithms used in this study. Then, we present the feature selection problem followed of the strategies considered in this work.

3.1 Classification Task

In classification tasks, let $\mathcal{T}$ be a set of n instances described by the pair $(\mathbf{x}_i, y_i)$, with $\mathbf{x}_i$ being a d-dimensional vector and y_i its associated class label. Then, the aim of classification is to induce a classifier $\mathcal{C}$ such that $\mathcal{C} : \mathbf{X} \rightarrow \mathcal{Y}$. The classifier $\mathcal{C}$ is a function that maps from a vector $\mathbf{X}$ to the class labels $\mathcal{Y}$. In this work we use the following classifiers: Bayesian Network (BN), Naive Bayes (NB) and Support Vector Machine (SVM).

Bayesian Network Classifier [8] (BNC) represents the joint probability distribution over a set of features by means of a directed acyclic graph (DAG). In this network a node represents a feature while the edges represent conditional dependencies between two features. The structure of the network is induced during the learning process of the classifier and, then, finds the set of conditional probability tables (CPTs) that best describes the probability distribution over the training data. This process is addressed as an optimization problem because it is computational very expensive. Therefore, it requires the use of heuristics and approximate algorithms. In this work we use K2; a greedy algorithm that attempts to find the network structure that maximizes the posterior probability of the network over the training data.

Naive Bayes (NB) [3] is another probability based classifier that uses, given an instance, the Bayes rule to compute the probability of a given class label. Given the class label $\mathcal{Y}$, the conditional probability of each feature X_i is learned from training data. It is worth noting that NB, despite the fact of assuming that all features are conditionally independent given the values of the class, the performance of this classifier is surprisingly good. NB is just a simple kind of BNC in which edges only connect each feature with the class label.

Finally, Support Vector Machine [12] (SVM) is a classifier based on the idea of finding a hyperplane that optimally separates the data into two categories by solving an optimization problem. Such hyperplane is the one that maximizes the margin between classes. In case of non-separable data, it optimizes a weighted combination of the misclassification rate and the distance of the decision boundary to any sample vector. In order to find non-linear decision boundaries it uses a kernel function for mapping the original feature space to some higher-dimensional feature space where the data set is separable.

3.2 Feature Selection

In a classification task the aim of feature selection is to reduce the dimensionality of the data while keeping or improving the predictive power of the classifier. Given an objective function $J(S)$, with S a feature subset of size $|S|$, the associated optimization problems seeks to find the smallest subset of features that keeps the information about the class. In this work we study the performance of two feature subsets: Fast Correlation Based Filter (FCBF) and Scatter Search (SS).

Fast Correlation Based Filter (FCBF) [14] is basically an efficient backward search strategy that uses Symmetrical Uncertainty (SU) measure as objective function $J(S)$ to estimate non-linear dependencies between features. For such purpose, it first performs an analysis of relevance to remove irrelevant features. A feature X_i is considered irrelevant if $SU(X_i, \mathcal{Y})$ is lower than a given threshold δ. There is no rule about the optimal value of the threshold since such value depends on the data. So, in this work we set $\delta = 0$ since a value of 0 indicates that the feature has no information about the class. After this step the algorithm remove redundant features according to the concept of Markov blanket.

Scatter Search (SS) [4] is an population-based metaheuristic that works with a reduced set of solutions that evolve by means of intensification and diversification mechanisms. Intensification is focused on finding high quality solutions while diversification on diverse solution to avoid the search get stuck in a local optimum. The initial population evolve as a result of combination and local search procedures until it converges to a local optimum. In order to evaluate the goodness of the feature subsets, we use as $J(S)$ the Correlation Feature Selection [5] (CFS) measure, which considers that good subsets include features highly correlated with the class while little correlated to each other.

4 Experiments

This section presents the analysis of the dataset. The aim of the experimentation The aims of the experiments are:

(a) First, we study if the different datasets from different regions of Paraguay contain similar information about the class by computing SU of each feature and analyzing the predictive power of the classifiers built on each sample.
(b) Second, we compare the performance of the classifier with and without dimensionality reduction.

The second experiments were conducted using cross-validation to assess the predictive performance of the classification models. The number of folds k considered is important since lower values of k lead to higher bias in the error estimates and less variance. Conversely increasing k decreases the bias and increases the variance. In our experiments we use Leave-one-out (LOOCV) cross validation which is a special case of k-fold cross validation with $k = n$, where n is the sample size.

The predictive models are evaluated by means of the classification error averaged over the folds. In the feature selection analysis, in addition to the error we also report the average number of features selected by each strategy and the robustness of each algorithm. The robustness, also called stability, assesses the variations of a feature selection algorithm when it is applied to different samples of a dataset. This issue is important because different subsets may yield to similar predictive models performance. Thus, this analysis improves the confidence in the analysis of the feature selection strategies.

To evaluate the robustness in the experiments we use the Jaccard index [9]. Such index measures the differences between two subsets of features. Given two subsets of features A and B such that $A, B \subseteq \mathcal{X}$. The Jaccard index $\mathcal{I}_J(A, B)$ is defined as follows

$$\mathcal{I}_J(A, B) = \frac{|A \cap B|}{|A \cup B|}.$$

In our experiments we evaluate the predictive model using LOOCV and, so, we need to estimate the robustness $\Sigma(S)$ for more than two subsets. In this case, we compute $\Sigma(S)$ by averaging the pairwise $\mathcal{I}_J(\cdot, \cdot)$ (Σ) similarities. Given a set of solutions $\mathcal{S} = \{S_1, \ldots, S_m\}$ the robustness among this set of solutions is defined

$$\Sigma(\mathcal{S}) = \frac{2}{m(m-1)} \sum_{i=1}^{m-1} \sum_{j=i+1}^{m} \mathcal{I}_J(S_i, S_j).$$

According to this measure, higher values correspond to more stable subsets.

4.1 Analysis of Data Samples

In this section we analyze the differences of the three samples collected from different regions of the country. This analysis is traditionally performed by using statistical test. However in our case we have multivariate data with numerical and categorical features. This issue make difficult this approach. Therefore, since we are interested in the amount of information that such features have about the class and how this information is distributed along the features.

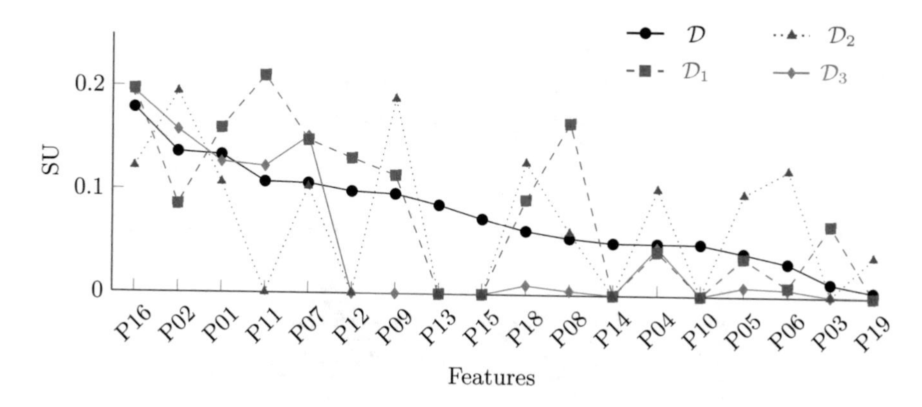

Fig. 2. Ranking of the features with the full data and SU values for each feature on each sample.

In order to compute the information of each feature, we use SU. Figure 2 show the ranking of the features from full data and the amount of information that each feature has from each sample $\mathcal{D}_i$ individually. We can see that each sample distribute in a different way the information about the class.

Now we study if despite the information about the class is differently distributed on each sample, the amount of information is similar or not. For this case, we study their predictive power by training a classifier on each sample $\mathcal{D}_i$ and test it on the remaining samples $\mathcal{D}_j,\ j \neq i$.

Table 2 presents the results with NB, BN and SVM classifiers. Although there is some variability in specific results, on average, the predictive power of all samples are close and lie between 67% and 72%. This variability may be partly due to the small sample size. Therefore, from now on, we continue our analysis on full data.

4.2 Data Analysis

We now proceed to analyze the performance of the classifiers with and without feature selection. Table 3 presents such results. The first three rows shoe the performance with each classifier on baseline and after applying FCBF and SS respectively. Finally the number of features used and the robustness of each feature selection techniques are in the last two rows.

Table 2. Performance of the classifiers when training on each sample $\mathcal{D}_i$.

$\mathcal{D}$	$\mathcal{C}$			avg
tr/te	NB	BN	SVM	
$\mathcal{D}_1/\mathcal{D}_2$	72.06	70.60	69.12	70.60
$\mathcal{D}_1/\mathcal{D}_3$	68.83	70.13	72.73	70.56
$\mathcal{D}_2/\mathcal{D}_1$	69.01	70.42	77.47	72.30
$\mathcal{D}_2/\mathcal{D}_3$	72.73	71.43	67.53	70.56
$\mathcal{D}_3/\mathcal{D}_1$	63.38	63.38	74.65	67.14
$\mathcal{D}_3/\mathcal{D}_2$	64.71	77.94	69.12	70.59

Table 3. Performance of the classifiers before and after feature selection.

$\mathcal{C}$	Baseline	FCBF	SS
NB	70.80 ± 6.41	68.94 ± 5.56	69.88 ± 5.58
BN	71.28 ± 4.02	70.81 ± 3.70	68.94 ± 6.34
SVM	71.26 ± 8.44	71.27 ± 4.98	72.15 ± 10.63
#feats.	19	5.60 ± 0.55	6.80 ± 2.60
$\Sigma(S)$	–	0.44	0.50

In general we can see that all models have a similar predictive power and, so, removing some feature lead to more simple models. On average FCBF finds smaller subsets than SS and reduces more the variability of the classifiers. However, SS is the strategy that presents a higher robustness.

Finally the top most frequently selected features with each feature selection strategy is in Fig. 3. We only show those features that were elected by the algorithm at least 50% of the time. On the left the features selected by FCBF while on the right those selected by SS. As it is expected due to its robustness, SS presents more features.

5 Discussion

Below we show the most relevant questions selected by the feature selection algorithms:

- P01. How do teachers perceive the concept of competence in mathematics?
- P02. Do teachers know how to solve problems using arithmetic operations?

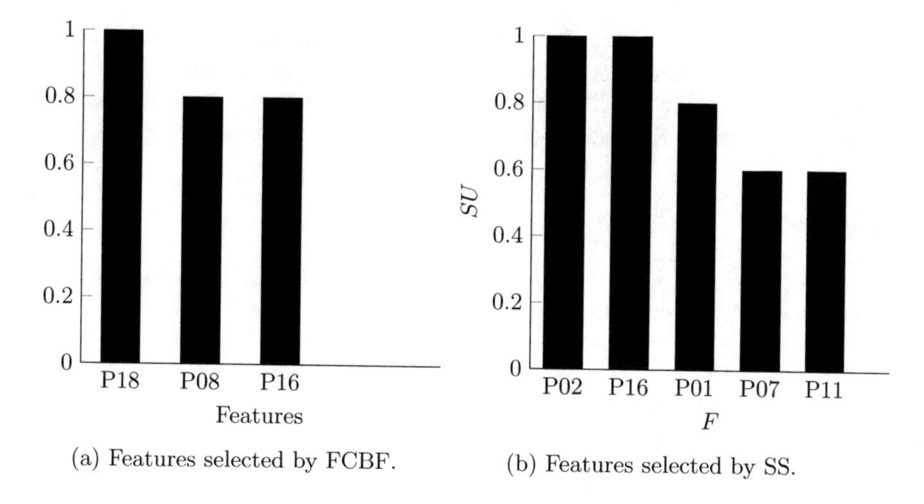

(a) Features selected by FCBF.

(b) Features selected by SS.

Fig. 3. The most frequently selected features

- P07. Do teachers know how to solve problems using geometry?
- P08. Is the van Hiele model used to teach geometry?
- P11. Do teachers know basic concepts from geometry?
- P16. Do teachers know how to assess mathematics competences?
- P18. Do teachers develop the lesson plan correctly?

As we can see, FCBF selects P18, P08 and P16. These results suggest the importance of a good scheduling. Furthermore, the correct assessment of skills related to the competences also arise relevant, just as the use of the Van Hiesel model for teaching geometry. On the other hand, SS Also finds relevant the correct evaluation of competences and includes the basic knowledge on arithmetic and the perception of the concept of competences. Finally, SS also selects the importance of geometry selecting P07 and P11.

6 Conclusions and Future Work

In this work we have conducted a survey based on the TERCE project results to asses the key competences of the mathematics teachers in Paraguay. Results of such survey suggest that teachers do not understand the different evaluation types structures. Therefore, to remedy this situation the government must perform action policies directed especially at teachers.

Although the different sample show different distributions of information about the class, the predictive power is, on average, similar in all cases. Finally it is worth noting that not all questions are equally relevant to answer our question. Therefore, feature selection is useful to assess the adequacy of the questions.

The samples used in this work were collected from different regions with different socio-economic conditions in Paraguay. So, this may yield to differences in the needs and, therefore, this may be reflected showing different distribution in the information. Hence, future work should be extended to identify the needs according to the conditions as well as study the relevance of the different questions selected by the feature selection techniques in a greater depth.

References

1. Asif, R., Merceron, A., Ali, S.A., Haider, N.G.: Analyzing undergraduate students' performance using educational data mining. Comput. Educ. **113**, 177–194 (2017)
2. Castro, F., Vellido, A., Nebot, À., Mugica, F.: Applying data mining techniques to e-learning problems. In: Evolution of Teaching and Learning Paradigms in Intelligent Environment, pp. 183–221. Springer (2007)
3. Duda, R.O., Hart, P.E.: Pattern Classification and Scene Analysis. Willey, New Yotk (1973)
4. García-López, F.C., García-Torres, M., Melián-Batista, B., Pérez, J.A.M., Moreno-Vega, J.M.: Solving the feature selection problem by a parallel scatter search. Eur. J. Oper. Res. **169**(2), 477–489 (2006)
5. Hall, M.A.: Correlation-based feature subset selection for machine learning. Ph.D. thesis, University of Waikato, Hamilton, New Zealand (1998)
6. Hegazi, M.O., Abugroon, M.A.: The state of the art on educational data mining in higher education. Int. J. Comput. Trends Technol. **31**(1), 46–56 (2016)
7. Kabra, R., Bichkar, R.: Performance prediction of engineering students using decision trees. Int. J. Comput. Appl. **36**(11), 8–12 (2011)
8. Pearl, J.: Probabilistic Reasoning in Intelligent Systems: Networks of Plausible Inference. Morgan Kaufmann Publishers Inc., San Francisco (1988)
9. Saeys, Y., Abeel, T., van de Peer, Y.: Robust feature selection using ensemble feature selection techniques. In: Proceedings of the European Conference on Machine Learning and Knowledge Discovery in Databases, vol. 5212. Lecture Notes In Artificial Intelligence, pp. 313–325 (2008)
10. Shahiri, A.M., Husain, W., et al.: A review on predicting student's performance using data mining techniques. Procedia Comput. Sci. **72**, 414–422 (2015)
11. UNESCO: TERCE: associated factors, executive summary (2015)
12. Vapnik, V.N.: Statistical Learning Theory. Wiley-Interscience, New York (1998)
13. Velmurugan, T., Anuradha, C.: Performance evaluation of feature selection algorithms in educational data mining. Int. J. Data Min. Tech. Appl. **5**(02), 131–139 (2016)
14. Yu, L., Liu, H.: Efficient feature selection via analysis of relevance and redundancy. J. Mach. Learn. Res. **5**, 1205–1224 (2004)
15. Zaffar, M., Hashmani, M.A., Savita, K.: Performance analysis of feature selection algorithm for educational data mining. In: 2017 IEEE Conference on Big Data and Analytics (ICBDA), pp. 7–12. IEEE (2017)

Cooperative Evaluation Using Moodle

Enrique Domínguez, Ezequiel López-Rubio, and Miguel A. Molina-Cabello(✉)

Department of Computer Science, University of Málaga,
Boulevar Louis Pasteur, 35, 29071 Málaga, Spain
{enriqued,ezeqlr,miguelangel}@lcc.uma.es

Abstract. Evaluation is an important task in the classroom and it conditions the effort of the students. Moreover, their marks are usually the main goal of the students. The traditional methodology about exams and the subsequent average of them to calculate the grade of the students is not the most suitable way in order to aim different facets like a significant learning or a critical thinking. Nowadays, several methods are indicated to improve this aspects among others, even the students attitude. Another evaluation methodology by cooperative assessment is shown in this paper.

Keywords: Cooperative assessment · Peer review evaluation · Web-based education · Higher education

1 Introduction

Cultured and knowledgeable students has been the objective of the higher education. Traditional methodologies are replaced by other modern approaches which produce improved skills in the students such as professional and personal abilities [7]. Techniques like problem-based or case-based learning are examples of these new proposals.

A methodology based on assessment can produce an extra information and a more efficient feedback. Nevertheless, the higher education presents a problem which is not in primary or secondary education: the number of students who are in class is higher. Thus, this approach in higher education exhibits the inconvenience of the increase of the teacher work.

In order to alleviate that work, online cooperative evaluation offers an excellent tool. Interest and skills of the student are increased by using this kind of methodology [1,4,5]. Its use provides additional and powerful benefits based on the teacher revision. For example, from the given teacher feedback, a student can feel a more clarified meanings, a learning of the way to solve future problems, etc. Peer assisted evaluation can be seen as a part of peer assisted learning [2,3,9].

Additionally, peer review is an effective method which produces an appreciated information. Tasks which employ student peer review help them by offering

© Springer Nature Switzerland AG 2020
F. Martínez Álvarez et al. (Eds.): CISIS 2019/ICEUTE 2019, AISC 951, pp. 295–301, 2020.
https://doi.org/10.1007/978-3-030-20005-3_30

a reinforcement learning. On the other hand, several inconveniences are presented in peer review. Problems such as students can be so pleasant with their mates and the opposite with non-friend students, the feedback explanation is not detailed with the necessary precision... increase the difficulty of grading in a fair way. Moreover, not only students are the issue; instructors are also involved on it. For example, a non clear rubric will produce the student confusion with the subsequent ineffective evaluation.

In spite of all the inconveniences which are exhibited by this method, educational scientists are sure of the goodness of it. A clear rubric and an adequate training can make easier better student reviewers. In this way, scoring rubrics are essential to reduce the student review complexity and diffused answers, by allowing to produce more specific feedbacks. The more evaluating exercises they do the more proficient reviewers they will be. This is due to that they internalise the judgement that the exercises they do embrace.

The objective of this work is to study how cooperative evaluation affects to the learning process and different relations between the marks given by the students and the teacher.

The paper is structured as follows. The academic context is shown in Sect. 2, while the evaluation methodologies are presented in Sect. 3. Then, Sect. 4 exhibits the project results. Finally, some conclusions are provided in Sect. 5.

2 Academic Context

The Universities Organic Law (UOL) regulates the Spanish higher education system. Furthermore, Spain is also a member of the World Conference on Higher Education, which is sponsored by the United Nations Educational, Scientific and Cultural Organization (UNESCO) and the Bologna process.

The most important challenges in higher education with their suggested indications are the principal objective of the World Conference on Higher Education. Previously, several conferences has been developed in order to prepare the work of the four commissions of the world conference: relevance, quality, management and financing, and international cooperation. In addition, several considerations such as the take of actions, to execute projects and support an international cooperation based on solidarity and the construction of an equal society thanks to research, community projects and specialist training.

Regarding to the Bologna process, many European countries are involved on it. This agreement has certain aims, such as the support of student mobility, the proposition of comparable teaching methodologies and evaluations, and the motivation of institutional cooperation. The Bologna process also establishes an comprehensible and comparable education system. This way, several considerations can be yielded easily such as the search of jobs. The degree system that is proposed is based on two levels. The first one is the undergraduate (BSc) and the second one is the graduate (MSc). An European Credit Transfer System (ECTS) is used commonly in order to organize those degrees. So that, student interchange among other considerations.

In Spain, universities can design their own degrees, but always under the existing laws. Nevertheless, the national common degree catalog from the previous system has been removed. At this moment, three different levels compose the Spanish higher education system: undergraduate (BSc), graduate (MSc) and doctorate (PhD).

In particular, an usual academic undergraduate or graduate year has 60 ECTS credits. Undergraduate programs have between 180 and 240 ECTS credits (there are some exceptions with 300 or 360 ECTS credits), while most graduate programs have between 60 and 120 ECTS credits. The usual undergraduate program has 240 ECTS credits and the usual graduate program has 60 ECTS credits.

The Bologna process has changed the organisation of the courses from the previous contents system to the competences system. With this change, students will be awarded with their degree if they acquire the corresponding competences, which have to be assessable.

The undergraduate Health Engineering degree at the Computer Science Technical School at the University of Málaga (Spain) is carrying out the current educational innovation project. Next sections exhibit the different results that have been obtained. The Intelligent Systems course, which is acquired during the fifth semester in the third academic year, is the course where the obtained results have been yielded from different tasks.

3 Evaluation Methodologies

Online peer assessment are supported by several resources. In our case, Moodle is the selected software since the our institution uses it as learning management system [8]. Moodle provides a wide range of tools, the workshop among them, which is the most powerful peer assessment mean [6]. Thus, the different tasks developed by the students are supported by the workshop tool.

For a given activity, its workflow is as follows. First of all, students do their work and submit it. After that, each student has to evaluate a fixed number of submissions from other students. A rubric is provided by the instructor and it is used by students in order to evaluate the fellow works. In this process the works are anonymous in order to achieve an impartial assessment. Finally, the mark of a student work is given by the teacher considering its own evaluation and the fellow student assessments.

The instructor has the possibility of selecting a simple grading or a complex one. In this case, and with the aim of a clear interpretation of the grading process, in this project a range between 0 and 100 has been used as a numerical evaluation.

4 Project Results

As it is commented previously, students have to carry out several tasks along the course and each one has to be done and assessed by their fellow students.

After that, the instructor also evaluates each task. Thus, a comparison can be done in order to study marks and reviews.

Each task is assessed by five fellow students in an anonymous way and the teacher. These reviews are guided by a rubric and the mark is a number in the range between 0 and 100. In the context of this project, 43 students and 1 instructor are involved in assessments. The task selected has been an exam.

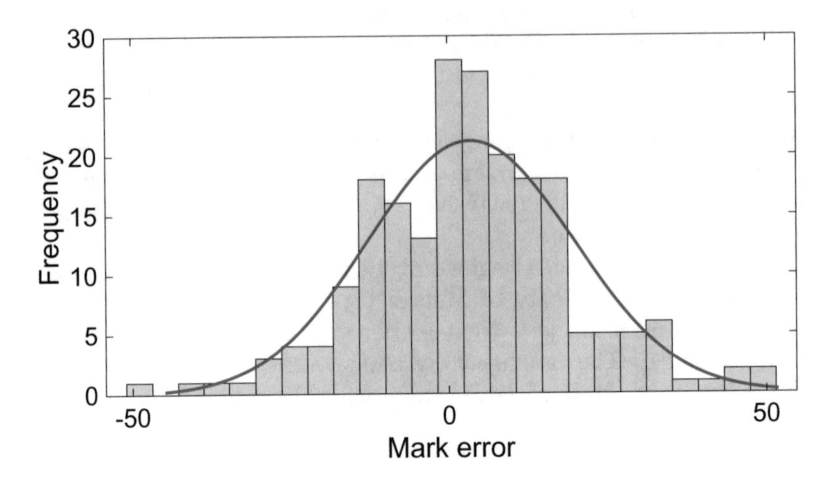

Fig. 1. Histogram of marks error and Gaussian best fit.

Figure 1 depicts the histogram of the marks error and the best fit Gaussian distribution according to the data. A negative marks error means that students have awarded in an optimistic way; on the other hand, a positive marks error means that students have awarded in an pessimistic way. As it can be observed, the marks error is approximately Gaussian, the student fellow marks are close to the instructor mark (which means marks error equals to 0) and slightly pessimistic.

The scatter plot of the marks error and the regression line according to the marks and the marks error are reported in Fig. 2. It shows the marks error are close to 0, so that, student evaluations are, in general, similar to the teacher evaluation. In addition, an optimistic mark is given by the students when the task that they are grading is close to the line which distinguishes between the fail and the pass (mark equals to 50). Furthermore, students are opposed to provide an excellent mark, so that, they grades pessimistically.

Figure 3 also shows the marks error and the regression line according to the marks and the marks error. However, in this case it is shown the mean and the median students marks error per mark. It is shown that those marks error are close to 0, so the mean and median students evaluation may be considered as a very similar mark in relation to the instructor mark. Moreover, the median students evaluation seems to work better than the mean one.

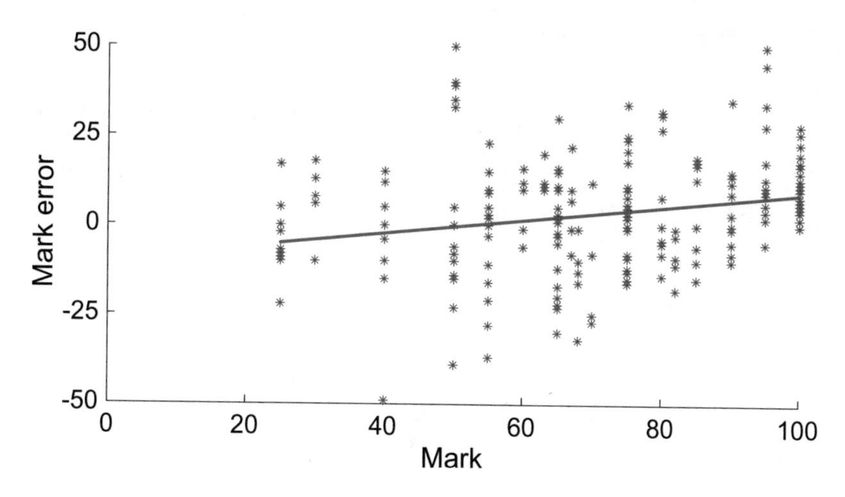

Fig. 2. Scatter plot of marks error.

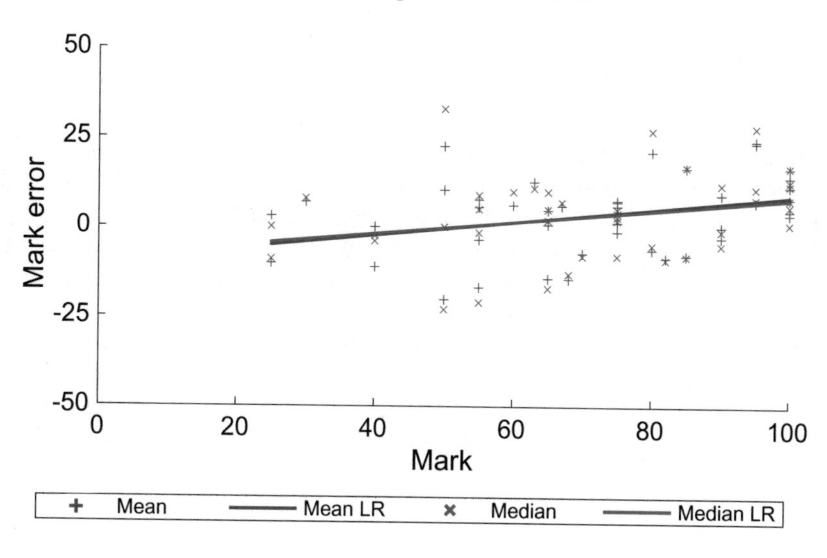

Fig. 3. Scatter plot of mean and median marks error.

Finally, the whisker plot of the marks error is reported in Fig. 4. Red crosses represent the outliers and they are pessimistic marks in four of the five outliers. Besides, students grades present a more heterogeneous value in tasks whose instructor mark is around the fail and the pass.

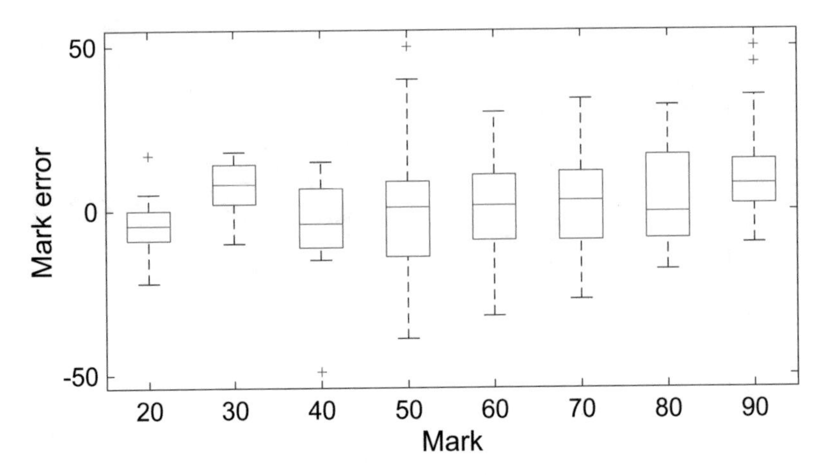

Fig. 4. Whisker plot of marks error.

5 Conclusions

The evaluation that students do is similar to the instructor one. Student are optimistic when marks are lower; on the other hand, they are pessimistic when marks are higher. In addition, tasks which are in the line which distinguishes between the fail and the pass exhibit a wide range of grades awarded by students.

Students consider peer assessment as a positive complement. Moreover, this activity has produced an increase of the discussions in class and the quality of the questions that they do is higher than the previous one. Furthermore, they understand with a higher level the evaluation criteria used by the instructor. Besides, they have increased their curiosity, and resulting interesting debates.

It can be said that students feel assessment as a learning process which is part of the course contents. It is because of that the possibility to apply this assessment peer review process to other courses is being considered. This way, more data may be obtained and compared in order to extract more relevant information about peer student assessment.

Acknowledgments. This work is partially supported by the University of Málaga (Spain) under the grant PIE17-162, project name New Cooperative Evaluation Techniques and Tools.

References

1. Chiu, C.Y., Chiang, F.C., Chung, S.C.: Effects of Online Cooperative Evaluation and Group Rewards on High School Students' Argumentative Writing Attitudes and Interaction Behaviors, pp. 243–268. De Gruyter Mouton, Boston (2011)
2. Glynn, L.G., MacFarlane, A., Kelly, M., Cantillon, P., Murphy, A.W.: Helping each other to learn - a process evaluation of peer assisted learning. BMC Med. Educ. **6**(1), 18 (2006)

3. Hodgson, Y., Benson, R., Brack, C.: Student conceptions of peer-assisted learning. J. Further High. Educ. **39**(4), 579–597 (2015)
4. Kim, C.J.: The effects of peer assessment and peer feedback in writing education for premedical students. Ewha Med. J. **40**(1), 41–49 (2017)
5. Montanero, M., Lucero, M., Fernández, M.J.: Iterative co-evaluation with a rubric of narrative texts in primary education. J. Study Educ. Dev. **37**(1), 184–220 (2014)
6. Nash, S.S.: Moodle 3.x Teaching Techniques. Packt Publishing Limited, Birmingham (2016)
7. Palomba, C.A., Banta, T.W.: Assessment Essentials: Planning, Implementing, and Improving Assessment in Higher Education. Higher and Adult Education Series. Jossey-Bass, San Francisco (1999)
8. Rice, W.: Moodle E-Learning Course Development. Packt Publishing Limited, Birmingham (2015)
9. Ross, M.T., Cameron, H.S.: Peer assisted learning: a planning and implementation framework: AMEE Guide no. 30. Med. Teach. **29**(6), 527–545 (2007)

Automatic Categorization of Introductory Programming Students

Miguel A. Rubio(✉)

University of Granada, Granada, Spain
marubio@ugr.es

Abstract. Learning to program can be quite difficult for CS1 students. They must master language syntax, programming theory and problem-solving techniques in a short period of time. Consequently, a significant percentage of students struggle to successfully complete CS1.

Several studies have shown that students that do poorly the first weeks tend to fail at the end of the course. We would be able to help these students if we could automatically infer their learning stages during the course. One powerful tool that we can use to this end is cluster analysis: a class of computational methods that has been proved effective in analyzing complex datasets, including novice programmers' learning trajectories.

Our aim in this study is to explore the feasibility of using exercises based on the neo-Piagetian model together with cluster analysis to infer the learning stage of the novice programmer. We have been able to automatically classify students at the end of an introductory programming course obtaining highly stable clusters compatible with Neo-Piagetian theory.

Keywords: Neo-Piagetian theory · Learning analytics · Cluster analysis · Introductory programming · Novice programmers

1 Introduction

Lecturers in charge of an introductory programming course face a complex challenge [12]. Students must master language syntax, programming theory and problem-solving techniques in a short period of time, something they usually struggle to do.

As a consequence introductory programming courses tend to show poor results and high dropout rates [20]. Although some progress has been made in the last years [9, 19] the following quote from Carter and Jenkins [4] is still valid:

> *"Few teachers of programming in higher education would claim that all their students reach a reasonable standard of competence by graduation. Indeed, most would confess that an alarmingly large proportion of graduates are unable to 'program' in any meaningful sense."*

In order to improve student learning in introductory programming courses, we can focus on identifying and helping those students at risk of failing.

One promising approach is based on recent research that suggests that there is a strong correlation between students' performance during the first weeks of the course and their final performance [7]. These results suggest an effective way to enhance

© Springer Nature Switzerland AG 2020
F. Martínez Álvarez et al. (Eds.): CISIS 2019/ICEUTE 2019, AISC 951, pp. 302–311, 2020.
https://doi.org/10.1007/978-3-030-20005-3_31

students' learning. We would only have to test our students early in the course to see which students are struggling and offer them remedial help.

This approach would be especially useful if it could be implemented in an automated way using machine learning methods [1]. Such a system could provide instructors the opportunity to intervene and support those students in need.

A possible first step is to select a model that describes the learning stages that introductory programming students present during their learning. One interesting approach proposed by Lister [8] is to use a model based on neo-Piagetian theory. This theory describes the cognitive development of a person that is acquiring knowledge in a specific domain.

We will describe the different stages proposed by the neo-Piagetian theory following closely the descriptions given by Lister [8] and Teague [15].

- Sensorimotor Stage: Sensorimotor is the first stage of development. A sensorimotor novice programmer has minimal language skills in the domain and is still learning to recognize syntax and distinguish between the various elements of code. At this stage the novice programmer requires considerable effort to trace code and only occasionally do they manage to do so accurately.
- Preoperational Stage: At the next stage the preoperational novice has mastered the basic programming concepts. They are capable of tracing code accurately in a consistent manner. Preoperational novices are not yet able to perform abstract reasoning about segments of code.
- Concrete Operational Stage: By the time a novice is at the concrete operational stage, their focus shifts from individual statements to small segments of code which allows them to consider the overall purpose of code. Their ability to reason at a more abstract level allows them to understand short pieces of code simply by reading that code. Two defining qualities of this stage are being able to perform transitive inference and reverse code.
- Formal Operational Stage: This is the most abstract of the Piagetian types of reasoning. A person reasoning at this level exhibits the thinking characteristics traditionally emphasized at university. It can be defined succinctly thus: formal operational reasoning is what competent programmers do, and what we'd like our students to do.

A set of programming questions based on the neo-Piagetian model can be a powerful tool to assess the learning stage of our students. Several studies have tried to assess the learning stage using exercises to assess preoperational skills [15] or concrete operational skills [16] in different programming languages.

If we want to integrate this model in an automated way, we need to develop a methodology to automatically assess the learning stages of our students. This is not an easy task as students can take different paths when moving from one learning stage to the next (Fig. 1). One possible approach is the use of clustering techniques: a class of computational methods that has been proved effective in analyzing complex datasets.

Several studies have successfully applied these techniques in the introductory programming context. Rubio et al. [13] used cluster analysis to assess the learning stages of introductory programming students in terms of their reading and writing code

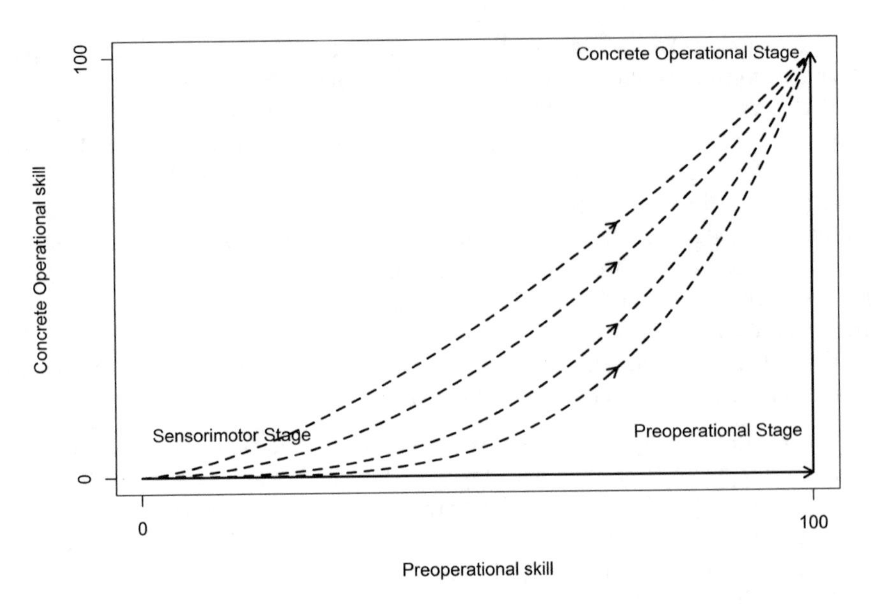

Fig. 1. Theoretical path from the sensorimotor stage to the concrete operational stage.

skills. Another relevant example is a study done by Lahtinen [6]. She used cluster analysis to assess the learning stage of the novice programmer using the Bloom's taxonomy as a reference.

The main goal of this study is to assess if it is feasible to use a set of exercises based on the neo-Piagetian approach combined with cluster analysis to automatically assess the learning stages of introductory programming students.

2 Methods

This study is motivated by the question of how to automatically categorize introductory programming students using machine learning techniques. To this end we are going to analyze the results obtained in an introductory programming course.

2.1 Research Questions

The study research questions are:

- RQ1: Is it possible to use a set of exercises based on neo-Piagetian theory to partition a course of introductory programming students in stable clusters?
- RQ2: Are the clusters obtained consistent with Neo-Piagetian theory?

2.2 Study Design

The goal of our study was to explore the possibility of inferring the students' neo-Piagetian stage using cluster analysis techniques. To this end we used an introductory programming course in a Biology degree at a research university. Students in this course learn basic computing skills and devote ten weeks to learn to program using MATLAB. Each week students attend two lectures of one hour and a two hours lab session.

The instructor in this study (not the author) used traditional teaching methods: PowerPoint slides and multimedia material were used to introduce theoretical concepts. In lab sessions students would work individually to solve different programming exercises. 53 students participated in the study. 29 were females and 24 males.

2.3 Assessment Tool

There is a notable lack of easily accessible and validated assessment tools in introductory programming [17] but a small set of tools have been recently developed. For example, Simon et al. [14] have proposed several questions that can be used to benchmark student performance in introductory programming courses at a wide range of institutions.

In our study we measured students' learning achievements by means of a set of exercises based on neo-Piagetian theory. The set of questions was selected from published studies. There are three questions designed to test skills associated to the preoperational stage and two questions designed to test skills associated to the concrete operational stage. One example of each type of question is presented in Fig. 2. The first concrete operational question requires the use of transitive inference skills and the second concrete operational question requires the use of inversion skills.

Describe the purpose of the following code using only one sentence.

```
%numbers is a numeric array
result = 0;
for pos=1:length(numbers)
    if numbers(pos) < 0
        result = result + 1;
    end
end
```

Describe the purpose of the following code using only one sentence. The variables y1, y2 and y3 contain numeric values. In each of the three boxes the appropriate code is provided.

```
if y1 < y2
  Swap the values in y1 and y2.
end

if y2 < y3
  Swap the values in y2 and y3.
end

if y1 < y2
  Swap the values in y3 and y1.
end
```

Fig. 2. Two exercises presented to the students. The question to the left assesses preoperational skills. The question to the right assesses concrete operational skills.

These exercises were presented to the students at the end of the course as part of their final exam. The use of formal examinations in assessing the reasoning of the novice programmer present several advantages: participation in the study represents no extra cost to the student, students are motivated to solve the exercises and data collection has a minimal cost to the researcher. The use of exam questions in this type of studies also present disadvantages: grading can be subjective and tends to be generous with the less proficient students [18].

All the questions were graded by the author following different guidelines published in other studies. To avoid bias in grading the author did not know the students and could not see their name on the exam.

2.4 Analysis Performed

We used the grades from the preoperational and concrete operational exercises to assess the students' learning stages. We added the grades of the three preoperational exercises to obtain the preoperational score and added the two concrete operational exercises to obtain the concrete operational score.

We analyzed this dataset using clustering techniques: a class of computational methods that has been proved effective in analyzing complex datasets. Several studies have successfully applied these techniques in the introductory programming context [3, 21]. We clustered the dataset using the K-Medoids technique, a variation of K-Means clustering where centroids are represented by the median. We have used the Partitioning Around Medoids (PAM) algorithm [11] implemented in R [10].

The first step in this method is to choose the correct number of clusters. To determine the optimal number of clusters we partitioned the dataset using different number of clusters (from k = 2 to k = 5) and assessed the stability of the clusters obtained.

Cluster stability was measured using the Jaccard coefficient following the methodology proposed by Hennig [5]. The Jaccard coefficient measures the similarity of a set of sets, and is defined as the size of the intersection divided by the size of the union of the sample sets. Jaccard coefficient values lie in the unit interval. A cluster is considered to be stable if its Jaccard coefficient is greater than 0.85.

Table 1 shows the results obtained for different partition going from k = 2 to k = 5. The partition with k = 3 was the only partition comprised of stable clusters. In this partition all the clusters are highly stable with Jaccard coefficients around 0.95. When the number of clusters is two (k = 2) only one cluster is stable. When k = 4 or k = 5 most clusters are not stable.

Table 1. Jaccard coefficients for different number of clusters.

Cluster	k = 2	k = 3	k = 4	k = 5
1	0.81	0.96	0.58	0.75
2	0.86	0.94	0.75	0.68
3	-	0.96	0.92	0.85
4	-	-	0.77	0.65
5	-	-	-	0.89

Clusters with Jaccard coefficients greater than 0.85 are stable.

This result is coherent with our theoretical framework. As we are only assessing preoperational and concrete operational skills we would expect the optimal number of clusters to be three: those that show preoperational and concrete operational skills, those that show preoperational skills and those that show neither of them.

3 Results

We assessed students' performance using a cluster analysis of the preoperational and concrete operational scores. We first determined the optimal number of clusters measuring the stability of the clusters obtained when using different number of clusters.

Following the previous results, we classified students in three different groups and labeled them according to their preoperational and concrete operational scores. This is shown in Fig. 3.

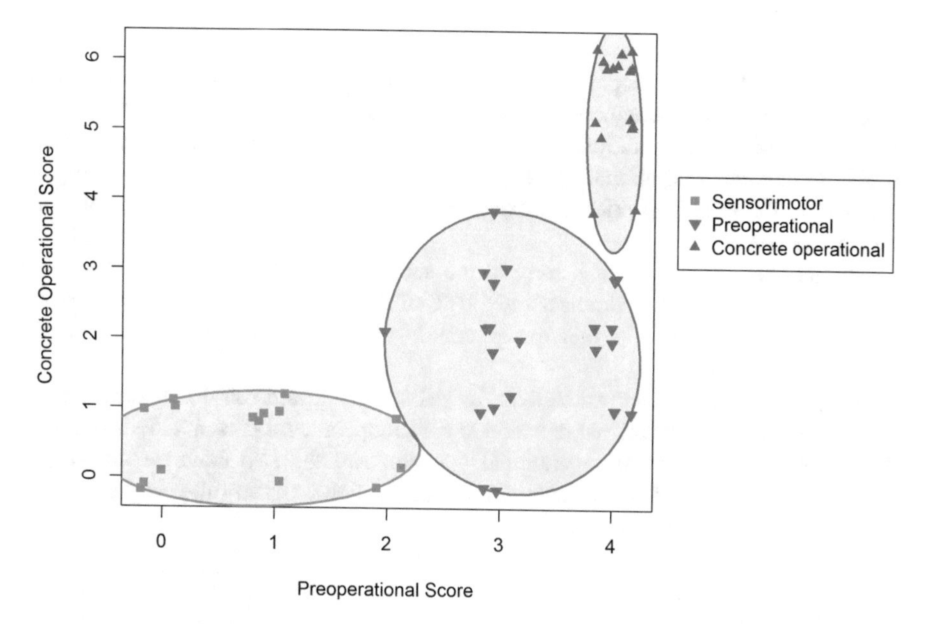

Fig. 3. Clustering of preoperational and concrete operational scores. A small amount of jitter has been added to reduce the points overlap.

We can observe that students that belong to the sensorimotor cluster present very low concrete operational scores and low preoperational scores. Students that belong to the concrete operational cluster show the maximum preoperational score and high concrete operational scores. The preoperational cluster comprises students that have high preoperational scores and low concrete operational scores.

One noticeable fact is the presence of an empty area at the top left corner of the graph. That indicates an absence of students with high concrete operational scores and low preoperational scores. This result supports the notion that the preoperational stage is previous and supports the concrete operational stage.

Table 2 contains the percentage of students who answered correctly each question broken down by cluster and the statistical difference (if any) between the clusters. Students that belong to the concrete operational cluster show a strong performance in all the exercises. All of them completed successfully the preoperational exercises and most of them completed successfully both the transitive inference and the reverse code exercise.

Table 2. Percentage of students who answered correctly each question broken down by cluster.

Cluster	Neo-Piagetian stage	Exercise 1	Exercise 2	Exercise 3	Exercise 4	Exercise 5
1	Concrete operational	100%	100%	100%	81.3%	93.8%
			**		**	**
2	Preoperational	90.9%	50%	90.9%	31.8%	0%
		**		**		
3	Sensorimotor	13.3%	6.7%	13.3%	0%	0%

Statistical significance is computed using Fisher's exact test. ** $p < 0.01$

Students within the preoperational cluster present more mixed results. All of them completed successfully preoperational exercises 1 and 3 but only half of them completed successfully preoperational exercise 2. They show a weaker performance when facing the concrete operational exercises: only a third of them managed to complete successfully the transitive inference exercise and none of them managed to solve the reverse code exercise.

Results from students that belong to the sensorimotor cluster are very poor. For each of the preoperational exercises only 10% of the students managed to complete it successfully. No student in this group could solve correctly any of the concrete operational exercises.

If we use Fisher's exact test to look for statistically significant differences, we can see that the concrete operational cluster and the preoperational cluster show significant differences in the second preoperational question and the two concrete operational questions. The preoperational cluster and the sensorimotor cluster differ in the first and the third preoperational question.

4 Discussion

In this study we have used a set of programming exercises to group the students in different clusters according to their scores. Our final aim is to associate these clusters to the different neo-Piagetian stages of the novice programmer.

We started determining the optimal number of clusters for our dataset. We used a bootstrapping method to determine the stability of the clusters and found that only when using three clusters we were able to obtain a stable configuration.

The three clusters obtained present characteristics that allow us to associate them to different neo-Piagetian stages. The questions success rate is significantly different among the clusters. The concrete operational and preoperational clusters show significantly different rates for the second preoperational question and the two concrete operational questions. The preoperational and sensorimotor clusters significantly differ in the first and third preoperational question.

These results indicate that students in each of the cluster show different skills. Students that belong to the concrete operational cluster are able to use of transitive inference and reverse code. Possessing these skills strongly support the idea that these students have attained the concrete operational stage.

Students in the preoperational cluster possess preoperational skills but lack concrete operational skills at a significant level and become natural preoperational stage candidates. Students in the sensorimotor cluster were able to solve neither the concrete operational exercises nor the preoperational exercises. These results place them in the sensorimotor stage.

Other authors have obtained similar results using other statistical methods. Alireza et al. [2] designed a set of exercises in Python and Java that was able to distinguish between students in different learning stages. They used the chi square test on the percentage of students that answered correctly each question. Teague and Lister [15] performed a think aloud study on a programming exercise and were able to associate different processes followed by the students to different neo-Piagetian stages.

The results described above allow us to answer the first research question: Is it possible to use a set of exercises based on neo-Piagetian theory to partition a course of introductory programming students in stable clusters?

The answer is affirmative. Applying cluster analysis to the students' scores we have been able to distinguish three different groups of students and these clusters are highly stable.

The answer to the second research question is also affirmative. We have shown that students in each cluster show statistically different behavior when completing the preoperational and concrete operational exercises. These differences can be used to link each cluster to a different Neo-Piagetian stage.

In our study we also found that all the students with high concrete operational skills have high preoperational skills. Similar results have been obtained by other authors [2]. This suggests that acquiring a minimum level of preoperational skills is a first step in the path of learning concrete operational skills.

Our study has several limitations. Probably the most fundamental one is that we are inferring the student neo-Piagetian stage using final answers to exercises and not the process followed by the student to solve it. This can be problematic as Teague and Lister showed in their study [15].

Another limitation is that the sample size is small as the study comprised only one course. We plan to extend this study to see if the results that we have obtained are reproducible when working with students from other degrees in different institutions.

5 Conclusions

In this study we have used a set of exercises based on the neo-Piagetian model combined with cluster analysis to automatically infer the learning stages of introductory programming students.

We have obtained a highly stable cluster configuration that is consistent with the Neo-Piagetian model. The clusters show significantly different completion rates for different type of questions. These differences allow us to link each cluster to a Neo-Piagetian stage.

These results suggest that the Neo-Piagetian model can be a useful tool to design exercises to automatically assess novice programmers learning throughout the course.

Acknowledgments. This work was partially supported by the University of Granada and by the MINECO under its Competitive Research Programme (DPI2015-69585-R).

References

1. Ahadi, A., et al.: Exploring machine learning methods to automatically identify students in need of assistance. In: Proceedings of the Eleventh Annual International Conference on International Computing Education Research, pp. 121–130. ACM, New York (2015)
2. Ahadi, A., et al.: Falling behind early and staying behind when learning to program. In: 25th Anniversary Psychology of Programming Annual Conference (2014)
3. Bumbacher, E., et al.: Student coding styles as predictors of help-seeking behavior. In: Chad, H.L., Kalina, Y., Jack, M., Philip, P. (eds.) Artificial Intelligence in Education, pp. 856–859. Springer, Heidelberg (2013)
4. Carter, J., Jenkins, T.: Gender and programming: what's going on? SIGCSE Bull. **31**(3), 1–4 (1999)
5. Hennig, C.: Cluster-wise assessment of cluster stability. Comput. Stat. Data Anal. **52**(1), 258–271 (2007)
6. Lahtinen, E.: A categorization of novice programmers: a cluster analysis study. In: Proceedings of the 19th Annual Workshop of the Psychology of Programming Interest Group, Joensuu, Finland, pp. 32–41 (2007)
7. Liao, S.N., et al.: Lightweight, early identification of at-risk CS1 students. In: Proceedings of the 2016 ACM Conference on International Computing Education Research, pp. 123–131. ACM, New York (2016)
8. Lister, R.: Concrete and other neo-Piagetian forms of reasoning in the novice programmer. In: Proceedings of the Thirteenth Australasian Computing Education Conference, vol. 114, pp. 9–18 (2011)
9. Porter, L., et al.: Success in introductory programming: what works? Commun. ACM **56**(8), 34–36 (2013)
10. R Core Team: R: A Language and Environment for Statistical Computing, Vienna, Austria (2014)
11. Reynolds, A.P., et al.: Clustering rules: a comparison of partitioning and hierarchical clustering algorithms. J. Math. Model. Algorithms **5**(4), 475–504 (2006)
12. Robins, A., et al.: Learning and teaching programming: a review and discussion. Comput. Sci. Educ. **13**(2), 137–172 (2003)
13. Rubio, M.A., et al.: Closing the gender gap in an introductory programming course. Comput. Educ. **82**, 409–420 (2015)
14. Simon et al.: Benchmarking introductory programming exams: How and Why. In: Proceedings of the 2016 ACM Conference on Innovation and Technology in Computer Science Education, pp. 154–159. ACM, New York (2016)
15. Teague, D., Lister, R.: Blinded by their plight: tracing and the preoperational programmer. In: 25th Anniversary Psychology of Programming Annual Conference (PPIG), Brighton, England, 25–27 June (2014)
16. Teague, D., Lister, R.: Programming: reading, writing and reversing. In: Proceedings of the 2014 Conference on Innovation & Technology in Computer Science Education, pp. 285–290. ACM, Uppsala, Sweden (2014)
17. Tew, A.E.: Assessing Fundamental Introductory Computing Concept Knowledge in a Language Independent Manner. Georgia Institute of Technology (2010)

18. Traynor, D., et al.: Automated assessment in CS1. In: Proceedings of the 8th Australasian Conference on Computing Education, vol. 52, pp. 223–228 (2006)
19. Utting, I., et al.: A fresh look at novice programmers' performance and their teachers' expectations. In: Proceedings of the ITiCSE Working Group Reports Conference on Innovation and Technology in Computer Science Education-Working Group Reports, pp. 15–32. ACM, New York (2013)
20. Watson, C., Li, F.W.B.: Failure rates in introductory programming revisited. In: Proceedings of the 2014 Conference on Innovation & Technology in Computer Science Education, pp. 39–44. ACM, Uppsala, Sweden (2014)
21. Worsley, M., Blikstein, P.: Programming pathways: a technique for analyzing novice programmers' learning trajectories. In: Chad, L.H., Kalina, Y., Jack, M., Philip, P. (eds.) Artificial Intelligence in Education, pp. 844–847. Springer, Heidelberg (2013)

Using Scratch to Improve Undergraduate Students' Skills on Artificial Intelligence

Julian Estevez(✉), Gorka Garate, Jose Manuel López-Guede, and Manuel Graña

University of the Basque Country (UPV-EHU), Donostia, Spain
julian.estevez@ehu.eus

Abstract. This paper presents several educational applications in Scratch that are proposed for the active participation of undergraduate students in contexts of Artificial Intelligence. The students are asked to understand the mathematics involved in an automatic clustering algorithm and two simple neural networks for data learning. They must practice with the implementation, following closely the short instructions and mathematical theory provided by teachers.

Keywords: Artificial Intelligence · Scratch · Neural network

1 Introduction

Scientific Method is key in the development of technologically advanced communities [6], and it is of great importance that our students include it in their curricula. As many other competences, it is convenient that some knowledge of scientific method is learned at the undergraduate stage, and any effort done in the Educational System to foster the learning of it undoubtedly pays.

For a better understanding of scientific method, the Educational Community widely recognizes that school curricula must move on from traditional expositive classes to more informal and collaborative contexts. The fact is, however, that active participation of the students is difficult to achieve at the classroom. One of the reasons should appeal to the interests of the students themselves; in order to achieve active involvement, the classes should make use of more attractive resources, such as fun, games, social interaction, observation of real problems, novelty, etc. This is specially clear in the case of undergraduate students. Surprisingly, there is a quite widespread mistrust of science in the post-truth era where we live. This makes it necessary to insist on a bigger effort to spread the benefits of science among younger students [1,2].

The present article presents the design and development of several simple educational exercises to promote understanding and learning of Artificial Intelligence (AI) at schools with the usage of Scratch, which is a graphic programming environment. By allowing novices to build programs by snapping together graphical blocks that control the actions of different actors and algorithms make programming fairly easy for beginners.

© Springer Nature Switzerland AG 2020
F. Martínez Álvarez et al. (Eds.): CISIS 2019/ICEUTE 2019, AISC 951, pp. 312–320, 2020.
https://doi.org/10.1007/978-3-030-20005-3_32

AI is one of the technologies that will transform our society, economy and jobs in a greater extent along the next decades. Some of the most known examples of AI are driverless cars, chatbots, voice assistants, internet search engines, robot traders, etc. These systems can be embedded in physical machines or in software, and the promising capacities of both architectures makes it necessary for society and politicians to regulate the functions and limits of these devices [8]. Despite the myth of destructive AI represented in films such as *Terminator* or *I, Robot*, the truth is that nowadays most usual smart algorithms consist of a series of simple rules applied to large series of numbers, and the result of that is called Artificial Intelligence.

British Parliament and other institutions recommend the education from high school of Artificial Intelligence [11], regardless of the pace of development of this technology, in order to cope future technological and social challenges.

Main reason is to improve technological understanding, enabling people to navigate an increasingly digital world, and inform the debate around how AI should, and should not, be used.

Thus, in this paper an educational proposal for teaching the basic mathematics behind simple AI algorithms is presented. The solution is developed in an open source platform, Scratch, and the main objective for the students is to be aware of the rules behind intelligent systems rather than learning or memorizing anything. The specific tasks to understand are automatic clustering, learning and prediction with AI. The software is designed for students of 16–18 years. The chosen algorithms to teach are specifically designed or adapted considering the mathematical background of those students.

The present article is divided in following sections. The mathematics that will be used in the software is going to be described in Sect. 2. Next, in Sect. 3, the Methodology that teachers will use with students is going to be detailed. Finally, in Sect. 4, conclusions of the experiment are going to be analyzed.

2 Content

This article will depict algorithms that can be practiced using Scratch [10] in a workshop with students 16–18 years old. After the workshop, the students should be able to understand, play and eventually code these algorithms.

Specific algorithms are:

- K-means
- Neural network

And the exercises that students will have to do with them are described next.

K-Means

The algorithm K-means, developed in 1967 by MacQueen [3], is one of the simplest unsupervised learning algorithms that solve the well known clustering problem. K-means is a clustering algorithm which tries to show the students how items in a big dataset are automatically classified, even while new items are being added. The adopted technique for creating as many clusters as we want, is the minimum square error rule.

Starting from a given cloud of points, the aim of this activity is that the student learns how to cluster the points into K groups, by programming an application in Scratch. In order to do the clustering, the points will be classified by colors, and each one will belong to the group that has the closest average or mass center.

Neural Network

The basic idea behind a neural network is to simulate lots of densely interconnected brain cells inside a computer so you can get it to learn things, recognize patterns, and make decisions in a human-like way. The main characteristic of this tool is that a neural network learns all by itself. The programmer just needs to design the physical structure (number of outputs, inputs, hidden layers) and set some very simple rules involving additions, multiplications and derivatives. They are based on perceptrons, which were developed in the 1950s and 1960s by the scientist Frank Rosenblatt [7], inspired by earlier work by Warren McCulloch and Walter Pitts [4]. It is important to note that neural networks are (generally) software simulations: they are made by programming very ordinary computers. Computer simulations are just collections of algebraic variables and mathematical equations linking.

Two different exercises are developed with neural networks. First, a simple neural network with two inputs and an output neuron, is trained with an AND logic gate in an iterative way. In each iteration, students will see the different weights that the neural network gets. Next, and OR logic gate will be used, and as a consequence, students will observe how adjusting parameters change with this new structure.

As a second neural network exercise, a more complex neural network exercise is presented. Considering a 3-2-1 neural network (three inputs, a hidden layer with two neurons and an output) the students will train the neural network with AND and OR logic gates again.

Simple Neural Network Operation. The neuron obtains an output Y_1 from the two inputs $Input_1$ and $Input_2$ using the corresponding weights and the bias.

$$N_1 = Input_1 \cdot W_1 + Input_2 \cdot W_2 \tag{1}$$

The error is defined as the difference between the desired output $Desired\_output$ and the obtained output

$$Error = Desired\_output - Y_1 \tag{2}$$

Each time the function is executed, the algorithm updates the weights and the bias using the backpropagation and gradient descent rule, until the output Y_1 converges to the desired output $Desired\_output$.

The new value of the first weight will be the sum of its previous value and the product of the first input, the learning rate (LR) and the error

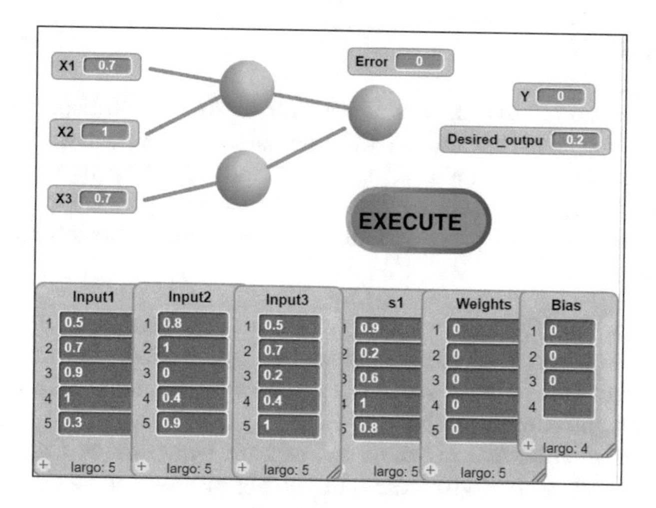

Fig. 1. Second neural network interface

$$W_{1,new} = W_1 + Input_1 \cdot LR \cdot Error \tag{3}$$

Similarly, the new value of the second weight will be the sum of its previous value and the product of the second input, the learning coefficient and the error.

Complex Neural Network Operation. The interface used is described in Fig. 1. There are three neurons: two at the input layer (N_1 and N_2) and one at the output (N_3). Their calculus depend on the inputs (X_1, X_2 and X_3) and weights (W_1, W_2, W_3, W_4, W_5) following next expressions:

$$\begin{cases} N_1 = B_1 + X_1 \cdot W_1 + X_2 \cdot W_2 \\ N_2 = B_2 + X_3 \cdot W_3 \\ N_3 = B_3 + N_1 \cdot W_4 + N_2 \cdot W_5 \end{cases}, \tag{4}$$

Employing again the backpropagation and gradient descent rule, the calculus for updating the weights and bias result in next equations:

$$\begin{cases} W_{1,new} = W_1 + LR \cdot Error \cdot X_1 \cdot W_4 \\ W_{2,new} = W_2 + LR \cdot Error \cdot X_2 \cdot W_4 \\ W_{3,new} = W_3 + LR \cdot Error \cdot X_3 \cdot W_5 \\ W_{4,new} = W_4 + LR \cdot Error \cdot N_1 \\ W_{5,new} = W_5 + LR \cdot Error \cdot N_2 \end{cases}, \tag{5}$$

$$\begin{cases} B_{1,new} = B_1 + LR \cdot Error \cdot W_4 \\ B_{2,new} = B_2 + LR \cdot Error \cdot W_5 \\ B_{3,new} = B_3 + LR \cdot Error \end{cases}, \tag{6}$$

Once the values of the error, weights and bias are calculated, the student has to store them on the corresponding neurons.

3 Methodology

In all algorithms presented in Sect. 2, the first task of the student is to fill in the white gaps left to him/her among the lines of code, marked with a comment. That is, the students do not need to create the algorithm or write the whole code themselves; the code will be provided for the most part.

Students will work in pairs and the workshop will be structured in the following steps:

- At the beginning, the teachers will give a short explanation of 15–20 min about AI and the objective of the workshop, with a twofold aim: first, to demystify Artificial Intelligence; second, to understand some simple mathematics underneath the computations.
- The students will work in couples. Teachers will provide them with some written theoretical background about the algorithms and the instructions for the exercises (K-means and neural networks). Moreover, teachers will explain in another 10 min of presentation the basics of the algorithms.
- Students will have one hour to finish the codes (20 min for K-means and 40 min for both parts of the neural networks) and eventually execute the applications to see whether they work properly. Teachers will be at hand to assist whenever necessary.
- Finally, at the end of the session, the teachers will provide the students with the finished proposed answers in Scratch software for the students to check them.

3.1 K-Means

The algorithm is developed using one main block and three function blocks of Scratch code. The function blocks are called `NewDataSet`, `KMeans` and `ColourPoints`.

The student must finish up the three function blocks: the block `NewDataSet`, block `KMeans` and block `ColourPoints`.

The main block will create the mass centers. The set of mass centers contains K random points with X coordinates between $(-230, 230)$ and Y coordinates between $(-170, 170)$. These K points will be the mass centers of the clusters (see Fig. 2).

The first task is to finish the programming of the block `NewDataSet` that will create the cloud of N points. The cloud must contain N points with X coordinates between $(-230, 230)$ and Y coordinates between $(-170, 170)$ (see Fig. 3).

Block `KMeans` has two parts: in the first, it stores the number of the mass center that will be assigned to each point. To do that, the student must code the calculation of the Euclidean distance from each of the points to each the mass centers in variable A. The program will then find the minimum distance of them all and fill up vector `Clusters`, which contains the numbers of the cluster to which each point belongs.

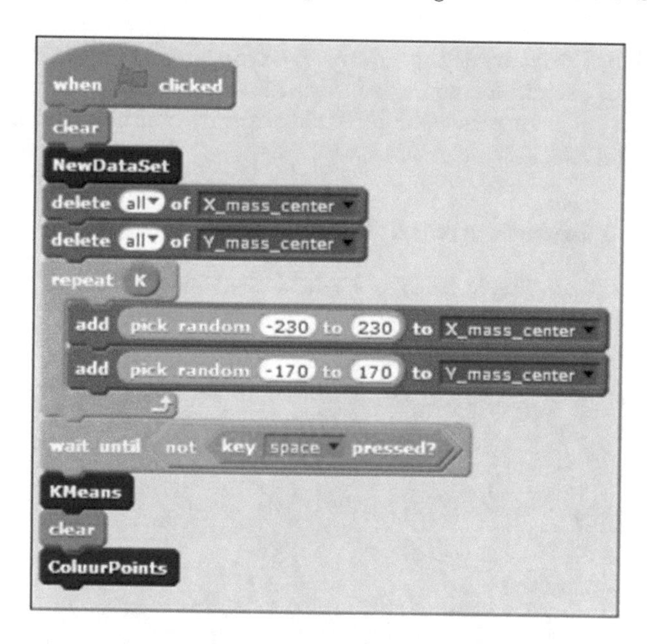

Fig. 2. Main block to create the mass centers

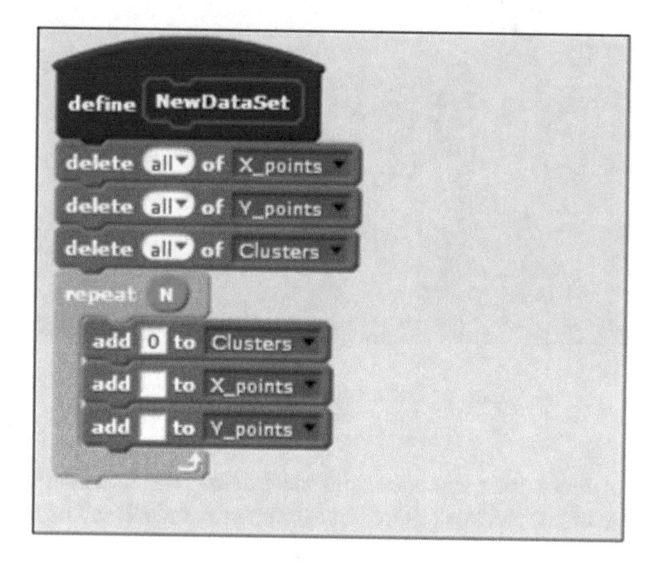

Fig. 3. Unfinished function to create the cloud of points

The second part of block `KMeans` changes the coordinates `X_mass_center` and `Y_mass_center` to the real centers of mass of the clusters. The student is not asked to do anything in this second part other than understanding the logic used.

Finally, block `ColourPoints` graphs the clouds of points and the cloud of centers of mass, each in a colour given by vector `Clusters`.

The algorithm here presented just tries to show automatic clustering, and not getting equivalent size clusters.

3.2 Neural Network: AND/OR Logic Gate

The algorithm is developed using a main block, which initializes the data, and two blocks, `Neuron` and `ExecuteButton`.

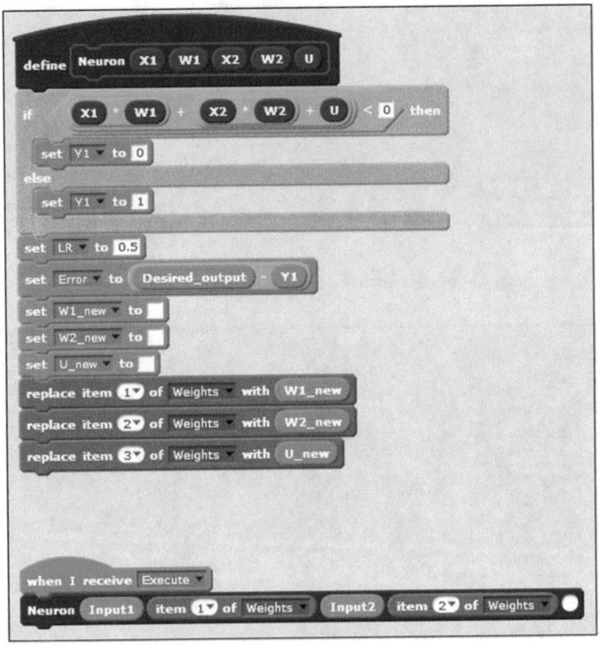

Fig. 4. Definition of block neuron

The student must code the equations that define the function that performs the only neuron of the network. The equations must calculate the new error, the new weights and the new bias. After having obtained the equations, the block must be finished (see Fig. 4).

The students must train the neuron using two sets of data: one set for the AND logic gate and another for the OR logic gate.

3.3 Complex Neural Network Operation

This exercise is based on training a three inputs neural network with AND and OR logic gates. As an additional complexity comparing to previous exercise, this neural network is multilayered, as can be seen in Fig. 1. It uses three neurons, `Neuron1`, `Neuron2` and `Neuron3`, and `ExecuteButton`.

The exercise is planned to fill the gaps of the following operations, as shown in Fig. 5:

- Net calculus of N_3
- Update of W_2 and W_5
- Update of B_2

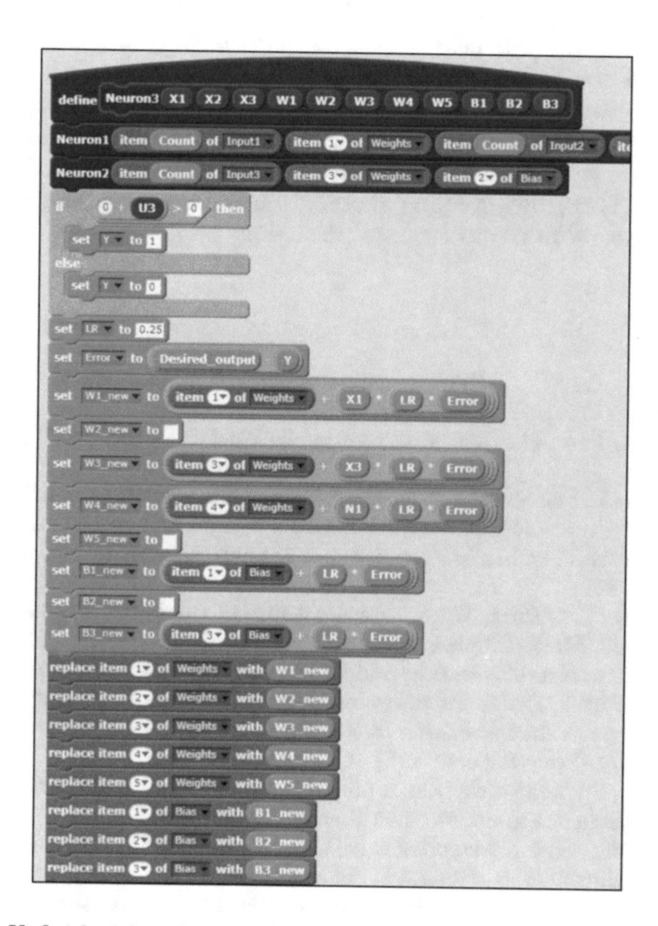

Fig. 5. Unfinished function to calculate net output, weights and bias update

4 Conclusions

The paper presents teacher-guided, easy to implant activities that can be performed at schools with Scratch. Moreover, the mathematics has been adapted for 16–18 year-old students mathematical background.

The tasks presented are scalable; students can delve into the maths involved in the mathematical iterations or into the Scratch code itself, or even propose new neural networks to deal with other problems.

The work can be extensible to students of different ages and more AI algorithms can be added to the system.

The easiness of the equations of K-means and neural networks permit their implementation in other formats, such as MS Excel, which in some occasions could be more familiar to students than Scratch.

Moreover, the students will realize that the mathematical knowledge acquired throughout the year will help them finishing the programming of automatic clustering and neuron networks, calculating AND and OR logic gates.

As a future work, first, the authors should organize the AI workshop several times and measure the grade of the objective achievement in collaboration of pedagogical researchers. Secondly, more programming languages should be explored (GeoGebra, HTML, ShynnyApps) in order to implement more exercises, such as the usage of neural networks for data prediction.

References

1. Arimoto, T., Sato, Y.: Rebuilding public trust in science for policy-making. Science **337**(6099), 1176–1177 (2012)
2. Haerlin, B., Parr, D.: How to restore public trust in science. Nature **400**(6744), 499 (1999)
3. MacQueen, J.: Some methods for classification and analysis of multivariate observations. In: Proceedings of the Fifth Berkeley Symposium on Mathematical Statistics and Probability. Statistics, vol. 1, pp. 281–297. University of California Press, Berkeley (1967)
4. McCulloch, W.S., Pitts, W.: A logical calculus of the ideas immanent in nervous activity. Bull. Math. Biophys. **5**(4), 115–133 (1943)
5. Nielsen, M.A.: Neural Networks and Deep Learning. Determination Press (2015)
6. OECD: Scientific advice for policy making: the role and responsibility of expert bodies and individual scientists. Technical report 21, OECD Science, Technology and Industry Policy Papers (2015)
7. Rosenblatt, F.: The perceptron: a probabilistic model for information storage and organization in the brain. Psychol. Rev. **65**(6), 386 (1958)
8. Stone, P., Brooks, R., Brynjolfsson, E., Calo, R., Etzioni, O., Hager, G., Hirschberg, J., Kalyanakrishnan, S., Kamar, E., Kraus, S., et al.: Artificial intelligence and life in 2030. One Hundred Year Study on Artificial Intelligence: Report of the 2015-2016 Study Panel (2016)
9. Woodford, C.: How neural networks work - a simple introduction (2016)
10. Resnick, M., Maloney, J., Monroy-Hernndez, A., Rusk, N., Eastmond, E., Brennan, K., Kafai, Y.: Scratch: programming for all. Commun. ACM **52**(11), 60–67 (2009)
11. AI in the UK: ready, willing and able? Technical report. Select Committee on Artificial Intelligence. House of Lords (2018)

Project-Based Methodology to Lecture on Web Frameworks Applied to the Management of Health-Related Data

Damián Fernández-Cerero(✉) and Alejandro Fernández-Montes

Departamento de Lenguajes y Sistemas Informticos,
Universidad de Sevilla, Sevilla, Spain
{damiancerero,afdez}@us.es
http://www.lsi.us.es

Abstract. The management and processing of the data generated by healthcare systems is getting more and more attention because of the digitalization of a traditionally analogical sector such as healthcare. Various data management frameworks and solutions have emerged to face some of the multiple challenges present in this environment, including: (a) Data privacy and security; (b) Inter-operability between heterogeneous systems; and (c) Usability and readability of health-related sensitive information.

In this paper, the authors share their experience on lecturing on how to address such issues by developing web-based software from several points of views: (a) Healthcare professional needs and requirements in terms of usability, accessibility and easiness; (b) Technical requirements and knowledge required to develop all the layers of a fully functional example of a micro health information system; and (c) Technical requirements to share and distribute the data required by several agents of the healthcare environment, focusing on the adoption of international standards such as HL7; (d) Perform of all operations in a secure, available and reliable way.

It is concluded that the application of Flipped Teaching among with Project-based Learning may have a very positive impact on both in grades and drop rates for late-years (junior and senior) courses within the Health Engineering bachelor's degrees.

Keywords: Project-based Learning · Flipped classroom · Health Engineering

1 Introduction

1.1 Health Engineering Degree

The University of Sevilla and the University of Málaga conforms the so called Andalucía-Tech campus. This is an aggregation of universities with the objective of develop and cooperate in this establishment of new and challenging degrees.

© Springer Nature Switzerland AG 2020
F. Martínez Álvarez et al. (Eds.): CISIS 2019/ICEUTE 2019, AISC 951, pp. 321–328, 2020.
https://doi.org/10.1007/978-3-030-20005-3_33

In this context, both universities agreed to offer a novel degree, not present in any other Spanish or European Universities: The Degree in Health Engineering. The main focus of this approach is to instruct students in the disciplines related to the application of engineering approaches to the Health Systems. The novelty of the approach is to incorporate three specializations: Clinic Informatics, Bioinformatics and Biomedical Engineering (which is the one present in several universities).

Clinics Informatics focuses on the application of Computer Science in clinic environments, especially for the interoperability of Healthcare Information Systems and for the definition of the Hospital Information Systems (HIS). In addition, various standards for the codification and transmission of health data are studied in this modality, including Health Level Seven (HL7), Digital Imaging and Communication in Medicine (DICOM). This specialization is only taught in the University of Sevilla.

Bioinformatics focuses on the scientific exploitation of clinic data to expand the knowledge on the relationship between them and the patients' pathologies. Moreover, the most advanced techniques to infer conclusions from patients' clinic data are studied in this modality. This specialization is only taught in the University of Málaga.

Biomedical Engineering focuses on the industrial design and manufacturing of clinical equipment for the various hospital environments, including cardiology units, operating rooms, general equipment, electrocardiogram equipment, and monitoring equipment.

1.2 Normative Context

The following verification files in Computer Science area have been taking into consideration to elaborate the lecturing project:

- The publication of the degree in Health Engineering Bachelor (Grado en Ingeniería de la Salud) present in the State official newsletter (BOE) number 5, 6th of August of 2012[1].
- Verification memory of the Health Engineering Bachelor (Grado en Ingeniería de la Salud)[2].
- Study program of the Health Engineering Bachelor (Grado en Ingeniería de la Salud)[3].

The concrete environment of the subjects taught in the Computing School (Escuela Técnica Superior de Ingeniería Informática) of the University of Sevilla and the working market have also taking into account for the elaboration of the study program.

[1] https://www.boe.es/boe/dias/2012/01/06/pdfs/BOE-A-2012-227.pdf.
[2] https://www.informatica.us.es/docs/estudios/ing_salud/memoria_verificacion.pdf.
[3] http://webapps.us.es/fichape/Doc/MVBOE/226_memVerBOE.pdf.

1.3 Project-Based Learning

Project-based Learning (PBL) is a model which organizes lecturing around projects. According to some definitions found in the literature, the projects are complex tasks based on problems or questions which are challenging. The students must perform design, problem-resolution, decision-making, and exploration tasks. Such tasks give students the opportunity to work with a relative autonomy for a long period of time. As a result, students may elaborate realistic products [1,5].

The characteristics found in the literature which define Project-based Learning include [4,7]:

- Original content.
- Teacher supervision but not direction.
- Clearly defined teaching objectives.
- Collaborative learning.
- Deliberation and incorporation of mature skills.

Project-based learning is considered the best tool to achieve multiple instructional goals. Among them: (a) Proposition and resolution of real-life challenges; (b) Students' involvement in the presentation of their projects, methodology and proposals; (c) Focus on the critical aspects of a project and prioritization of tasks; (d) Development of team-managing and communication skills to solve challenging issues; (e) Creativity, reflection, critique and revision on the project execution are some of the required collaborative skills to achieve a successful goal.

This combination of the aforementioned aspects of this teaching model provides students with the tools needed to successfully face the challenges that they will meet in the future in the working market.

In this environment, students are asked to perform a project which final goal consists on the development of subsystems of a Hospital Information System (HIS), including: clinic management system, administration system, and interoperation systems.

PBL can be very useful in the context of the bachelor's degrees taught in the School of Computing (ETSII), and that is why it is proposed as one of the learning techniques available to teachers that lecture subjects in the aforementioned degrees. Specifically, for Computer Science subjects, several studies have validated its effectiveness [6], showing many of the benefits obtained. Among these benefits, we highlight: (a) students are able to apply their knowledge; (b) acquisition of practical skills in programming; (c) involvement in teamwork processes; and (d) understanding of the factors that influence project management.

One of the most influential elements on the success of the use of PBL is the origin of the project idea. Three of the most commonly used scenarios include: (a) the students propose the topic of the project with complete freedom (requires validation by the faculty, which evaluates the complexity, viability and fulfilment of the teaching objectives); (b) the students choose from a set of projects proposed by the faculty; and finally (c) A single project is proposed for all the students.

Each approach has its own advantages and disadvantages: when students have the option to choose it, that decision usually translates into greater motivation. However, this decision freedom requires a highly involved teacher with a great ability to adapt to changing problems. In addition, this option requires also more complex evaluation and validation techniques. Plus, students can sometimes perceive that the difficulty, supervision and evaluation processes among students and projects are not homogeneous. On the other hand, when a single project is proposed to the students, such heterogeneity in terms of evaluation is minimized and requires less effort on the faculty side, since they only require knowledge on one problem domain and on one set of determined solutions. However, this approach presents a risk of lack of motivation from the students caused by the proposed project, meaning that apathy may grow, and the results could be less satisfactory.

Students may be oriented to carry their project out in groups or individually. The choice will depend on the teaching objectives, fundamentally if the faculty is eager to promote the group skills, and if the typology of the proposed projects is usually faced as a team or individually in the professional environment.

The orientation and the period of time given the students to complete their projects also marks the difficulty and extension of the projects, and therefore the teaching staff must propose or validate projects that are viable and at the same time hard enough for the objectives set.

The specific characteristics of the PBL that will be implemented in the course will be determined by: (a) the teaching objectives of each subject; (b) the size of the class groups; and (c) teaching staff and resources available.

1.4 Flipped Teaching

Flipped Teaching is a model that swaps the traditional lecturing process, where the teacher is the central source of knowledge which is spread among the students following a lecturing style. Students are not usually highly engaged nor motivated, since their role is framed within the absorption of the knowledge and asking the lecturer for some guidance on the lessons. The teacher can also suggest students to perform some activities and to solve problems/tasks in an autonomous way that could be corrected during following lessons in order to reinforce the knowledge transmitted.

In Flipped Teaching, unidirectional learning activities are moved to outside class hours thanks to online resources, forums, and personal research [3]. On the other hand, activities usually classified as homework are pushed into the classroom to have the real-time support and guidance of the teacher [8]. In this model, the learner become the center of the learning process, since he can adjust the knowledge-acquisition rate and depth and can enquiry lecturers when the knowledge is already present, strengthening the learning process. The guidance provided by teachers in this model is usually more personalized and interactive, achieving higher rates of motivation and engagement [2].

Flipped Teaching can optimally fit into the final courses of bachelor's degrees taught in the School of Computing (ETSII), where students need to be put as the center of the learning process to reinforce important skills before arriving to work market. Among these skills, flipped teaching include: (a) autonomous research; (b) self-management of time; (c) collaboration to achieve difficult goals; and (d) understanding of the importance of continuous work and organization.

Even if the most important factors for a successful learning process in Flipped Teaching falls on the students themselves, some decisions made by teaching staff may highly impact on the results. Among them: (a) the quality of the supporting material; (b) the quality of online resources for collaboration, communication and research; (c) the type of activities proposed in the classroom; and (d) the willingness of the teachers to go beyond of what's expected.

2 Subject and Student Population

The subject where the previously mentioned teaching methodologies were applied is a third course subject from the Degree in Health Engineering. The subject title is "Codificacin y gestin de la informacin sanitaria" (Codification and management of health data) and is included in one of the three specializations of the degree: Mencin en Informática Clínica (Clinic informatics).

The student population under study covers students from five academic courses, which sum up 57 students. Demographic information about students was not included as part of this study, but gender, age and other demographics were similar between both populations. Academic courses of 2013 and 2014 a project-based learning methodology was applied, mixed with conventional lessons conducted by academic staff. A total of 27 students participated during these courses. Academic courses of 2015, 2016 and 2017 a flipped teaching methodology was applied, and no conventional lessons were conducted by academic staff. A total of 30 students participated during these three courses.

Figure 1 shows the individual scoring for each academic year. The period of time under consideration comprises the following years: 2013, 2014, 2015, 2016 and 2017 (which sum up a total of 5 academic years). Academic courses of 2013, 2014 a project-based learning methodology was applied mixed with a more conventional lessons conducted by teachers. By the academic courses of 2015, 2016 and 2017 a flipped teaching + project-based learning methodology was applied (no conventional lessons conducted by teachers).

3 Statistical Analysis

Results from these five courses are presented in Table 1. In this section we performed various statistical analysis in order to check the goodness of the proposed teaching methodology based on flipped teaching.

It can be noticed that the proposed teaching methodology based on flipped teaching achieves its goals of putting the students as the core and center of the learning process, method and activities. As a consequence of this new approach several Key Performance Indicators (KPI) were improved in the three academic years where the flipped teaching methodology was successfully applied. This Key Performance Indicators include:

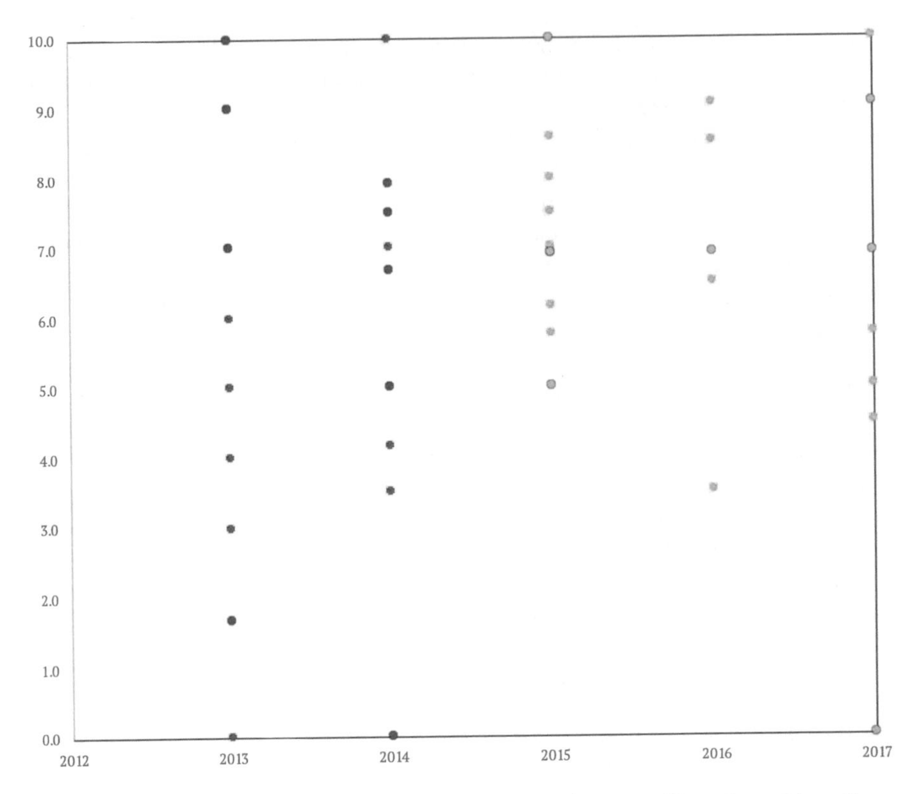

Fig. 1. Individual scoring grouped by academic year. (Red: no flipped-teaching, Green: flipped-teaching)

- **%passed.** Ratio of students that passed the subject.
- **%failed.** Ratio of students that failed the subject.
- **%drop-out.** Ratio of students that abandoned the subject.
- **Average score.** Mean value of the scores (from 0 to 10) achieved by students.
- **Variance.** Variance of the scores achieved by students.

From these results and comparison between the two groups we notice that the ratio of students that passed the subject was increased by 20% and at the same time the ratio of students that decided to drop-out the subject was decreased.

It is important to notice that the variance between the scores achieved by students was also decreased (from 9.23 to 5.81) which indicates that there was a more homogeneous subject understanding by students.

The p-value test was performed to demonstrate the significance of the improvement in the score results between students grouped by the learning methodology. The one-tailed study is described below:

Table 1. Statistical analysis summary results.

Measure	Traditional method	Flipped teaching	Diff
Total Students	27.00	30.00	3.00
Passed	18.00	26.00	8.00
%passed	0.67	0.87	0.20
Failed	9.00	4.00	−5.00
%failed	0.33	0.13	−0.20
Drop-out	3.00	2.00	−1.00
%drop-out	0.11	0.07	−0.04
Average	5.64	6.64	1.01
Mode	7.00	6.92	−0.08
Deviation	3.10	2.45	−0.65
Variance	9.23	5.81	−3.43

Difference Scores Calculations

Subset 1 (no flipped teaching method):

$$N_1 : 27$$

$$df_1 = N - 1 = 27 - 1 = 26$$

$$M_1 : 5.64$$

$$SS_1 : 248.69$$

$$s_1^2 = SS_1/(N - 1) = 248.69/(27 - 1) = 9.56$$

Subset 2 (flipped teaching method):

$$N_2 : 30$$

$$df_2 = N - 1 = 30 - 1 = 29$$

$$M_2 : 6.64$$

$$SS_2 : 173.45$$

$$s_2^2 = SS_2/(N - 1) = 173.45/(30 - 1) = 5.98$$

T-value Calculation:

$$s_p^2 = ((df_1/(df_1 + df_2)) * s_1^2) + ((df_2/(df_2 + d_f2)) * s_2^2)$$
$$= ((26/55) * 9.56) + ((29/55) * 5.98) = 7.68$$

$$s_{M_1}^2 = s_p^2/N_1 = 7.68/27 = 0.28$$
$$s_{M_2}^2 = s_p^2/N_2 = 7.68/30 = 0.26$$

$$t = (M_1 - M_2)/\sqrt{(s_{M_1}^2 + s_{M_2}^2)} = -1/\sqrt{0.54} = -1.36$$

The t-value is −1.35515. The p-value is .090455. The result is significant at $p < 0.10$.

4 Conclusions

Authors shared their experiences on lecturing on how to address issues related to healthcare information management and codification by developing web-based software from several points of views: (a) Healthcare professional needs and requirements in terms of usability, accessibility and easiness; (b) Technical requirements and knowledge required to develop all the layers of a fully functional example of a micro health information system; and (c) Technical requirements to share and distribute the data required by several agents of the healthcare environment, focusing on the adoption of international standards such as HL7; (d) Perform of all operations in a secure, available and reliable way.

Evidences made us conclude that the application of Flipped Teaching among with Project-based Learning may have a very positive impact on both scores and drop rates for late-years (junior and senior) courses within the Health Engineering bachelor's degrees. It is shown that flipped teaching reduced the number of drop-outs from students, and at the same time reduces the differences between the scores achieves by them.

References

1. Davcev, D., Stojkoska, B., Kalajdziski, S., Trivodaliev, K.: Project based learning of embedded systems. arXiv preprint arXiv:1606.07498 (2016)
2. Herreid, C.F., Schiller, N.A.: Case studies and the flipped classroom. J. Coll. Sci. Teach. **42**(5), 62–66 (2013)
3. Jinlei, Z., Ying, W., Baohui, Z.: Introducing a new teaching model: flipped classroom. J. Distance Educ. **4**(8), 46–51 (2012)
4. Kokotsaki, D., Menzies, V., Wiggins, A.: Project-based learning: a review of the literature. Improv. Sch. **19**(3), 267–277 (2016)
5. Larson, J.S., Farnsworth, K., Folkestad, L.S., Tirkolaei, H.K., Glazewski, K., Savenye, W.: Using problem-based learning to enable application of foundation engineering knowledge in a real-world problem. In: 2018 IEEE International Conference on Teaching, Assessment, and Learning for Engineering (TALE), pp. 500–506. IEEE (2018)
6. Pucher, R., Lehner, M.: Project based learning in computer science-a review of more than 500 projects. Procedia Soc. Behav. Sci. **29**, 1561–1566 (2011)
7. Savery, J.R.: Overview of problem-based learning: definitions and distinctions. In: Essential Readings in Problem-Based Learning: Exploring and Extending the Legacy of Howard S. Barrows, vol. 9, pp. 5–15 (2015)
8. Xinglong, Z.: The design of teaching mode based on knowledge construction in flipped classroom. Mod. Distance Educ. Res. **2**, 55–61 (2014)

Game-Based Student Response System Applied to a Multidisciplinary Teaching Context

José Lázaro Amaro-Mellado[1(✉)], Daniel Antón[2,3], Macarena Pérez-Suárez[4], and Francisco Martínez-Álvarez[5]

[1] Department of Graphic Engineering, University of Seville, Seville, Spain
jamaro@us.es

[2] Department of Graphic Expression and Building Engineering, University of Seville, Seville, Spain
danton@us.es

[3] The Creative and Virtual Technologies Research Laboratory, School of Architecture, Design and the Built Environment, Nottingham Trent University, Nottingham, UK

[4] Department of Applied Economy III, University of Seville, Seville, Spain
mperez32@us.es

[5] Data Science & Big Data Lab, Pablo de Olavide University, ES-41013 Seville, Spain
fmaralv@upo.es

Abstract. The prevailing need to promote the application of active methodologies to develop the teaching-learning process in an effective way led to this work. The aim was to evidence the efficiency of an Information and Communication Technologies (ICT) resource for the generation of a collaborative and multidisciplinary context in the university teaching activity within dissimilar fields of knowledge. A Game-based Student Response System (GSRS) was implemented as a common tool for different teachers of non-related fields of knowledge with a view to enhance critical thinking and engagement in the students. The outcomes showed that this methodology increased the involvement and reduced part of the degree of abstraction that complex contents entail. Thus, the environment varied significantly with respect to the traditional relationship between students and teachers, since the latter are in charge of generating student-centred experiences based on interaction. As a result, feedback information was gathered in order to rearrange contents and consider novel strategies. In addition, this work eased the acquisition of knowledge, competence development and the achievement of teaching objectives by the students.

Keywords: Game-based Student Response System (GSRS) · Multidisciplinary context · Teaching innovation · Information and Communication Technologies (ICT)

D. Antón—Visiting Research Fellow.

© Springer Nature Switzerland AG 2020
F. Martínez Álvarez et al. (Eds.): CISIS 2019/ICEUTE 2019, AISC 951, pp. 329–339, 2020.
https://doi.org/10.1007/978-3-030-20005-3_34

1 Introduction

Nowadays, research on teaching is both a current need and a complex task, especially in the integration of new technologies in university education. The effective application of Information and Communication Technologies (ICT) in real teaching contexts requires teachers to endeavour to assimilate and put them into practice so that students take advantage of this sort of experiences [13]. This, applied to a multidisciplinary field, reveals reflection and collaboration among teachers as essential, since knowledge should not be limited. Specific subjects and apparently dissimilar areas of knowledge can find common and complementary didactics to meet their respective teaching objectives [18]. There is a positive relationship between the position of teachers regarding the integration of ICT in the curricula and the use of these technologies in the classroom [8]. Thus, the benefits of ICT in teaching must be recognised.

In this sense, multiple research works [15, 16, 23, 24] constitute demonstrations based on this type of experience. These authors ensure that the integration of ICT in the teaching-learning process entails challenges for all participants in it, and that the simple use of ICT is associated with the most successful teaching practices [22].

A new paradigm based on pedagogical socio-constructivism in education emerges as a result of applying the ICT, thus guaranteeing competences in the use of these technologies, as well as transversal competences along with curricular contents [12].

Currently, the educational technology is integrated into a broad multidisciplinary context of study, although it is not possible to foresee the real impact of new technologies on complex, well-established institutions [7]. According to [3], the educational technology refers to constant emerging technologies that complete their theoretical foundations in any field. The volume of these technologies has increased progressively. Under this statement, the disciplinary diversity can be found in different contexts, for instance in information technology [21]; mathematics [27]; health sciences [14], engineering drawing [20], as well as in the efforts against illiteracy, for which the effectiveness of the use of smart devices has been proven [9].

Finally, there are other empirical demonstrations related to the use of the Student Response System (SRS) and games in teaching. On the one hand, [25] addresses the issue of how to integrate the development of games in education. Concerning the use of response systems, [4] analyses the function of pushbuttons and their repercussion in the teaching of the subject of biology, then providing a series of guidelines on their use. On the other hand, [10] summarise the benefits of SRS employment in three areas: classroom environment, teaching, and evaluation. The classroom environment results in more class attendance, likewise the students experience both greater concentration and participation. In terms of learning, it improves the interaction and constructive discussion between colleagues to build knowledge and increase the feedback the teachers receive which allows them to rearrange contents and questions. Regarding the evaluation, the constant awareness of the degree of understanding of the students, both by the teacher and their own peers, is a powerful tool for educational progress. It is worth mentioning the research carried out by [26] who analyses how interest falls on the part of the students when the use of SRS based on games (Game-based SRS, GSRS) becomes part of the habitual dynamics in the lessons throughout a complete academic course.

This framework, considering different fields of knowledge, led this work to address the implementation of ICT as a teaching resource with a view to promote student participation and collaboration among teachers. Thus, the use of games based on ICT in which students participate using their own devices (phones and tablets) is intended to leave the traditional educational model behind and agree with what has been demonstrated by [5], who state that the laptop user profile in the classrooms is different from the tablet user. Later, the results from this deployment are shown and analysed, and the suitability of this type of technological tools in teaching is supported.

The general objective is to demonstrate the efficiency of an ICT resource in the generation of a multidisciplinary and collaborative context in the university teaching activity from different areas of knowledge.

In this way, the specific objectives are: (1) to verify the capacity of these different areas of knowledge to interact and design practical and dynamic activities; (2) to know the degree of acceptance by students of the inclusion of game-based learning; and (3), to estimate the benefits of the educational resource implemented by the teacher.

In this sense, this research practices an inductive method by applying an ICT teaching innovation resource to a sample of students; that is, an empirical approach subject to greater future extrapolation is explained.

2 Methodology

This work is the result of the simultaneous implementation of a specific didactic resource into two different subjects in university teaching: (1) on local policies development (Degree in Labour Relations and Human Resources); and (2) on graphic expression and cartography (Degree in Agricultural Engineering). In this section, the methodology of both this research and the training given are firstly described.

Regarding the teaching methodology, the following parts are approached: introduction; identification and approach of the problem through key questions; theoretical notions; activity performance by the students and the use of the ICT tool, through which a diagnosis of the evolution of the process is conducted and, if necessary, the theoretical basis for solving the problem is emphasised; summarisation, discussion and evaluation of the achievement and interest by the students in the experience developed.

For this study an ICT tool is implemented in a context of five university groups of the aforementioned subjects with a total of 171 students. This ICT tool must be able to link reality with the contents of these subjects. The experience put into practice is based on the paradigm "Bring Your Own Device" (BYOD), which facilitates the participation of students. Thus, a GSRS called *Kahoot!* [2], is used, where the teacher is the moderator, whereas the students are responsible for answering questions interactively. Previous experiences in this field were based on the Lecture Quiz prototype, which additionally required to install Java and its graphics library. Currently, it is worth highlighting the positive characteristics of *Kahoot!*: there is no need to install programs to use it, it is freeware, and its integration with social networks is easier.

Regarding the creation of questions, these can be presented in questionnaire (*quiz*), debate (*discussion*) or *survey* mode. In the questionnaires of objective questions, which can be proposed as a contest, there can be between two and four options, of which one or more may be correct. There is a limited time for the response, between 5 and 120 s, and the option of whether students can get points or not with the answer. The question can be either textual, or include an image or video to illustrate it (this is essential for purely graphical subjects). In the quiz mode, the teacher displays the questions on the screen or in the classroom projector and the students proceed to answer as correctly and quickly as possible using their own device. After each question, the program shows global statistics of how the students have responded, so the teacher receives a feedback immediately, which allows him to act accordingly. In addition, the scores and nicknames of the students with the best results are shown. At the end of the game, the winner, the second and the third places are revealed.

The peculiarity of *Kahoot!* is that it is specially oriented to social networks, using a concept that is similar to the videogame Buzz! (Sony PlayStation). At the same time, there are other platforms that somehow respond to the SRS paradigm [26]: Socrative (similar to *Kahoot!*); Quizlet (it is not a SRS, but a web-based learning tool with flashcards); i > clicker; Learning Catalytics, etc.

As established by [11], to do something fun to learn you can establish three categories: the challenge (objectives with uncertain results); fantasy (captivate through intrinsic or extrinsic fantasy); and curiosity (sensory, through graphics and sound or cognitive, where the player must solve a problem). *Kahoot!* follows this model to persuade the students; thus, it has been considered in this research.

As explained above, one of the main advantages of using *Kahoot!* is that it eases to share the theoretical-practical questionnaires, either directly with other colleagues or through social networks. Thus, the creation and design of questionnaires is not strictly necessary, since there may already be one on the subject addressed. The creation tool allows reorganising the sequence of questions, making them public or private, adding metadata, such as language, main recipients or difficulty. Table 1 briefly shows some relevant aspects of this teaching-learning experience carried out.

Table 1. ICT teaching resource implemented.

Educational resource	GSRS (Game-based Student Response System)
Literature	[1, 6, 19, 25, 26]
Description	Game-based Student Response where an access to packages of questions is designed by the teacher or the students with different themes and levels of difficulty, establishing an entertaining environment for the participants
Competence incidence	Information related to knowledge, opinion, evolution and learning outcomes is recorded
Activity	The students answer questionnaires designed by the teachers in *Kahoot!* in a more relaxed and friendly way, and can visualise the correct answers and the scores they are obtaining. In addition, they can be encouraged to create questions and share them with the rest of students. These questionnaires are specifically based on: (a) local development policies: institutional framework; (b) orthographic parallel projection representation system: basic concepts

(*continued*)

Table 1. (*continued*)

Educational resource	GSRS (Game-based Student Response System)
Observations	This teaching resource leaves the usual and unidirectional classroom dynamics behind through reviewing ideas. Also, the acquisition of knowledge is promoted from the success-error par (immediate feedback). Eventually, both interaction and motivation in the competition mode are enhanced

For this experience, ten objective questions for each subject have been raised, considering four alternatives and a single correct answer, as well as a response time from 20 to 60 s. For the answers, the Likert scale of four points (1–4) was used according to the degree of agreement with the corresponding statement (1 Agreement - 4 Disagreement). In order to make the most of the students feel comfortable with the game, each one's nickname is as anonymous as they prefer. It should also be noted that the questionnaires have been designed by the authors of this work.

With a view to show the interface of *Kahoot!*, Figs. 1 and 2 illustrate the series of prototype 'unknowns' raised to students. On the left side, what is displayed on screen in the classroom; on the right, what the students experience using their devices: they have to tap on the colour to choose their answers. After results of this part were registered, a second part was conducted equally for the total sample of the students being responsible for preparing the questions along with their objective answers. Once both parts have been completed, the collected data are analysed following a descriptive and selective statistical approach to ascertain the behaviour of the students' responses.

Fig. 1. Implementation of the *Kahoot!* ICT teaching resource in the subject Local Development Policies. Source: Authors.

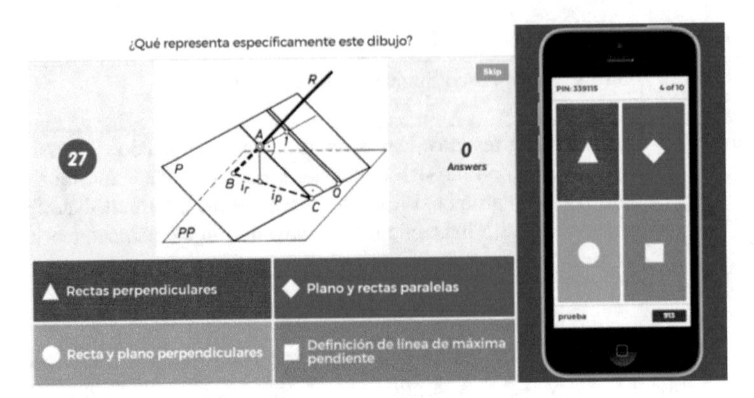

Fig. 2. Implementation of the *Kahoot!* ICT teaching resource in the subject Graphic Expression and Cartography. Source: Authors.

3 Results and Discussion

Initially, it can be observed that the classroom environment relaxes, with which the predisposition to learning increases. This may be due to the more advanced, immediate and spontaneous nature of BYOD against the traditional SRSs such as remote push buttons (clickers), keyboards (key-pads), headphones with microphone (handsets), etc. [4]. Certainly, the social circumstances have changed, since most of the students usually attend classes with smart devices. This avoids expenses by the Government on both acquisition and management of the devices and licenses [26].

According to the specialised scientific literature, there are different ways to integrate games in education [25]. Firstly, the exercises or tasks are replaced by allowing the students to play motivating games, thus giving the teacher an opportunity to monitor the students' progress in real time; secondly, the development of games can be used to teach other subjects such as pattern design, literacy, software architecture, computer science, mathematics, physics, etc.; finally, games can be an integrated part of a traditional class to improve motivation and participation. Consequently, the use of *Kahoot!* is a suitable starting point.

Table 2 shows the results of an analysis of the success percentage for each question by groups. Those questions with the highest and lowest percentage of success are highlighted in italics.

Table 2. Degree of success according to the unknown of Group 1 (G1) and Group 2 (G2).

Success rate	<25%	25–50%	51–75%	>75%
Unknown # (G1)	*P4*	P5, P8	P2, P7, P10	P1, P3, *P6*, P9
Unknown# (G2)	P2, *P6*	P1, P3, P7, P10	P4, P8, P9	*P5*

The overall success rate for G1 is 65% and 44% for G2. It is worth noting that the success percentage values should be put in context depending on the subject. That is, they should be considered according to the general level of the class, estimated as the percentage of global success for each subject.

The results of this research eased the assessment of the understanding degree for each question and subject. Table 3 shows correct answers (%) by groups, where significant differences are found in Group 2, which presents a lower number of correct answers. However, questions 1, 3 and 4 have major percentage in Group 2. The record of successes seems to decrease as the number of uncertainties raised increases. It is worth highlighting the debatable data existing when both groups are compared.

Table 3. Responses of the first part by groups.

	Group 1		Group 2	
	Answers	Success rate	Answers	Success rate
Unknown #1	2	1.29%	2	12.50%
Unknown #2	2	1.29%	0	0.00%
Unknown #3	2	1.29%	4	25.00%
Unknown #4	11	7.10%	7	43.75%
Unknown #5	24	15.48%	1	6.25%
Unknown #6	36	23.23%	1	6.25%
Unknown #7	43	27.74%	1	6.25%
Unknown #8	27	17.42%	0	0.00%
Unknown #9	8	5.16%	0	0.00%
Total	155	100.00%	16	100.00%

Table 4 indicates that the responses by Group 1 were different from those by Group 2, since there are only two students in Group 2 who, due to the way they responded, could belong to Group 1. Therefore, a differentiated response pattern by group of origin is found. The bias given by the area of knowledge may be more determinant than initially estimated.

Table 4. First part analysis after the teaching resource implementation.

		Group for analysis	
		G1	G2
		Check	Check
Source	Group 1	155	0
	Group 2	2	14

It can be noted that there are no students achieving 100% success rate (Table 5).

Table 5. Total responses (both groups) in the first part of the teaching resource implementation.

	Responses	Success rate
Unknown #1	4	2.34%
Unknown #2	2	1.17%
Unknown #3	6	3.51%
Unknown #4	18	10.53%
Unknown #5	25	14.62%
Unknown #6	37	21.64%
Unknown #7	44	25.73%
Unknown #8	27	15.79%
Unknown #9	8	4.68%

Also, students are encouraged to prepare questions with answers for the same content discussed in the previous part. This becomes a measurement pilot exploration, since only approximately 20% (34) out of 171 students participated in that proposal.

The results confirm the teachers' intention when designing the experience. The degree of success is proportionate; therefore the direction taken is adequate, and must be continued to ensure that the students become more motivated to acquire skills according to the teaching planning.

From the purely physiological point of view, the average time in which human attention is high does not exceed 20 min, and then the use of an SRS can help to restart that attention clock [4], which is even more evident using GSRS.

It should be highlighted that the use of ICT tools is closely linked to the teachers' technological skills and pedagogical training in teaching strategies development [17].

The BYOD paradigm becomes a well rated tool by the students within a powerful learning environment, since it basically lets them use their own devices, which are mainly intended for leisure. Thus, the method is revealed as effective and sufficient for the parties involved. In this way, teachers from different areas are able to find a common tool that is capable of being applied in each case in order to increase the motivation in learning and strengthen the concepts involved in the process.

Finally, based on observations, the experience produced has been satisfactory, both for students and teachers. Beyond the usefulness of recording the students' responses for their consequent analysis, the main investigations confirm the suitability of the teaching methodology described in this work.

4 Conclusions

The outcomes of this work are diverse and involve the system as a whole, as well as teachers, students and other agents of the teaching-learning process.

The main finding is the potential of this new framework for bi-directional and technological learning, in which the teacher's disadvantage against the students in the use of these ICT tools is generally evident. In this way, teachers from different fields of knowledge are able to find a collaborative tool that can be applied in each case in order to increase motivation in learning and strengthen the concepts involved in the process. The didactic resource based on GSRS and SRS tests a pattern of differential response in learning. This study indicates that Group 1 responded in a different way to Group 2. There is no student who responds correctly to the total of unknowns raised. As the students' active participation increases, the degree of success and suitability decreases. Regarding the impact on the students, several studies have shown that the use of GSRS and SRS involves students in really high percentages, greater than 85% [4, 26]. The values estimated in this experience confirm this statement in form, but not in the acquisition of knowledge.

Next, possible future lines of research are established. GSRS may be used to flip the situation: the total number of students is required to create a questionnaire (quiz) on the same topic for other classmates to answer, in such a way that the commitments to learning reach levels almost unthinkable otherwise. This would require an intense knowledge of the subject to be able to pose interesting questions and answers. In this way, a flipped lesson supervised by the teacher could be undertaken, thus constituting a real and powerful method. According to the data gathered, the students would be willing to participate, although it should be established an objective reward (extra score in the evaluation) to encourage them to take part.

The surprise factor of *Kahoot!* is noticeable at a first glance by the students, although there are few scientific studies on the verification of its regularity, including [26]. It would be useful to inquire into the regularity of the students' interest in the use of ICT teaching resources, as well as the methodological improvement of increasing the sample and the number of areas of knowledge. Also, the 360° evaluation of the results could be addressed. In any case, this experience is revealed as an enjoyable way to survey the degree of understanding of the fundamental concepts of practically any area of knowledge. A significant percentage of students do not usually express their level of understanding, either because of shyness or embarrassment, among other reasons. Nevertheless, the students can feel more comfortable with using tools such as *Kahoot!*, given the possibility of participating in a pseudo-anonymous way.

To conclude, this research finds ICT tools innovating and encouraging to be implemented in teaching. Thus, the students' predisposition to learn is enhanced, experiencing a seemingly minor effort, but certainly more efficient than following traditional methods.

Acknowledgments. The authors would like to thank the University of Seville for funding the Second Own Plan for Teaching Innovation (2016) which has supported this research.

References

1. Barreras, M.A.: Experiencia de la clase inversa en didáctica de las lenguas extranjeras. Educatio Siglo XXI **34**(1), 173–196 (2016)
2. Brand, J., Brooker, J., Furuseth, A., Versvik, M.: "Kahoot!". Oslo (2019). https://kahoot.com/. Accessed Jan 2019

3. Cabero-Almenara, J.: ¿Qué debemos aprender de las pasadas investigaciones en Tecnología Educativa? RIITE. Revista Interuniversitaria de Investigación en Tecnología Educativa. 0, 23–33 (2016)
4. Caldwell, J.E.: Clickers in the large classroom: current research and best-practice tips. CBE Life Sci. Educ. **6**(1), 9–20 (2007)
5. Castillo-Manzano, J.I., Castro-Nuño, M., López-Valpuesta, L., Sanz-Díaz, M.T., Yñíguez, R.: To take or not to take the laptop or tablet to classes, that is the question. Comput. Hum. Behav. **68**, 326–333 (2017)
6. Fotaris, P., Mastoras, T., Leinfellner, R., Rosunally, Y.: Climbing up the leaderboard: an empirical study of applying gamification techniques to a computer programming class. Electron. J. e-Learning **14**(2), 94–110 (2016)
7. Halverson, R., Smith, A.: How new technologies have (and have not) changed teaching and learning in schools. J. Comput. Teacher Educ. **26**(2), 49–54 (2014)
8. Hue, L.T., Ab Jalil, H.: Attitudes towards ICT integration into curriculum and usage among university lectures in Vietnam. Int. J. Instr. **6**(2), 53–66 (2016)
9. Jiménez-García, M., Martínez-Ortega, M.A.: El Uso de una Aplicación Móvil en la Enseñanza de la Lectura. Inf. Tecnol. **28**(1), 151–160 (2017)
10. Kay, R.H., LeSage, A.: Examining the benefits and challenges of using audience response systems: a review of the literature. Comput. Educ. **53**(3), 819–827 (2009)
11. Malone, T.W.: What makes things fun to learn? Heuristics for designing instructional computer games. In: The 3rd ACM SIGSMALL Symposium and the First SIGPC Symposium on Small Systems. ACM Press, Palo Alto (1980)
12. Marqués, P.: Impacto de las TIC en la educación: funciones y limitaciones. 3c TIC: Cuadernos de Desarrollo Aplicados a las TIC, vol. 3, pp. 1–15 (2013)
13. Martínez, P., Pérez, J., Martínez, M.: Las TICS y el entorno virtual para la tutoría universitaria. Educación XXI: Revista de la Facultad de Educación **19**(1), 287–310 (2016)
14. Martínez-Galiano, J.M., Amaro, P., Gálvez-Toro, A., Delgado-Rodríguez, M.: Metodología basada en tecnología de la información y la comunicación para resolver los nuevos retos en la formación de los profesionales de la salud. Educación Médica **17**(1), 20–24 (2016)
15. Martins Dos Santos, G., Mourão, R.G.: ICT in Education: personal learning environments in perspectives and practices of young people. Educação e Pesquisa AHEAD **21**, 1–20 (2017)
16. Mirate, A.B., García, F.A.: Rendimiento académico y TIC. Una experiencia con webs didácticas en la Universidad de Murcia. Pixel-Bit. Revista de Medios y Educación **44**, 169–183 (2014)
17. Muñoz-Carril, P., Fuentes, E.J., González-Sanmamed, M.: Necesidades formativas del profesorado universitario en infografía y multimedia. Revista de Investigación Educativa. **30**(2), 303–321 (2012)
18. Pérez-Suárez, M., Antón, D., Amaro-Mellado, J.L.: Las TIC como recurso docente de convergencia entre distintas áreas de conocimiento. In: TIC Actualizadas Para Una Nueva Docencia Universitaria, pp. 611–629. McGraw-Hill Interamericana de España, S.L. (2016)
19. Pintor, E., Gargantilla, P., Herreros, B., López, M.: Kahoot en docencia: una alternativa practica a los clickers. In: XI Jornadas Internacionales de Innovación Universitaria. Educar para transformar, Madrid, 7–8 July 2014
20. Ramírez-Juidías, E., Tejero-Manzanares, J., Amaro-Mellado, J.L., Ridao-Ceballos, L.: Developing experimental learning in a graphical course using Thrustone's law of comparative judgment. Eng. Lett. **25**(1), 61–67 (2017)
21. Révészová, L.: Development of information competencies via process modelling and ICT integration. In: 36th International Convention IEEE Information and Communication, Technology, Electronics and Microelectronics (MIPRO), 20–24 May, Opatija (2013)

22. Ricoy, M.C., Couto, M.J.V.S.: Best practices with ICT and the value attributed by the students newly integrated in university. Educação e Pesquisa **40**(4), 897–912 (2014)
23. Ricoy, M.C., Fernández, J.: Contributions and controversies generated by the use of ICT in higher education: a case study. Revista de Educación **360**, 509–532 (2013)
24. Rubio-Escudero, C., Asencio-Cortés, G., Martínez-Álvarez, F., Troncoso, A., Riquelme, J.C.: Impact of auto-evaluation tests as part of the continuous evaluation in programming courses. In: Advances in Intelligent Systems and Computing, vol. 771, pp. 553–561 (2018)
25. Wang, A.I.: Extensive evaluation of using a game project in a software architecture course. ACM Trans. on Comput. Educ. (TOCE) **11**(1), 1–28 (2011)
26. Wang, A.I.: The wear out effect of a game-based student response system. Comput. Educ. **82**, 217–227 (2015)
27. Yildiz, B., Usluel, Y.: A model proposal on ICT integration for effective mathematics instruction. Hacettepe Egitim Dergisi **31**(1), 14–33 (2016)

Implementation of an Internal Quality Assurance System at Pablo de Olavide University of Seville: Improving Computer Science Students Skills

C. Rubio-Escudero[1](✉), F. Martínez-Álvarez[2], E. Atencia-Gil[3], and A. Troncoso[2]

[1] Department of Computer Science, University of Seville, Seville, Spain
crubioescudero@us.es
[2] Data Science and Big Data Lab, University Pablo de Olavide, 41013 Seville, Spain
{fmaralv,atrolor}@upo.es
[3] University Pablo de Olavide, 41013 Seville, Spain
eategil@admon.upo.es

Abstract. This work describes how an internal quality assurance system is deployed at Pablo de Olavide University of Seville, Spain, in order to follow up all the existing degrees among the faculties and schools, seven centers in total, and how the teaching-learning process is improved. In the first place, the quality management structure existing in all the centers and degrees of the university is described. Additionally, all the actions related to the quality and improvement of the degrees of a center are reported. Unlike in other Universities, in the Pablo de Olavide University there is no specific procedure for monitoring degrees, but the strategic procedure *PE04: Measurement, analysis and improvement* of the Internal Quality Assurance System is used to carry out such a procedure. Therefore, the procedure is detailed specifying the different phases it consists of and those responsible for each of them. Once this procedure has been implemented, the centers have a follow-up report for each of their degrees, which also includes an improvement plan to be developed during the next course. The case of the degree of Computer Science in Information Systems, included in the School of Engineering, is analyzed over time in order to show how the implementation of such a system improves the overall performance of students.

Keywords: Quality · Innovation · Education · Students skills · Computer science

1 Introduction

The internal quality assurance is of utmost importance in educational centers because it relates to their trust [11]. The design of an Open System of Internal Quality Assurance (OIQAS) has been proved to be useful to improve the overall

© Springer Nature Switzerland AG 2020
F. Martínez Álvarez et al. (Eds.): CISIS 2019/ICEUTE 2019, AISC 951, pp. 340–348, 2020.
https://doi.org/10.1007/978-3-030-20005-3_35

process of learning [9]. The OIQAS can be described as the systematic, structured and continuous attention to quality with the aim of its maintenance and improvement. Within the framework of the training policies and processes that are developed in the universities, the OIQAS must allow such institutions to demonstrate that they take seriously the quality of their qualifications as well as their commitment to ensure and demonstrate this quality, which is not opposed to reach high research standards [10].

In the particular case of the Pablo de Olavide University (UPO), its seven centers (Faculty of Business, Faculty of Experimental Sciences, Faculty of Social Sciences, Faculty of Sport, Faculty of Law, Faculty of Humanities, School of Engineering) and degrees have been submitted to an external evaluation of the National Agency for Quality Assessment and Accreditation of Spain or ANECA, by its Spanish abbreviation (www.aneca.es/eng/ANECA), under the AUDIT program [3]. Furthermore, they have been submitted to an external evaluation of the Andalusian University Evaluation and Accreditation Agency, or AGAE by its Spanish acronym (deva.aac.es/?id=acreditacion), under the VERIFICA program. After its implementation in both centers and degrees, it was submitted to the ANECA for a certification process of the Center, and to the AGAE for an accreditation process of the degrees. Once this objective was achieved, the availability of a certified IQAS in each center will facilitate the verification of future university degrees.

The development of the OIQAS, as in any other Spanish university or within the European Space, requires an adequate balance between the actions promoted by the institution itself (that is, the IQAS) and the external evaluation, audit [7] and certification procedures carried out by the Agencies (which constitutes the External Quality Guarantee) on which we depend, in particular the European Association for Quality Assurance (ENQA) for the European area, the ANECA for the national territory and the AGAE for the Region of Andalusia. In this sense, the agencies promote an articulation between the Internal Guarantee and the External Guarantee, also protecting, as one of their tasks, the implementation of the IQAS, closely linked to specific actions and programs.

In this context, and being aware of the importance of this articulation, the UPO attended in 2007 the ANECA's AUDIT Program call, and signing on October 25th 20007 an agreement by which it was committed to the implementation of IQAS in all its centers, and consequently, in all its degrees.

The UPO has prepared the manual with the guidelines to be followed by the IQAS of each center and its degrees, according to the AGAE procedures [1], which was eventually approved by Governing Council approved on July 7th 2008. The design of these systems includes actions that allow:

1. To determine the needs and expectations of students, employers and other groups of interest in relation to the training offered in the university.
2. To establish the IQAS objectives and scope in relation to the training programs.
3. To determine the internal quality assurance criteria.

The IQAS design forms an essential element in the policy and training activities of the UPO centers and degrees. For this reason, the centers must set in advance the objectives they expect to achieve as a result of their implementation. These objectives have been established autonomously by the them, but they are particular cases of the common objective that the UPO aims to achieve with the implementation of these systems, adapted to every center: to guarantee in a responsible way the quality of all the graduate and postgraduate degrees, reviewing and improving, whenever it is considered necessary, their training programs. This must be based on the needs and expectations of their stakeholders, to which they must keep informed promptly following an Institutional Communication Plan for the performance of accounts, also keeping permanently updated the own IQAS of each center and its degrees. With this as ultimate goal, it is expected:

1. Respond to the UPO's commitment with the satisfaction of the needs and expectations generated by society.
2. Order their teaching initiatives in a systematic way so that they contribute effectively to the quality assurance.
3. Facilitate the process of accreditation of the degrees implemented in the UPO centers.
4. Incorporate strategies for continuous improvement [12,14].
5. Offer the transparency required within the framework of the European Higher Education Area [6].

The rest of the paper is structured as follows. Section 2 describes the quality management structure at UPO. The follow-up protocol is detailed in Sect. 3. A results analysis can be found in Sect. 4. Finally, the conclusions drawn are reported in Sect. 5.

2 Quality Management Structure

The quality management structure [13] is composed of several members, as listed below:

1. The quality committee.
2. The delegate of the governing council, which is the body that approves its composition and regulations.
3. The monitoring and control committee for the strategic plan, whose composition and regulations depend on the senate.
4. The person in charge of quality and planning of the directors board, who is the vice-rector for quality and planning and acts by delegation of the rector.
5. The internal quality assurance committee (hereinafter, IQAC) of the centers, a consultative body under the quality committee, constituted by the vice-Rector for quality and planning, the quality and planning representatives of each center, the president of the student council and a representative of the planning, analysis and quality area.

6. The IQAC for Services, a consultative body under the quality committee, constituted by the vice-rector of quality and planning, the quality and planning members of each service, the president of the student council and a member from the planning, analysis and quality area.

Each center implements, in accordance with the UPO statutes, a quality management structure, which will be responsible for the center's IQAS and its degrees. This structure is composed of a quality and planning center manager and an center's IQAC, a quality and planning manager for each degree and an IQAC for each degree. The head of quality and planning of each center is appointed by the rector on the proposal of the dean or director of the center and the quality and planning manager of each degree, as well as the center's IQAC. The center and each IQAC's degree are appointed by the board of the center, which determines their skills in the elaboration, development, monitoring and improvement of the center's IQAS and its degrees.

In addition to that, the departments have their own quality management structure, composed of a quality committee chaired by the director of the department. The quality committee is appointed by the department council, which determines its skills in its internal operating regulations.

On the other hand, the Center for Postgraduate Studies (hereinafter CEDEP, as abbreviated in Spanish) has a quality management structure responsible for its IQAS and its degrees. This structure is composed of:

1. A head of quality and planning of the center, appointed by the postgraduate commission, proposed by the postgraduate vice-rector.
2. An IQAC for each macro area, present at the UPO, of those recognized by the Ministry responsible for higher education.
3. A person in charge of quality and planning for each degree.
4. An academic committee of each degree.

The head of quality and planning of each degree and the IQAC by macroarea and the academic committee for each degree are appointed by the graduate commission, which determines their skills in the preparation, development, follow-up and improvement of the IQAS of the CEDEP and its degrees.

In the case of degrees jointly coordinated between two universities (Pablo de Olavide University and University of Seville), the structure of quality management will be determined as stipulated in the respective collaboration agreement. Otherwise, it will be adapted to the general structure described in this section and, in any case, the different groups of interest must be represented.

3 Follow-Up Protocol

The Spanish royal decrees 1393/2007 [4] and 861/2010 [5], by which the organization of official university education is established, offers the framework of the quality assurance of the training programs.

The quality assurance systems, which are part of the new curricula, are also the basis for the new teaching organization to function efficiently and to create the confidence on which the process of accreditation of degrees relies on.

In these royal decrees, the autonomy in the design of the degree is combined with an adequate system of evaluation and accreditation, which will allow to supervise the effective execution of the teachings and inform society about the quality of it. The concretion of the system of verification and accreditation will allow the balance between a greater capacity of the universities to design the degrees and the accountability oriented to guarantee the quality and to improve the information of the society on the characteristics of the university offer.

All the training programs leading to bachelor and postgraduate degrees, since their implementation proposal, have an IQAS. The centers possess a robust and powerful IQAS that allows them to monitor and control all their degrees, with the aim of guaranteeing the quality of the training programs for which they are responsible for.

The centers have mechanisms that allow them to maintain and renew their training offer and develop methodologies for the approval, control and periodic review of their programs. To this end, and in its different organizational levels:

1. Determine the groups of interest, bodies and procedures involved in the design, control, planning, development and periodic review of the degrees, their objectives and associated competencies.
2. They have information collection and analysis systems (including information from the national and international environment) that allow them to assess the maintenance of their training offer, its updating or renewal.
3. They have mechanisms that regulate the decision-making process regarding the training offer and the design of the degrees and their objectives.
4. They ensure that the necessary mechanisms are developed to implement the improvements derived from the process of periodic review of the degrees.
5. They determine the way (how, who, when) in which the groups of interest are held accountable for the quality of the teachings.
6. They define the criteria for the eventual suspension of the degree.

In all cases, and in the actions to guarantee the quality of the training programs of a center, the centers have criteria of quality in relation to:

1. The relevance of the justification of the degree and the needs of the groups of interest.
2. The relevance of the general objectives and skills.
3. The clarity and sufficiency of the systems that regulate the access and admission of students.
4. The coherence of the planned planning.
5. The adequacy of the academic and support staff, as well as material resources and services.
6. The expected efficiency in relation to the expected results.
7. The IQAS of the Center and its degrees.

And, finally, the centers take into account that training in any professional activity should contribute to the knowledge and development of the human rights, democratic principles, principles of equality between women and men, solidarity, environmental protection, universal accessibility and design for all, and promotion of the culture of peace, as contemplated by the royal decrees that regulate the university education.

In order to carry out the monitoring of university education, the AGAE designs a protocol and a specific procedure [2] on which the *PE04 procedure is based: Measurement, analysis and continuous improvement* of the OIQAS of the centers and degrees of UPO, with which the UPO's training programs are monitored.

4 IQAS Analysis

This section is divided into three subsections. In particular, Sect. 4.1 analyzes the results derived from the IQAS. Later in Sect. 4.2 the IQAS reports analyses are discussed. Finally, some quality indicators directly related to the teaching-learning process, from the degree of computer science and information systems, are assessed in order to determine how the quality of a degree is improved in Sect. 4.3.

4.1 Results Analysis

The IQAC for Bachelor's degrees, and the Academic Committee, for Master's degrees, annually follow up the improvement plans approved in the previous revision of the IQAS and of the policy and objectives of quality of the degree. In addition, they analyze the indicators [8] related to the enrollment of students (*PC03: access, admission and enrollment of students of the centers*), the entrance/exit profiles (*PC04: Profiles of entrance/exit and recruitment of students*), adequacy of academic staff (*PA03: Recruitment and selection of academic and administrative staff and services, PA05: Evaluation of academic and administrative staff and services*), student mobility (*PC08: Management and review of the mobility of students*), external internships (*PC09: Management and Review of external internships*), job placement (*PC11: Management of job placement*), academic results (*PC12: Analysis of the results of the learning*) and the satisfaction of the groups of interest (*PA09: Satisfaction, needs and expectations of the groups of interest*).

The center's IQAC in the case of Bachelor's degrees and the Head of Quality and Planning of CEDEP in the case of Master's Degrees analyze the incidents, claims and suggestions of the center every year, respectively. They also analyze the general indicators related to the management of degrees (*PA02: File management and processing of degrees*) and adaptation of material resources (*PA06: Management of material resources*).

4.2 IQAS Reports Analysis

After the analysis carried out described in the previous section, the following reports are prepared:

1. Monitoring report on the center's quality objectives (Bachelors degree only).
2. Follow-up report on the quality objectives of the degree.
3. Annual monitoring report of the degree that includes:
   (a) Results of the Improvement Plan established in the monitoring of the previous year.
   (b) Summary of incidents, claims and suggestions.
   (c) Assessment of the indicators included in the procedures mentioned in Sect. 3.
   (d) Annual improvement plan for next year.

The annual follow-up report must be approved by the Center Board, in the case of Bachelor's degrees, and must be reviewed by the IQAC of the degree, and approved by the Graduate Commission, in the case of Master's degrees.

Likewise, each Center (except CEDEP) signs a Contract-Program of conditional financing, based on the objectives of the Plans, with the Vice-Rectorate of Quality and Planning and the Board of Directors of the UPO. Annual improvements included in the monitoring reports of the center's degrees, and whose follow-up will be carried out by the Planning, Analysis and Quality Department of the University.

4.3 Assessing Quality Indicators

As case study, we show the quality indicators associated to the School of Engineering and, in particular, to its Degree in Computer Science and Information Systems. Although many indicators are included in the degree annual report for quality, only those related to the students performance are here discussed. All data can be found at the IQAS UPO website, after authentication (www.upo.es/calidad).

Table 1 refers to the evolution of the selected indicators over time, since the degree was implemented. The indicators are defined below:

1. Performance rate. It is defined as the total amount of credits passed divided by the total amount of credits enrolled.
2. Efficiency rate. It is defined as the amount of credits that a cohort of students should have been enrolled in divided by the actual credits enrolled.
3. Graduation rate. It is defined as the number of a students within a cohort that finish the degree in n or $n + 1$ years (being n the number of years a degree is planned to be done) divided by the number of students that formed such a cohort during their first academic year.
4. Success rate. It is defined as the total number of credits passed divided by the total number the students are examined.

Table 1. Quality indicators evolution over time for the Degree in Computer Science and Information Systems. All values are expressed in %.

Year	Performance	Efficiency	Graduation	Success
2010–2011	48.24	N/A	N/A	68.42
2011–2012	47.91	N/A	N/A	77.06
2012–2013	61.43	N/A	N/A	81.25
2013–2014	58.35	93.41	N/A	81.02
2014–2015	60.69	90.66	16.95	84.30
2015–2016	62.25	80.04	18.33	84.26
2016–2017	57.25	88.78	11.11	77.14
2017–2018	61.33	87.60	14.81	78.89

Please note that some values of both efficiency and graduation rates are not available (N/A), since they could not be measured until the first cohort ended the degree.

As it can be seen, the performance and success indicators exhibit a very relevant increase from the first year (with no IQAS still implemented) to the last year measured (from 48.24% to 61.33% and from 68.42% to 78.89%, respectively). As for the efficiency rate, also slightly lower, it remains almost constant over time with a very high value for 2018 (87.60%). Finally, the variation of the graduation rate is hardly −2%.

5 Conclusions

It is well-known that the implementation of an internal quality assurance system in public universities reports multiple benefits for both the institution and the students. In this paper, we have introduced how such a system has been implemented at Pablo de Olavide University. Thus, all roles and commissions are detailed. The Degree in Computer Science and Systems Information, belonging to the School of Engineering, is used as study case. The analysis of several quality indicators shows how this system has improved the teaching-learning process.

Acknowledgments. The authors want to thank the financial support given by the Spanish Ministry of Economy and Competitiveness project TIN2017-88209-C2-R. Also, this analysis has been conducted under the Innovation Teaching Project helps (2015–2018) by the University of Seville and Pablo de Olavide University.

References

1. Agencia Andaluza de Evaluación de la Calidad y Acreditación Universitaria (AGAE). Orientaciones practicas para el establecimiento de un Sistema de Garantia de Calidad de titulos universitarios oficiales de Grado (2008). http://deva.aac.es/?id=verificacion
2. Agencia Andaluza de Evaluación de la Calidad y Acreditación Universitaria (AGAE). Procedimiento para el seguimiento de los títulos oficiales (Grado y Máster) (2010). http://deva.aac.es/?id=seguimiento
3. Agencia Nacional de Evaluación de la Calidad y Acreditación (ANECA). Programas AUDIT y VERIFICA (2008). http://www.aneca.es/Programas-de-evaluacion/Evaluacion-institucional/AUDIT
4. Boletín Oficial del Estado (BOE). Real Decreto 1393/2007, de 29 de octubre, por el que se establece la ordenación de las ensenanzas universitarias oficiales (2007). https://www.boe.es/buscar/act.php?id=BOE-A-2007-18770
5. Boletín Oficial del Estado (BOE). Real Decreto 861/2010, de 2 de julio, por el que se modifica el Real Decreto 1393/2007 de 29 de octubre, por el que se establece la ordenación de las ensenanzas universitarias oficiales (2010). https://www.boe.es/buscar/doc.php?id=BOE-A-2010-10542
6. Bologna Declaration. The European Higher Education Area. Joint Declaration of the European Ministers of Education 19 (1999)
7. Dickins, D., Johnson-Snyder, A.J., Reisch, J.T.: Selecting an auditor for Bradco using indicators of audit quality. J. Account. Educ. **45**, 32–44 (2018)
8. Gojkov, G., Stojanović, A., Rajić, A.G.: Critical thinking of students – indicator of quality in higher education. Procedia Soc. Behav. Sci. **191**, 591–596 (2015)
9. Huet, I., Figueiredo, C., Abreu, O., Oliveira, J.M., Costa, N., Rafael, J.A., Vieira, C.: Linking a research dimension to an internal quality assurance system to enhance teaching and learning in higher education. Procedia Soc. Behav. Sci. **29**, 947–956 (2011)
10. Palali, A., van Elk, R., Bolhaar, J., Rud, I.: Are good researchers also good teachers? The relationship between research quality and teaching quality. Econ. Educ. Rev. **64**, 40–49 (2018)
11. Praraksa, P., Sroinam, S., Inthusamith, M., Pawarinyanon, M.: A model of factors influencing internal quality assurance operational effectiveness of the small sized primary schools in Northeast Thailand. Procedia Soc. Behav. Sci. **197**, 1586–1590 (2015)
12. Ramírez-Juidias, E., Tejero-Manzanares, J., Amaro-Mellado, J.L., Ridao-Ceballos, L.: Developing experimental learning in a graphical course using Thurstone's Law of comparative judgment. Eng. Lett. **25**, 61–67 (2017)
13. Rezeanu, O.M.: The implementation of quality management in higher education. Procedia Soc. Behav. Sci. **15**, 1046–1050 (2011)
14. Rubio-Escudero, C., Asencio-Cortés, G., Martínez-Álvarez, F., Troncoso, A., Riquelme, J.C.: Impact of Auto-evaluation Tests as Part of the Continuous Evaluation in Programming Courses. Adv. Intell. Syst. Comput. **771**, 553–561 (2018)

Author Index

© Springer Nature Switzerland AG 2020
F. Martínez Álvarez et al. (Eds.): CISIS 2019/ICEUTE 2019, AISC 951, pp. 349–350, 2020.
https://doi.org/10.1007/978-3-030-20005-3

Printed in the United States
By Bookmasters